Experimental Methods in
Organic Fluorine Chemistry

Experimental Methods in Organic Fluorine Chemistry

Tomoya Kitazume
Tokyo Institute of Technology, Yokohama, Japan

Takashi Yamazaki
Tokyo Institute of Technology, Yokohama, Japan

Kodansha   1998

Tokyo

Gordon and Breach Science Publishers

Australia·Canada·China·France·Germany·India
Japan·Luxembourg·Malaysia·The Netherlands·Russia
Singapore·Switzerland

Copublished by
KODANSHA LTD.
12-21 Otowa 2-chome, Bunkyo-ku, Tokyo 112-8001, Japan
and
GORDON AND BREACH SCIENCE PUBLISHERS
Amsteldijk 166, 1st Floor, 1079 LH Amsterdam, The Netherlands

British Library Cataloguing in Publication Data

A catalogue record for this book is available from the British Library.

ISBN 90-5699-122 -1 (Gordon and Breach Science Publishers)
ISBN 4-06-209353-7 (Kodansha)

Printed in Japan.

Contents

Preface

Fluorinated organic materials have been gaining significant importance among the biologically active and functional materials. Even though several reagents and reaction sequences have been developed for the construction of fluorinated materials in the last decade, the topic remains a specialized field unfamiliar to most chemists. There are a number of problems to be solved, including the handling of the materials, availability of reagents, and selectivity (stereo-, regio-, and/or chemo-). This volume aims to help the nonspecialist in preparing fluorinated materials and reactive intermediates as well as update the specialist on the reagents and reaction sequences in fluorine science.

The task of selecting as diverse references as possible for this volume was very difficult. Since the recent trend is to report important results from prestigious journals as a short communication, we sometimes cannot trace interesting procedures due to very little experimental information in a strictly limited number of printed pages. So, from the standpoint of our initial concept for this volume, we mainly collected the experimental methods from full papers, which usually contain ampler, more detailed, and more practical knowledge, even if the original work was published as a communication. Moreover, the editors only collected information on aliphatic compounds with a few fluorine atoms because works on the preparation of aromatic and heterocyclic materials have already been published.

We are grateful to the kind cooperation of a number of authors who have kindly sent us their detailed procedures or physical properties upon request, or promptly answered our questions. We also thank Mr. Ippei Ohta of Kodansha Scientific Ltd. for his enthusiastic support for this book.

November 1998

Tomoya Kitazume
Takashi Yamazaki

General Remarks

Chapters 2 through 5 deal with the experimental details of the formation of each specific material. Basically, such experimental procedures are described as originally published, but minor changes are sometimes made for consistency throughout the text. One example is the preparation of the famous and widely applicable non-nucleophilic strong base, lithium diisopropylamide (LDA). Because this volume contains many reaction schemes using this versatile base, each procedure only includes a brief description such as "to a solution of LDA (1 mmol) ···", and readers are referred to the following two representative procedures.

Method A: A three-necked, 500-mL, round-bottomed flask is fitted with a nitrogen inlet, a rubber septum, and a 125-mL dropping funnel. The flask is flame-dried, flushed with nitrogen and charged with diisopropylamine (22.0 mL, 0.16 mol) and 100 mL of anhydrous THF. This solution is cooled in an ice bath, and the dropping funnel is charged with a solution of n-BuLi (80.0 mL of a 1.88 M solution in hexane, 0.15 mol), which is added dropwise over 15 min.

Boeckman, Jr., R. K., *et al. Org. Synth. Coll. VIII* **1993**, 192.

Method B: An oven-dried, 250-mL, round-bottomed flask equipped with a stirring bar and a rubber septum is charged with 100 mL of anhydrous THF and 5.56 g (55 mmol) of anhydrous diisopropylamine. The flask is flushed with argon via a needle inlet-outlet and cooled to -78 °C with a dry ice-isopropyl alcohol bath. To this stirred solution 30 mL (54 mmol) of a 1.8 M solution of n-BuLi in hexane is added dropwise with a syringe, and stirring is continued at that temperature for 30 min.

Spitzner, D.; Engler, A. *Org. Synth. Coll. VIII* **1993**, 219.

In the case of NMR spectroscopic data, chemical shifts of ^{1}H and ^{13}C NMR in the indicated solvent are reported using tetramethylsilane (TMS) as the internal standard, and the chemical shifts of ^{19}F NMR are tabulated using trichlorofluoromethane ($CFCl_3$) as the reference unless otherwise noted. While downfield shifts in ^{19}F NMR are sometimes designated as negative, their signs are changed by the authors to keep the uniform description of downfield shifts positive.

When a procedure requires the use of poisonous or dangerous materials, readers will find a "*Caution*" section containing remarks on how to handle such compounds. Additional information for readers' convenience is included under the heading *Note*.

Abbreviations

[α]	specific rotation
Ac	acetyl
Ac$_2$O	acetic anhydride
AcOEt	ethyl acetate
AcOH	acetic acid
AIBN	2,2'-azobisisobutyronitrile (=2,2'-azobis(2-methylpropionitrile))
Allyl	prop-2-en-1-yl (=C$_3$H$_7$)
BINOL	1,1'-bi-2-naphthol
Bn	benzyl (=C$_6$H$_5$CH$_2$)
bp	boiling point
br	broad
Bu	butyl (=C$_4$H$_9$)
d	day(s); doublet
DAST	diethylaminosulfur trifluoride
DBH	1,3-dibromo-5,5-dimethylhydantoin
DBU	1,8-diazabicyclo[5.4.0]undec-7-ene
	(=2,3,4,6,7,8,9,10-octahydropyrimido[1,2-a]azepin)
de	diastereomeric excess
DIBAL-H	diisobutylaluminum hydride (=DIBAH or DIBALH)
DIP-Cl	diisopinocampheylboron chloride
DMAP	4-(dimethylamino)pyridine
DMF	N,N-dimethylformamide
DMPU	1,3-dimethyl-3,4,5,6-tetrahydro-2(1H)-pyrimidinone
	(=N,N'-dimethylpropyleneurea)
DMSO	dimethyl sulfoxide
ee	enantiomeric excess
Et	ethyl (=C$_2$H$_5$)
Et$_2$O	ether (=Diethyl ether)
EtOH	ethyl alcohol (=Ethanol)
F-TEDA-BF$_4$	1-(chloromethyl)-4-fluoro-1,4-diazabicyclo[2.2.2]octane
bis(tetrafluoroborate) (=Slectfluor™)	
h	hour(s)
HMPA	hexamethylphosphoramide
Hz	hertz
i-	iso
IR	infrared spectrum
LCIA	lithium cyclohexylisopropylamide
LDA	lithium diisopropylamide
LHMDS	lithium hexamethyldisilazide or lithium bis(trimethylsilyl)amide
m	multiplet

*m*CPBA	*m*-chloroperbenzoic acid
Me	methyl (=CH$_3$)
MeCN	acetonitrile
MEM	(2-methoxyethoxy)methyl
MeOH	methyl alcohol (Methanol)
min	minute(s)
MMPP	magnesium monoperoxyphthalate
MOM	methoxymethyl
mp	melting point
MS	molecular sieves
Ms	methanesulfonyl (=mesyl)
n-	normal
NaHMDS	sodium hexamethyldisilazide or sodium bis(trimethylsilyl)amide
Naph	naphthyl
NBS	*N*-bromosuccinimide
NIS	*N*-iodosuccinimide
NMR	nuclear magnetic resonance
PCC	pyridinium chlorochromate
Ph	phenyl (=C$_6$H$_5$)
PMP	*p*-methoxyphenyl
ppm	parts per million
Pr	propyl (=C$_3$H$_7$)
q	quartet
quint	quintet
Red-Al	sodium bis(2-methoxyethoxy)aluminum (=Vitride, SMEAH)
Rf	retention factor
R$_f$	perfluoro or polyfluoroalkyl groups
s	singlet
sec-	secondary
sept	septet
sex	sextet
t	triplet
TASF	tris(dimethylamino)sulfur (trimethylsilyl)difluoride
TBAF	tetra-*n*-butylammonium fluoride
TBS	*tert*-butyldimethylsilyl
tert-	tertiary
Tf	trifluoromethanesulfonyl (=triflyl)
THF	tetrahydrofuran
TMEDA	*N,N,N',N'*-tetramethylethylenediamine
TMS	trimethylsilyl; tetramethylsilane
Ts	*p*-toluenesulfonyl (=tosyl)

1 Introduction

When Henri Moissan first isolated elemental fluorine in 1886,[1] could he have imagined the explosive development of fluorine chemistry over these past decades? Recently, fluorine chemistry has been recognized as one of the most significant fields in chemistry. This is because the special properties resulting from the incorporation of fluorine sometimes adds an unexpected quality to organic molecules. The exceptional importance of fluorine chemistry can be readily understood from the fact that many books[2] and reviews in journals[3,4] have been published in this field, especially in the 1990s. On the other hand, its special nature, in particular regarding reactivity and toxicity, can be considered to be one of the main reasons keeping general organic chemists away from this field. However, if such a barrier could be removed, further development of fluorine chemistry with wide application in diverse fields can be anticipated. The primary aim of this publication is to provide as much useful procedural information as possible for easy access to materials readers would like to construct. The various chapters have been compiled from published articles which contain procedures *that do not require any special apparatus or technique.* This is why the present volume contains no scheme directly using F_2 gas, anhydrous HF, SF_4, or similar compounds.

This chapter briefly describes the basic chemistry of fluorine-containing substances discussing physical property changes and effects on the reactivity caused by the introduction of the fluorine atom as well as the care required when handling fluorinated materials.

1.1 Effect of Fluorine on Physical Properties

1.1.1 Mimic and Block Effects

Why has this particular atom been drawing so much attention in recent years? The main reasons stem from the following three special characteristics. Fluorine is

1) the most electronegative element of all,
2) the second smallest element next to hydrogen, and
3) the atom that can form a stronger bond with carbon than hydrogen.

Table 1.1 on the following page contains selected data on this atom together with those on a

Table 1.1   Representative physical properties of selected atoms

Atom	Electronegativity	van der Waals radius (Bondi) (Å)	Bond length (CH_3-X) (Å)	Dissociation energy (CH_3-X) (kcal/mol)
H	2.2	1.20	1.09	99
F	4.0	1.47	1.39	110
Cl	3.0	1.75	1.77	85
Br	2.8	1.85	1.93	71
C	2.5	1.70	1.54	83
O	3.5	1.52	1.43	86

few other elements for comparison. Looking at the first column of this table, the strong electronegativity of the fluorine atom is readily understood when compared with the other halogens as well as oxygen. Because of the location of fluorine on the right side of the periodic table (VIIb group), its small size is intuitively and qualitatively expected. Actually, van der Waals (vdW) radius improved by Bondi[5] numerically demonstrated the rather small value of fluorine, only about 20% larger than that of hydrogen (an even smaller difference of *ca.* 10% was previously estimated based on the vdW values by Pauling,[6] H: 1.20 Å, F: 1.35 Å). This element also forms a firm bond with the methyl carbon atom, which is evaluated to be 11 kcal/mol stronger than the methane C-H bond.

These are the basic properties fluorine inherently possesses. The special effects of fluorinated materials brought about by the combination of the above three characteristics are discussed below.

When a fluorine atom is introduced into a molecule, one sometimes expects the enhancement or modification of the original biological activity. Actually, there are a number of cases where just one fluorine causes a tremendous change in the parent physical properties, the most famous and important example being the profound effect of fluoroacetic acid in the original acetic acid metabolic system, the TCA cycle. Compound **1b**,[7] one of the few naturally occurring fluorinated compounds,[3h] is "mistakenly" recognized as acetic acid **1a** by the enzyme due to its already described close steric similarity to hydrogen and is converted to fluorocitric acid (2*R*,3*R*)-**4** by the same biological manipulation of the corresponding fluoroacetyl-CoA **2b** and oxaloacetic acid **3**, as shown in Fig. 1.1. However, different from its prototypic case further transforming the resultant citric acid **5** to isocitric acid **7** *via* the aconitase-catalyzed dehydration, (2*R*,3*R*)-**4** cannot follow the usual process because fluorine is located at the position of the usually abstracted hydrogen atom and the stronger C-F bond cleavage is apparently more difficult, leading to the inhibition of the TCA cycle and accumulation of this fluorinated material. It is quite interesting to note that citrate synthase stereospecifically constructs (2*R*,3*R*)-**4** out of four possible diastereomers and, in spite of the extremely poisonous nature of this isomer, causing convulsions and ventricular fibrillation, the three other stereoisomers are not toxic. Thus, this is expressed as the "lethal synthesis" by citrate synthase.[8] Moreover, use of ω-fluoroalkanoic acid derivatives (F-$(CH_2)_n$-CO_2R) requires similar significant attention when n is an odd number since these substances are metabolically transformed into fluoroacetic acid by the repetitive degradation *via* β-oxidation as a key step.

This is the typical "mimic effect" emerging from the combination of the second and third characteristics of fluorine stated above, which is usually applicable to compounds with

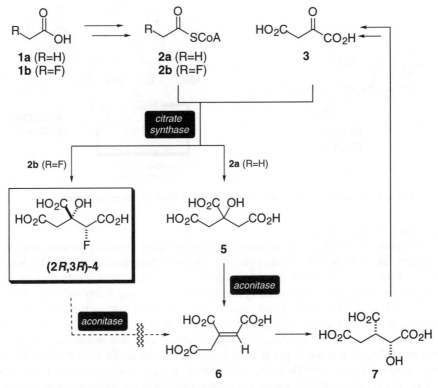

Fig. 1.1 Citric acid cycle with fluoroacetic acid **1b** as a substrate.

only one fluorine atom, but there are a few examples of the mimic effect for di-[9] as well as trifluoromethylated compounds.[10]

Apart from the metabolic pathway and from the synthetic point of view, the appearance of toxicity of (2R,3R)-**4** in Fig. 1.1 is also interpreted as a possible result of the introduction of a fluorine atom to **5** instead of the enzymatically labile hydrogen. Such replacement constitutes one of the important strategies effectively inhibiting the usual conversion pathways. This is sometimes classified as the "block effect." The slower oxidative metabolic degradation of p-fluorophenylalanine, about one-sixth of the parent phenyl-alanine,[11] was explained in a similar fashion, which suggests increased durability by the incorporation of a fluorine atom.

1.1.2 Effect of the Electron-withdrawing Nature of Fluorine

1.1.2.1 Deactivation Effect

It has been proved that the second and third characteristics play important roles for the mimic and block effects. Moreover, the electronegative quality of fluorine has also been widely utilized for alteration of the original physical properties along with the inherent

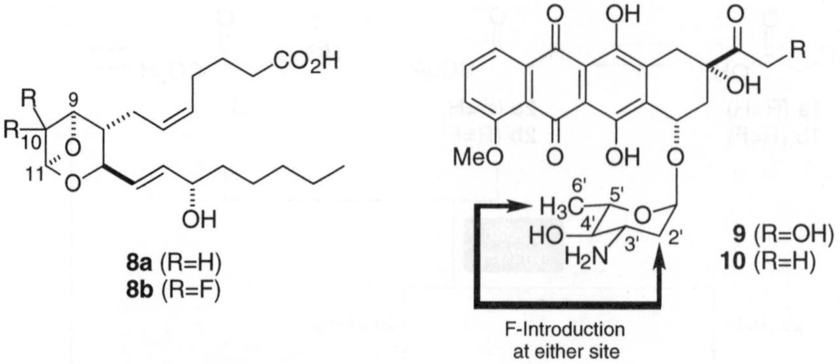

8a (R=H)
8b (R=F)

9 (R=OH)
10 (R=H)

F-Introduction
at either site

Fig. 1.2   Thromboxane A$_2$ and anthracyclines as fluorine-modified targets.

increase in its lipophilic nature.

For example, Fried and co-workers reported excellent resistance to hydrolysis when two fluorine atoms were introduced at the 10 position of thromboxane A$_2$ (TXA$_2$, **8a**, Fig. 1.2).[12] Their study stemmed from the necessity of finding more stable analogs of **8a** because of its powerful vasoconstricting and platelet aggregating ability but only 30 sec of half life at pH=7.4. However, Fried came to the conclusion that such inherent lability of oxetane acetal could be overcome by these two fluorines, which would effectively decrease the electron density of the neighboring oxygen atoms and also destabilize the possible formation of the cation at the 11 position on hydrolysis under acidic conditions (*vide infra*). This was indeed the case and **8b** attained dramatically greater chemical stability (*a half life of 270 years at physiological pH!*) retaining the same or even better biological activity than the parent **8a**.[13]

A similar concept was applied to increase the stability of anthracyclines such as doxorubicin **9** or daunorubicin **10** by Tsuchiya's group (Fig. 1.2).[14,15,16] Based on the fact that these compounds are also susceptible to acid hydrolysis and that the resultant aglycones themselves do not show any antitumor activity, incorporation of one or two fluorine(s) at the 2' position[14a,b] or three fluorines at the 6' position[14c] was investigated

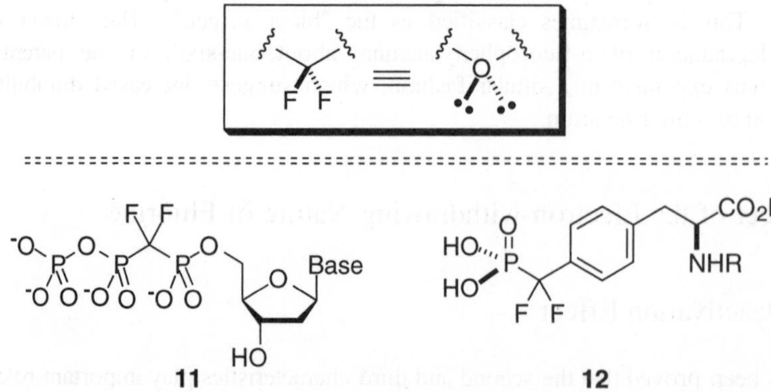

11   **12**

Fig. 1.3   Representative difluorophosphonates.

along with the modification of functional groups at the 3' and/or 4' positions to successfully suppress the unfavorable cleavage of the sugar moieties while maintaining their excellent original activities.

Substitution of biologically attractive phosphate intermediates for the corresponding phosphonates becomes the major protection strategy from undesirable hydrolysis. Recently, difluoromethylene phosphonates rather than the nonfluorinated counterparts have drawn significant attention due to the isosteric as well as isoelectronic nature of a CF_2 moiety to the ether oxygen with elimination of the hydrolysis susceptibility,[17] and such analogs lead to attainment of second pK_a values much closer to those of the parent materials.[18] Based on this concept, Blackburn demonstrated[19] the construction of difluoromethylene-substituted 2'-deoxyadenosine and 2'-deoxythymidine 5'-triphosphates **11** along with the corresponding monofluorinated counterparts as the DNA polymerase inhibitors (Fig. 1.3). On the other hand, two research groups[20] manipulated 4-(phosphonodifluoromethyl)phenylalanine **12** as the mimetics of the original phosphate as the phosphotyrosine phosphatase inhibitor.[21,22] In addition to the pK_a values, another advantage of oxygen substitution for a CF_2 group is the retention of the hydrogen bonding acceptance ability, which is completely lost in the corresponding CH_2 derivatives.

1.1.2.2 Activation Effect

The above constitutes the typical application of a difluoromethylene moiety as the mimic of oxygen, and the deactivation by the electronic acceptance from the neighboring functionalities results in the final stabilization of molecules. In sharp contrast to this concept, there exists another strategy activating the original substrates by incorporation of fluorine.

How then are functionalities in the vicinity of fluorine or fluorine-containing groups activated? The solution was suggested by a comparison of two model compounds, acetone **13** and 1,1,1-trifluoroacetone **14**, based on *ab initio* molecular orbital calculations (Fig. 1.4).[23] It might be intuitively understood that the carbonyl carbon atom of the latter is electronically more deficient than the one of **13** due to the direct attachment of the strongly electron-withdrawing CF_3 group. However, MO calculations by Linderman's group[23] unambiguously manifested the opposite tendency, *i.e.* the carbonyl carbon of **14** is less cationic than the one of **13**, and this theoretical result is experimentally supported by ^{13}C NMR chemical shifts of these carbon atoms in question, 206.0 ppm for **13** and 188.8 ppm for **14**. Instead of the partial charge, fluorines are considered to be responsible for about 1.4 eV lowering of the LUMO energy level, producing the quite different electrophilic

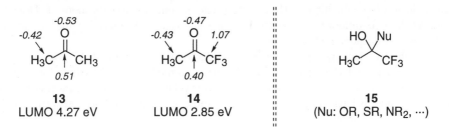

Fig. 1.4 *Ab initio* (HF/6-31G**) analysis of acetone and trifluoroacetone.
(*italic number*: point charge)

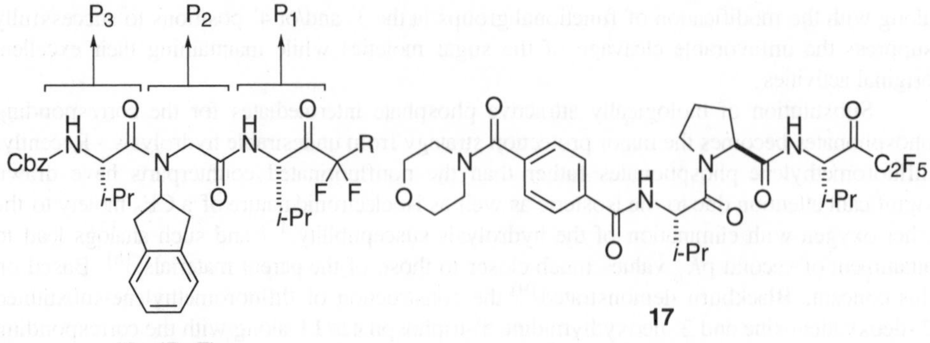

16a (R=F)
16b (R=CO₂Et)

Fig. 1.5   Protease inhibitors with fluorinated ketone functions.

circumstance between **13** and **14**.

This quality allows **14** to very easily react with a variety of nucleophiles to form the adduct **15**. Although this type of product is less stable than the starting carbonyl compound when such materials do not contain any fluorine atoms, a tetrahedral intermediate like **15** is extraordinarily stable for substances with two or three fluorines next to the original carbonyl group. This stability originates from the significant reluctance to form carbocations α to the CF_3 group and also the strong stability of the alkoxide from **15** for the electrostatic reason (*vide infra*), both inhibiting its hydrolysis under acidic and basic conditions, respectively. It is well known that trifluoroacetaldehyde CF_3CHO demonstrates a characteristic behavior: only its hydrate ($CF_3CH(OH)_2$) or hemiacetal ($CF_3CH(OH)(OR)$) forms are available from commercial suppliers, and *both forms must be heated to 120 °C or more in concentrated H_2SO_4 for the liberation of the desired aldehyde!*[24)]

Application of this unique activation concept based on the stability of a tetrahedral intermediate like **15** has been realized in a recent intensive investigation of protease inhibitors which mimic the tetrahedral hydrolysis transition states by the stable hemiketal formation with the hydroxy group of the active site Ser[195].[25)] The representative examples are described in Fig. 1.5. Skiles and his co-workers reported[26)] the preparation of **16a** as the human leukocyte elastase (HLE) inhibitor and the effectiveness of the achiral indan substituent on nitrogen instead of L-proline at the P_2 site as well as the importance of the valine-derived trifluoromethyl ketone structure at P_1 for the accomplishment of the acceptable *in vitro* inhibition. They further carried out the modification of **16a**[27)] to construct the α,α-difluoro-β-ketoester **16b** and succeeded in enhancing the activity to a level about twice that of cephalosporins.[28)]

Another research group studied[29)] a similar type of compound **17** for the human neutrophil elastase (HNE) inhibitors. NMR study clarified the ready hydrate formation of pentafluoroethyl and the corresponding trifluoromethyl ketones (the latter is more active than the former) and it is interesting to note that even substrates with five fluorines were found to be effective as the inhibitor.

1.1.3 Conformational Preference Including Hydrogen Bonding Ability

There is another special characteristic of fluorine sometimes determining molecular conformations which is attributable to its strong electron-withdrawing nature and small steric size, and is called as the *gauche* effect or (extended) anomeric effect.[30] This phenomenon is explained by using 1,2-difluoroethane **18** as the model compound (Fig. 1.6).

It is apparent that **18** possesses two stable conformations, *gauche* and *anti(periplanar)* forms as depicted in Fig. 1.6. As shown in Table 1.1, comparison of physical properties of fluorine with hydrogen indicates differences in vdW radii of 0.27 Å and in electronegative nature, both of these possibly allowing us to conclude that *anti*-**18** is energetically more stable than *gauche*-**18** due to less steric congestion, electrostatic as well as dipolar (C-F: 1.82 Debye, C-H: 0 Debye) repulsion. However, this is actually not the case and the latter *gauche* conformer was demonstrated to be approximately 1 kcal/mol more favorable than the former by various analytical methods (IR and Raman,[31] NMR,[32] and electron diffraction[33]) as well as by theoretical calculation.[34]

The preference for *gauche*-**18** would be accounted for by the attractive interaction between the electron-donating σ_{C-H} and the electron-accepting σ^*_{C-F} molecular orbitals. Thus, *gauche*-**18** possesses two such interactions, but the other conformer *anti*-**18** having the energetically lowered σ_{C-F} at the *antiperiplanar* position of the σ^*_{C-F} orbital obtains less energetic stabilization, and this orbital interaction, not significantly strong, is enough to compensate for the undesirable steric and electronic factors. While the energy level of σ^*_{C-X} (thus the stabilization energy when interacted with the σ_{C-H} orbital) theoretically decreases in the order I→Br→Cl→F, dominance of the steric factor is regarded to be the main reason for the other 1,2-dihalogenoethanes mainly or exclusively existing as the *anti* form.[35]

Such a conformational trend is also found for 2-fluoroethanol. With the above discussion and consideration of the close energy level of σ^*_{C-O} to σ^*_{C-F}, it is not difficult to assume that conformers with the *gauche* relationship between C-O and C-F bonds are more stable than the others. Moreover, electronic repulsion of fluorine and oxygen lone pairs leads to the conclusion that **gg'-19** would be the most preferable of all, since it also possesses intramolecular hydrogen bonding property (Fig. 1.7).

This interpretation is, at least, qualitatively correct, and irrespective of the analytical means[36] employed or theoretical calculation methods,[37] this **gg'-19** is actually obtained as the energetically most favorable structure. Bakke and his co-workers pointed out[37d] that the preference for this conformation stems from the above *gauche* effect as well as the electronic

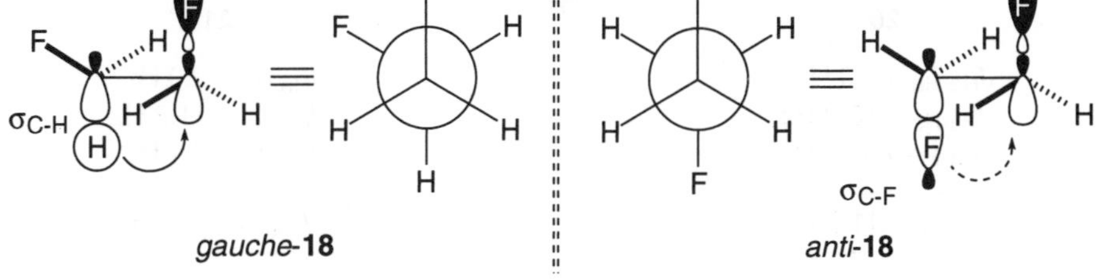

Fig. 1.6 *Gauche* and *anti* conformations of 1,2-difluoroethane.

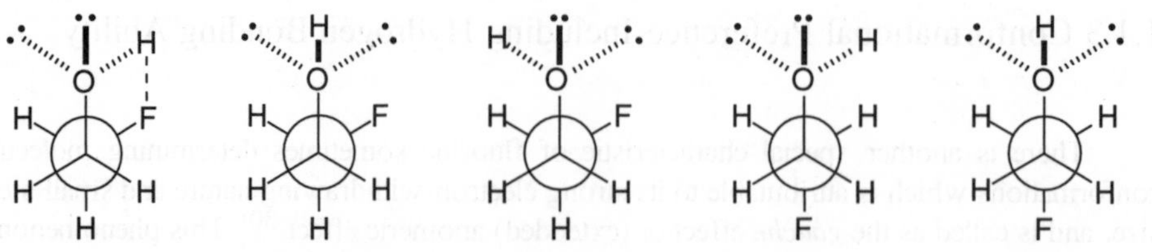

Fig. 1.7 Conformational isomers of 2-fluoroethanol **19**.

repulsion, not from the formation of intramolecular hydrogen bonding. Their suggestion is supported by their NMR experiments for **20** and **21** whose ring conformations are fixed by the bulky *tert*-butyl group (see the coupling constants related to H^a in both materials) and thus, these compounds can be regarded as the "rigid model" of 2-fluoroethanol and the corresponding methyl ether, respectively (Fig. 1.8).   In the case of **20**, the coupling constant J_{Ha-Hd} of 11.7 Hz estimated the isomeric composition to be *gauche:anti* = 6:94 around the H^a-C-O-H^d bond, supporting the energetically stable situation of the hydrogen bonding conformer.   On the other hand, replacement of the hydroxy hydrogen atom in **20** with a methyl moiety imposes additional steric congestion and, at the same time, eliminates the ability to properly interact with fluorine.   However, the coupling constant J_{Ha-Ca} of 4.0 Hz from the ^{13}C NMR allowed them to calculate that the composition of the conformer with a dihedral angle H^a-C-O-C^a of 180°, the same as the hydrogen bonding conformation of **20**, was still 40%.   Additional results on the theoretical calculation of 1-fluoro-2-methoxyethane anticipated the higher probability of *gauche* isomers of *ca.* 60%.

Møllendal's microwave spectral study on 2,2-difluoroethanol[38)] clarified the importance of the electronic repulsion between fluorine and oxygen lone pairs, and theoretical calculations in general anticipated the weaker intramolecular OH⋯F hydrogen bonding in a range of 2 to 3 kcal/mol[37,39,40)] (about half the strength of the corresponding OH⋯O interaction[41)]).   These results along with the OH⋯F nonbonding distance of 2.46 Å for **gg'-19** at 240 °C (the sum of van der Waals radii of 2.67 Å) would lead to the

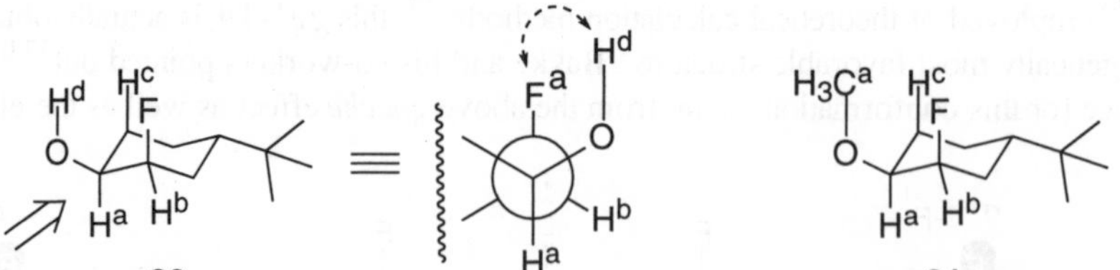

20

J_{Ha-Hb} = 2.0 Hz
J_{Ha-Hc} = 13.2 Hz
J_{Ha-F} = 29.9 Hz

J_{Ha-Hd} = 11.7 Hz

21

J_{Ha-Hb} = 2.0 Hz
J_{Ha-Hc} = 11.8 Hz
J_{Ha-F} = 29.1 Hz

J_{Ha-Ca} = 4.0 Hz

Fig. 1.8 Conformationally locked cyclohexanol and its methyl ether.

conclusion that the conformational determinant of 2-fluoroethanol is the combination of the *gauche* effect and the electronic repulsion, and, in consequence, the three-dimensional location of a hydroxy group and fluorine is observed *as if they formed intramolecular hydrogen bonding.*[42]

1.1.4 Steric Size of Fluorine-containing Groups

As noted above, fluorine itself is the second smallest of all atoms. Then, how large are the fluorine-containing groups? Table 1.2 shows the Taft steric parameters[43] and their modified values,[44] which clearly illustrate the steric similarity of fluorine to a hydroxy group rather than to hydrogen, as also anticipated from the vdW radius values in Table 1.1. Fluorinated methyl groups are relatively large, and it is interesting to note that a monofluoro-methyl (CH_2F) group is sterically more demanding than an ethyl group: thus, simple comparison of both moieties leads us to conclude that the fluorine atom, when introduced to a methyl group, is larger than the methyl group in spite of the explicitly inverse trend of their Es or Es' values. Taking Taft's Es values into consideration, the size of a difluoromethyl group is between that of isopropyl and isobutyl groups, and a CF_3 moiety is larger than them but not as large as a *tert*-Bu group. However, Dubois[44] suggested a smaller size for the fluorinated methyl moieties (see Es' values).

Alternatively, the steric bulkiness of a CF_3 moiety has been qualitatively determined from a comparison of the rotational barriers of appropriate aromatic substrates with this group (Fig. 1.9). For example, Virgili and his co-workers estimated the activation energy for the rotation about C^9-C^{11} of compounds **22** based on dynamic ^1H NMR data and pointed out the energetic similarity of **22c** to **22b**.[45] An analogous tendency was also observed for compounds **23**, whose ^1H NMR-derived activation energy difference $\Delta\Delta G^{\neq}$ between **23b** and **23c** was only 1 kcal/mol in favor of the latter,[46] and gas chromatographic measurement in the simpler biphenyl system also proved a 0.6 kcal/mol difference between **24a** and **24b**.[47] These results assume that the trifluoromethyl group is quite similar in size to the nonfluorinated isopropyl moiety and much larger than the methyl group in spite of the fact that fluorine has only a 20% larger vdW radius than hydrogen.

Table 1.2 Representative steric parameters[43,44]

Group	H	F	Cl	Br	I	OH	CH_3
Es	0.00	-0.46	-0.97	-1.16	-1.40	-0.55	-1.24
Es'	0.00	-0.55	-1.14	-1.34	-1.62		-1.12

Group	CH_2F	CHF_2	CF_3	Et	*i*-Pr	*i*-Bu	*tert*-Bu
Es	-1.48	-1.91	-2.40	-1.31	-1.71	-2.17	-2.78
Es'	-1.32	-1.47	-1.90	-1.20	-1.60	-2.05	-2.55

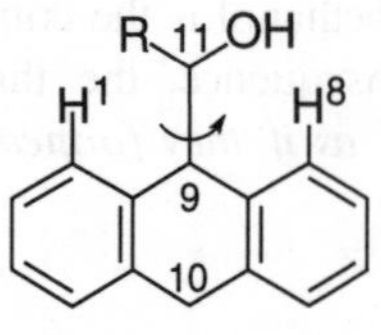

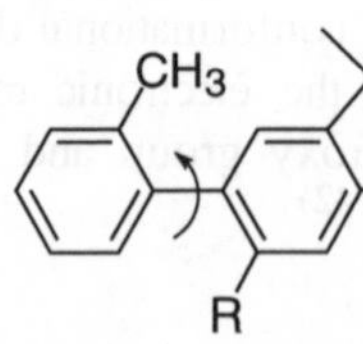

22a (R=Me), ΔG$^{\neq}$=11.0 kcal/mol   **23a** (R=Me), ΔG$^{\neq}$=80.9 kcal/mol
22b (R=i-Pr), ΔG$^{\neq}$=14.0 kcal/mol   **23b** (R=i-Pr), ΔG$^{\neq}$=93.0 kcal/mol
22c (R=CF$_3$), ΔG$^{\neq}$=14.5 kcal/mol   **23c** (R=CF$_3$), ΔG$^{\neq}$=92.0 kcal/mol

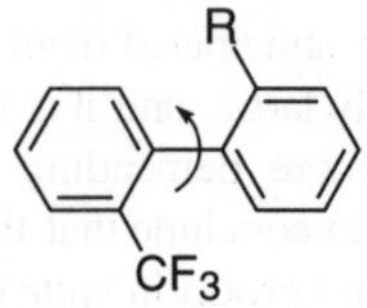

24a (R=i-Pr), ΔG$^{\neq}$=109.8 kcal/mol
24b (R=CF$_3$), ΔG$^{\neq}$=109.2 kcal/mol

Fig. 1.9   Various coalescence energies.

1.2 Effect of Fluorine on Chemical Reactivity

1.2.1 Stability of Cationic or Anionic Species Next to Fluorine or Fluoroalkyl Groups

The above discussion should facilitate the reader's understanding of the profound and interesting alteration of the original physical properties due to the introduction of fluorine(s). Next, we will briefly survey the effect of such qualitative changes on the original reactivity using various chemical species generated at the carbon atom next to a fluorine atom **25** or a trifluoromethyl group **26**.

In the case of a cation like **25-cat**, fluorine stabilizes such species by lone pair electron donation rather than the inductive effect working in the opposite way (Fig. 1.10). This discussion also helps one to understand that the anion **25-an** is definitely destabilized by the electrostatic repulsion.   Unlike the planar methyl radical,[48] the introduction of a fluorine atom to the carbon radical gives rise to partial pyramidalization, thus CF$_3$• becomes almost tetrahedral in shape[49] with a significant inversion barrier.[50] Because such deviation from planarity causes less effective resonance stabilization by lone pairs, **25-rad** is more stable than any other fluorine-containing methyl radical.   Moreover, based on the C-H bond

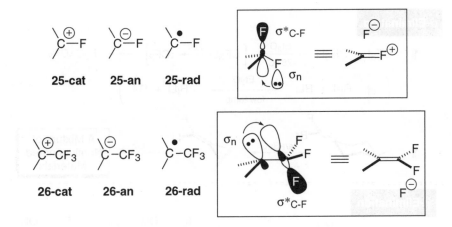

Fig. 1.10   Representative fluorinated reactive species.

dissociation energy as the traditional reference for radical stability, CH_2F-H and CHF_2-H dissociation energies are 3.6 and 1.6 kcal/mol less than that of CH_3-H respectively, but the CF_3-H bond is 1.9 kcal/mol stronger: thus, the stability order $CH_2F\bullet > CHF_2\bullet > CH_3\bullet > CF_3\bullet$ is clearly indicated.[51]

On the other hand, the opposite trend in stability between anion and cation was demonstrated by fluorinated alkyl groups (a CF_3 moiety is considered here as the representative example), where their strong electron-withdrawing nature plays a key role in the stabilization and destabilization[52] of **26-an** and **26-cat**, respectively.   Moreover, negative hyperconjugation (NHC)[53] also works as the dominant factor for the stabilization of this intermediate, which, similar to the already discussed *gauche* effect, is the interaction between the energetically lowered σ^*_{C-F} and the strongly electron-donating carbanion σ_n orbitals.   NHC between σ^*_{C-F} and fluorine lone pairs is also employed for the explanation of the consistent decrease in the observed C-F bond lengths in fluorine-containing methanes, $CH_{4-n}F_n$, which constantly decreases from 1.383 Å (n=1) to 1.320 Å (n=4) in spite of the almost same C-H distance in the range 1.093 to 1.100 Å.[54]   On the other hand, 0.048 Å elongation of the C-F bond *anti* to the carbanion in 2,2,2-trifluoroethyl anion was calculated relative to the C-F bond *gauche* to the anion, the result of the attractive overlap between carbanion σ_n orbitals and σ^*_{C-F}.[53]   β-Fluorinated (α-trifluorinated) radicals like **26-rad** are inductively destabilized, and the bond dissociation energies of CX_3CH_2-H simply increase as the number of fluorine increases: CH_2F, 1.9; CHF_2, 3.6; CF_3, 4.3 kcal/mol relative to the case of ethane (X=H).[51,55]

However, one must note the ability of the fluorine atom to act as the leaving group. As already pointed out, the relatively high stability of **26-an** would affect the smooth generation of such intermediates, but then fluorine is ready to depart from the original molecules in a β-elimination mode to furnish alkenes (**25-an** would produce carbenes by way of the α-elimination mechanism).   This seems to be especially the case for species possessing counter cationic metals with high affinity for this atom,[42,56] and sometimes almost spontaneous metal fluoride elimination is observed.

Figure 1.11 depicts selected examples of both α-[57] and β-elimination reactions. Although it has already been observed that the metal-halogen exchange occurs smoothly at

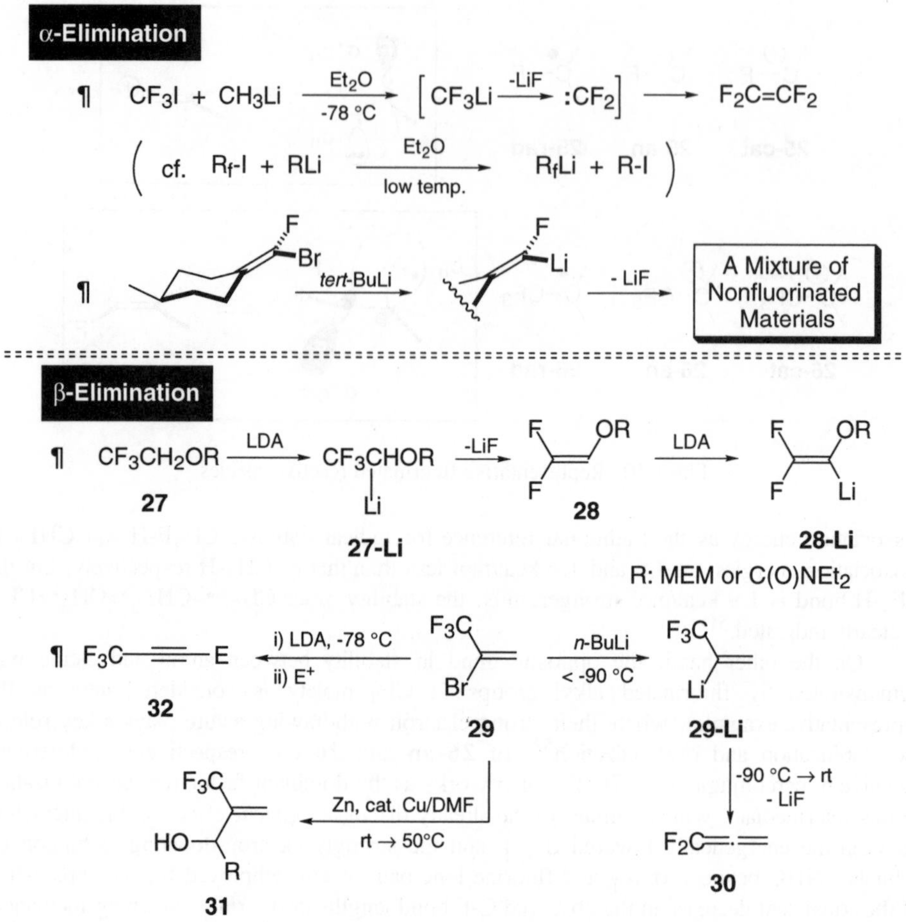

α-Elimination

¶ CF_3I + CH_3Li $\xrightarrow[-78\ °C]{Et_2O}$ $\left[CF_3Li \xrightarrow{-LiF} :CF_2 \right]$ ⟶ $F_2C=CF_2$

$\left(\text{cf.}\quad R_f\text{-I} + RLi \xrightarrow[\text{low temp.}]{Et_2O} R_fLi + R\text{-I} \right)$

¶ $\xrightarrow{\textit{tert}\text{-BuLi}}$ $\xrightarrow{-\ LiF}$

A Mixture of Nonfluorinated Materials

β-Elimination

¶ CF_3CH_2OR $\xrightarrow{LDA}$ CF_3CHOR $\xrightarrow{-LiF}$ $\xrightarrow{LDA}$

27   Li

                    27-Li   28   28-Li

R: MEM or C(O)NEt$_2$

¶ $F_3C\!\!\equiv\!\!E$ $\xleftarrow[\text{ii) } E^+]{\text{i) LDA, -78 °C}}$ $\xrightarrow[< -90\ °C]{n\text{-BuLi}}$

32   29   29-Li

$\xleftarrow[\text{rt} \to 50°C]{\text{Zn, cat. Cu/DMF}}$ │ $\xrightarrow[\substack{-90\ °C\ \to\ \text{rt} \\ -\ LiF}]{}$ $F_2C\!\!=\!\!\cdot\!\!=$

31   30

Fig. 1.11 Examples of fluoride ion elimination.

low temperature when methyllithium is reacted with perfluoroalkyl iodide,[58] trifluoromethyl iodide is the exception, and the decomposition *via* the α-elimination of LiF from the produced CF_3Li species would yield difluorocarbene, which further dimerizes to tetrafluoro-ethylene.[3f,59] This type of α-elimination also occurred for the sp^2-bound fluorine.[60]

As expected, deprotonation of hydroxy-protected 2,2,2-trifluoroethanol **27** is easily carried out by lithium diisopropylamide (LDA). However, the very facile fluoride ion departure proceeds from the resultant **27-Li** to give the terminally difluorinated enol ether **28**. This tendency would be, at least in part, supported by the experimental fact that the use of a limited amount of LDA led to the mixture formation of the starting 2,2,2-trifluoro-ethanol derivative **27** and **28-Li**.[61] Despite the presence of protective groups with chelation ability to lithium, it was necessary to keep the lithiated species formed, **28-Li**, at a temperature below -65 °C for safe survival.

On the other hand, the following is an instance in which a slight modification can give rise to significant change in reactivity. Thus, 2-bromo-3,3,3-trifluoropropene **29** was reported to follow lithium-halogen exchange by the action of *n*-BuLi and the resultant **29-Li**, which starts to decompose at temperatures above -90 °C, furnished difluoroallene **30** by warming to room temperature.[62] However, subjection of **29** to a solution of zinc and a catalytic amount of copper(I) chloride produced a highly efficient reaction with aldehydes to form adducts **31** in good yields even at 50 °C.[63] It is worth noting that the non-nucleophilic strong base, LDA, abstracted only the terminal proton from **29**, and the adduct **32** was obtained *via* β-elimination of bromide, followed by further lithium acetylide formation.[64]

Another special feature is found for materials with fluorine(s) at the sp²-hybridized carbon atom. In this case, participation of fluorine lone pairs plays an important role for the construction of the resonance structure as depicted in Fig. 1.12, while classically these lone pairs have been considered to be responsible for the strong electronic repulsive interaction, giving rise to the distinct localization of the π electron on the non-fluorinated carbon atom. Both effects eventually lead to the same conclusion that the fluorine-possessing sp² carbon is more positively charged than the other olefinic carbon atom and this tendency allows polyfluorinated alkenes to *smoothly react with nucleophiles* in quite sharp contrast to the preferential electrophilic attack occurring for the usual nonfluorinated alkenes.

For example, industrial raw material hexafluoropropene **34** was known to spontaneously form the intermediary adduct with appropriate secondary amines in such a smooth manner that bubbling of this gas into the amine solution does not give any unreacted gaseous material and thus, observation of exhaust from the outlet is a good indicator of the termination of the reaction. The intermediate obtained then follows the elimination of fluoride or capture of proton to furnish **35** as an *E*, *Z* mixture or **36**, respectively (Fig.

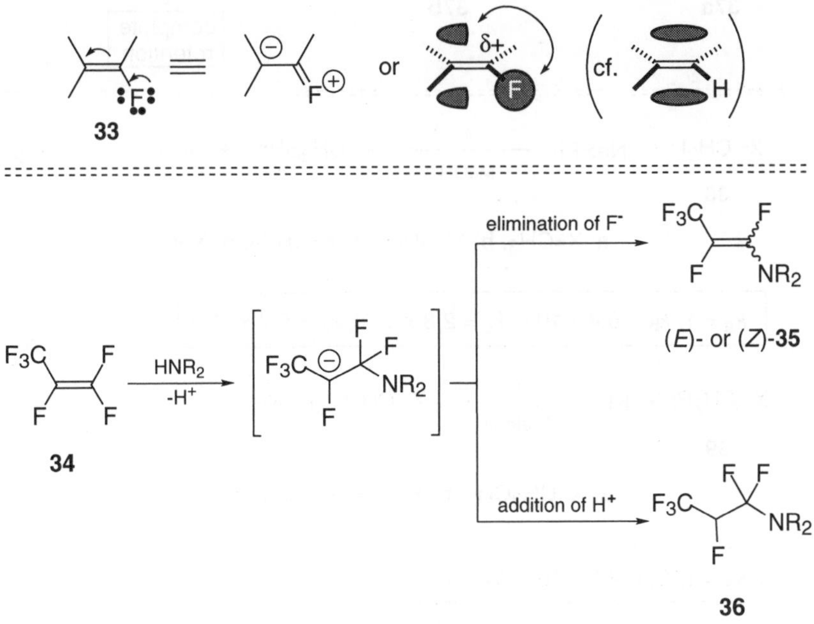

Fig. 1.12 Reaction of hexafluoropropene **34** with secondary amines.

1.12). The mixture of **35** and **36**, known as Ishikawa's reagent,[65] can be obtained from a commercial supplier and used as a promising fluorinating reagent, mainly for alcohols.

1.2.2 S_N1 and S_N2 Reactivity of Fluorinated Materials

This section discusses the qualitative analysis of the introduction effect of a CF_3 group (as the representative fluoroalkyl group) on the S_N1 or S_N2 type reactions.

The aforementioned instability of cationic species next to the trifluoromethyl group is typically exemplified by the S_N1 solvolysis of $PhC(CF_3)_2OTs$ *proceeding 10^{18} times slower than its corresponding nonfluorinated counterpart.*[66] Myhre *et al.* studied[67] the solvolysis of tosylate **37b** from 1,1,1-trifluoropropan-2-ol **37a** in a strongly acidic medium and found complete retention of the inherent stereochemistry even in 98% H_2SO_4; the mechanistic details were clarified by the use of the ^{18}O labeling technique as the result of the O-S bond cleavage instead of the usually anticipated O-C bond during the reaction (eq. 1 in Fig. 1.13).

Then, what about the S_N2 type reaction at the same site? In the literature,[68] incorporation of fluorine at the vicinity of the reaction center caused significant retardation. In spite of the expectation that strong electronic induction would contribute to enhance S_N2 reactivity by developing the more positive charge at the neighboring carbon atom, a CF_3

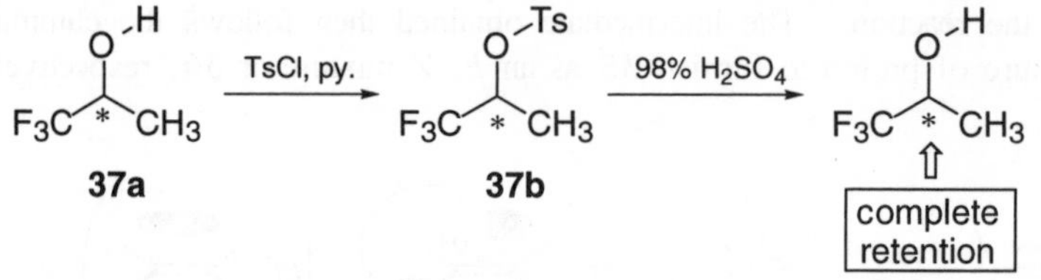

¶ $\quad$ (eq. 1)

$$\underset{\textbf{37a}}{\overset{\overset{\displaystyle H}{\underset{\displaystyle |}{O}}}{F_3C \overset{*}{\underset{}{\,}} CH_3}} \xrightarrow{\text{TsCl, py.}} \underset{\textbf{37b}}{\overset{\overset{\displaystyle Ts}{\underset{\displaystyle |}{O}}}{F_3C \overset{*}{\underset{}{\,}} CH_3}} \xrightarrow{\text{98\% } H_2SO_4} \underset{\uparrow}{\overset{\overset{\displaystyle H}{\underset{\displaystyle |}{O}}}{F_3C \overset{*}{\underset{}{\,}} CH_3}}$$

complete retention

===

¶ $\quad$ X–CH$_2$I + NaSPh $\xrightarrow[\text{MeOH}]{k}$ X–CH$_2$SPh + NaI $\qquad$ (eq. 2)

38

a: X=CH$_3$, b: X=CH$_2$F, c: X=CHF$_2$, d: X=CF$_3$

$k_a = 1$, $k_b = 6.4 \times 10^{-2}$, $k_c = 2.8 \times 10^{-3}$, $k_d = 5.7 \times 10^{-5}$

¶ $\quad$ X–CH$_2$Br + KI $\xrightarrow[\text{Acetone}]{k}$ X–CH$_2$I + KBr $\qquad$ (eq. 3)

39

a: X=CH$_3$CH$_2$, b: X=CF$_3$CH$_2$, c: X=CF$_3$

$k_a = 1$, $k_b = 2.0 \times 10^{-1}$, $k_c = 1.6 \times 10^{-4}$

Fig. 1.13 Effects of a CF_3 group on S_N1 or S_N2 reactions.

group was proved to play a strong role as the deactivating functionality and the nonfluorinated counterpart **38a** or **39a** was observed to react 1.8×10^4 or 6.3×10^3 times faster than the corresponding CF_3-containing **38d** or **39c**, respectively (eq. $2^{68a)}$ or $3^{68b)}$ in Fig. 1.13). Iodination of 2,2,2-trifluoroethyl tosylate proceeded at less than half the velocity of **39c**.[68c] Severe electronic as well as steric repulsion between fluoroalkyl groups and the incoming nucleophiles is considered to be an important factor for such reluctance. Quite recently, the behavior of 1-alkyl-2,2,2-trifluoroethyl mesylates or triflates was investigated in detail,[69] and it was found that the use of heteroatom nucleophiles like benzoic acid (oxygen), thiophenol (sulfur), or phthalimide (nitrogen) in the presence of KF or CsF furnished substitution products with virtually complete inversion of stereochemistry. This opened novel application routes of such chiral alcohols as **37a**, readily accessible both *via* enzymatic[70] and chemical[71] transformations, but a number of problems remain concerning the use of carbon nucleophiles.[72]

Brosylates **41b** and **40b** derived respectively from 4,4,4-trifluorobut-2-en-1-ol **41a** and 1,1,1-trifluorobut-3-en-2-ol **40a** are known[73] to be 113 and 1100 times as reactive as the corresponding saturated forms under the Finkelstein reaction conditions with KI in acetone. Moreover, the reaction was nine times more accelerated when **41e** was compared with the nonfluorinated crotyl chloride, which is consistent with the effective lowering of the LUMO energy level as discussed in section 1.1.2.2. On the other hand, the reaction rate difference between **41b** and **40b** was proved to be significant with the former reacting 7.8×10^4 times faster *via* the S_N2 mechanism in spite of the latter undergoing only the S_N2' process, both substrates leading to the formation of the same product **42** (Fig. 1.14). A similar tendency was reported by Kobayashi's group for the Pd(0)-catalyzed alkylation of **40** or **41** possessing the leaving group of acetate (**c**), phosphate (**d**), or carbonate with sodium malonate.[74]

On the other hand, we have recently demonstrated[75] the addition reactions of Grignard reagents to substrates such as **41c** catalyzed by copper(I) cyanide where TMS-Cl worked as the key component for the decrease of the transition state energy barrier by the coordination of chloride onto copper.[76] It is quite interesting to note that this reaction occurred in an S_N2' fashion irrespective of the substrates even when starting from **40**, and

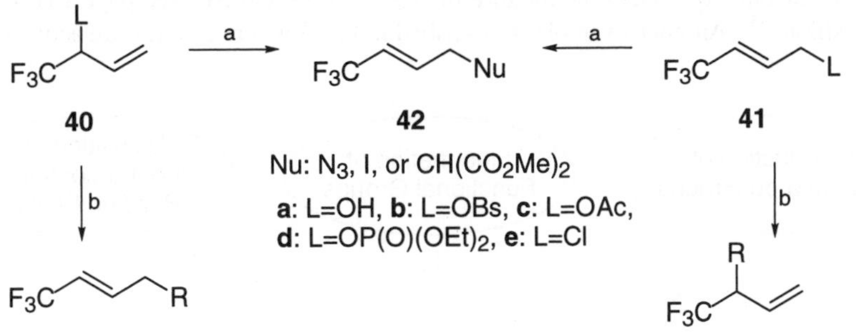

Fig. 1.14 Nucleophilic reaction of CF_3-containing allylic alcohols.

the stereochemical course was verified in detail by the chiral substrates as an almost complete $anti$-S_N2' mechanism.

1.3 Preparation of Fluorine-containing Materials

In this section, we will briefly look at methods for constructing the desired fluorine-containing compounds, which, as noted earlier, do not call for any special technique. Procedures not readily accessible are sometimes referred to only for comparison.

Roughly speaking, there are three methods for the synthetic attainment of the desired compounds, as shown in Fig. 1.15. The first pathway is to convert appropriate functional groups by the action of fluorinating reagents, and this sometimes serves as the easiest as well as the shortest preparative scheme as long as the requisite substrates are available. Direct incorporation of fluorine-containing groups is also possible, and the major part of the process is the reaction of perfluorinated halides with carbonyl compounds mediated by some specific metals, which is again recognized as a very straightforward preparation method. Versatility of various building blocks has been well documented in the literature,[1,2,3] and they are especially useful for the manipulation of complex molecules. The importance of the last route seems to be quite apparent for obtaining molecules with a CF_3 group, which would be elucidated at least in part as the result of the increasing difficulty of their construction by way of simpler fluorination techniques as the number of fluorine included in molecules increases. The following is a brief overview of these methods. However, the building block pathways are not considered here because of their great diversity.

For the preparation of monofluorinated substances, the metal fluoride (usually KF or CsF)-promoted reactions[77] or HF-related reagents[78] towards compounds with such common leaving groups as sulfonates, phosphates, halides, and so forth would be the most familiar route (eq. 1 in Fig. 1.16). Diethylaminosulfur trifluoride (DAST),[79,80] one of the most typical fluorinating agent converting alcohols into the corresponding fluorides, is also a convenient reagent (eq. 2) and the mixture of (E)-, (Z)-**35** and **36** (see page 13) works in a similar fashion.[65] Another example is described in eq. 3 where alkenes are activated by the

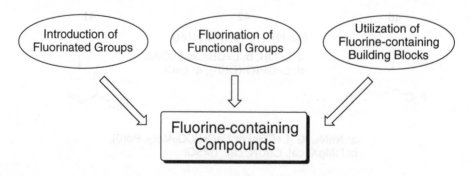

Fig. 1.15 Schemes to access the target molecules.

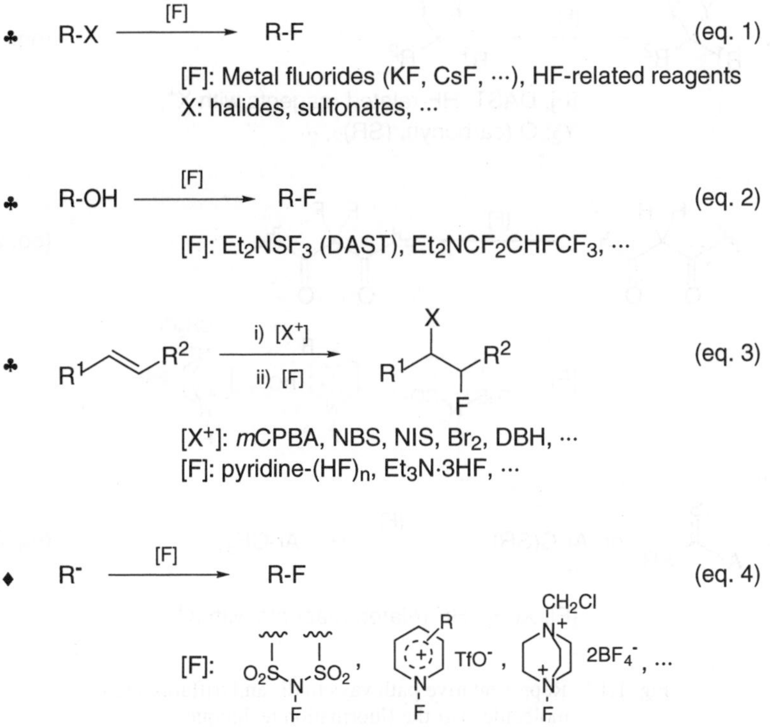

♣ R-X $\xrightarrow{\text{[F]}}$ R-F (eq. 1)

[F]: Metal fluorides (KF, CsF, ⋯), HF-related reagents
X: halides, sulfonates, ⋯

♣ R-OH $\xrightarrow{\text{[F]}}$ R-F (eq. 2)

[F]: Et_2NSF_3 (DAST), $Et_2NCF_2CHFCF_3$, ⋯

♣ $R^1\diagup\!\!\diagdown R^2$ $\xrightarrow[\text{ii) [F]}]{\text{i) [X$^+$]}}$ (eq. 3)

[X$^+$]: mCPBA, NBS, NIS, Br_2, DBH, ⋯
[F]: pyridine-$(HF)_n$, $Et_3N\cdot3HF$, ⋯

◆ R$^-$ $\xrightarrow{\text{[F]}}$ R-F (eq. 4)

[F]:

Fig. 1.16 Representative pathways to monofluorinated molecules
via the fluorination technique.

action of the halonium ions[81] or peroxides[82] (via epoxides), followed by the reaction with the fluoride source to furnish the corresponding halogenated or hydroxylated alkyl fluorides, respectively. Pyridinium poly(hydrogen fluoride), sometimes abbreviated as pyridine-$(HF)_n$ or PPHF, is known to be a very versatile reagent, and can be broadly applied as the fluorinating reagent in eqs. 1 to 3.[78] The commercially available stable solid material, xenon difluoride, is also employed for various types of fluorination with ease.[83] Usually, these anionic substitution reactions proceed via inversion of stereochemistry,[84] but participation of proximate functionalities is sometimes observed.[85]

These are typical examples of nucleophilic fluorination, but recently extensive investigation has been conducted for the development of electrophilic fluorinating reagents, generating a formal "F$^+$" species. Their general structure includes fluorine on a highly electron-deficient nitrogen atom, realizing sufficient stability for handling and facile fluorine incorporation into various enolates or organometallic reagents (eq. 4).[86] A similar type of reaction with enolates has traditionally been carried out by F_2 gas[87] or F_2-based reagents like CF_3OF,[88] or by an electrochemical pathway under the action of $Et_3N\cdot3HF$.[89] Optically active fluorinating reagents have been prepared from SF_4, affording the modified DAST structure but generally the enantiomeric excess values obtained are not high.[90]

Figure 1.17 summarizes the preparation methods of difluoro- and trifluoromethyl moieties via the fluorination technique. As discussed above, DAST[79a,91] is recognized as

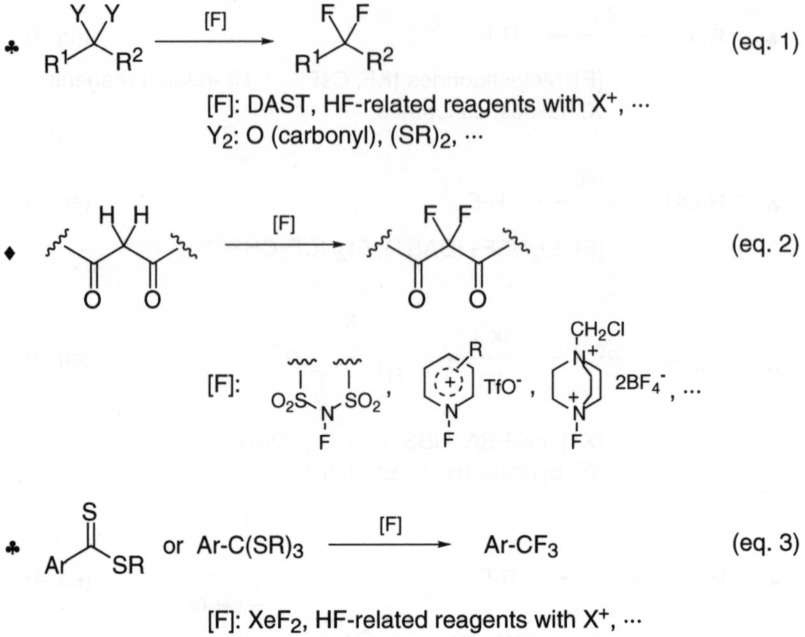

                                                                                    (eq. 1)

[F]: DAST, HF-related reagents with X⁺, ···
Y₂: O (carbonyl), (SR)₂, ···

                                                                                    (eq. 2)

[F]:

or Ar-C(SR)₃ $\xrightarrow{[F]}$ Ar-CF₃   (eq. 3)

[F]: XeF₂, HF-related reagents with X⁺, ···

Fig. 1.17 Representative pathways to di- and trifluorinated
molecules *via* the fluorination technique.

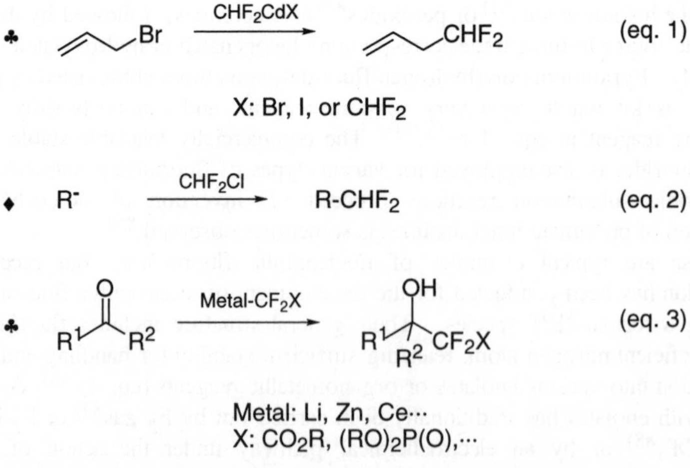

$\xrightarrow{CHF_2CdX}$ (eq. 1)

X: Br, I, or CHF₂

R⁻ $\xrightarrow{CHF_2Cl}$ R-CHF₂   (eq. 2)

$\xrightarrow{Metal-CF_2X}$ (eq. 3)

Metal: Li, Zn, Ce···
X: CO₂R, (RO)₂P(O),···

Fig. 1.18 Representative pathways to difluorinated molecules
via direct introduction pathways.

one of the most promising reagents for the conversion of carbonyl groups into difluoromethylene moieties, and tetra-*n*-butylammonium dihydrogen trifluoride ($n\text{-}Bu_4NH_2F_3$) in the presence of halonium ions (Br^+ or I^+ generated from *N*-bromo- or iodosuccinimide, respectively) also enables the difluorination of the corresponding dithioacetals at -78 °C in high yields.[92)] Moreover, smooth transformation of active methylene groups to the 2,2-difluoro-1,3-dicarbonyl structure was reported by utilization of *N*-fluorosulfonimide reagents.[86d,93)] On the other hand, the construction of a CF_3 group *via* fluorination methods, usually carried out by the very strong but toxic (*comparable to the toxicity of phosgene!*) fluorinating reagent SF_4,[94)] is quite difficult by conventional laboratory techniques, and eq. 3 in Fig. 1.17 would be the only feasible route at present.[95)]

Various types of direct incorporation methods of fluorine-containing groups are summarized in Fig. 1.18 (difluoro-) and 1.19 (trifluoromethyl), where the above architectural disadvantage by fluorination has been overcome, particularly in the case of CF_3 materials. CHF_2CdX in eq. 1 was readily prepared from bromo- or iododifluoromethane and cadmium powder at room temperature, smoothly forming the adducts with allylic halides preferentially at the less hindered site (Fig. 1.18).[96)] The second scheme is on the utilization of carbenoid derived from chlorodifluoromethane with the reaction of an anionic species (eq. 2).[57,97)] Eq. 3 describes the smooth introduction of difluorinated functionalities such as

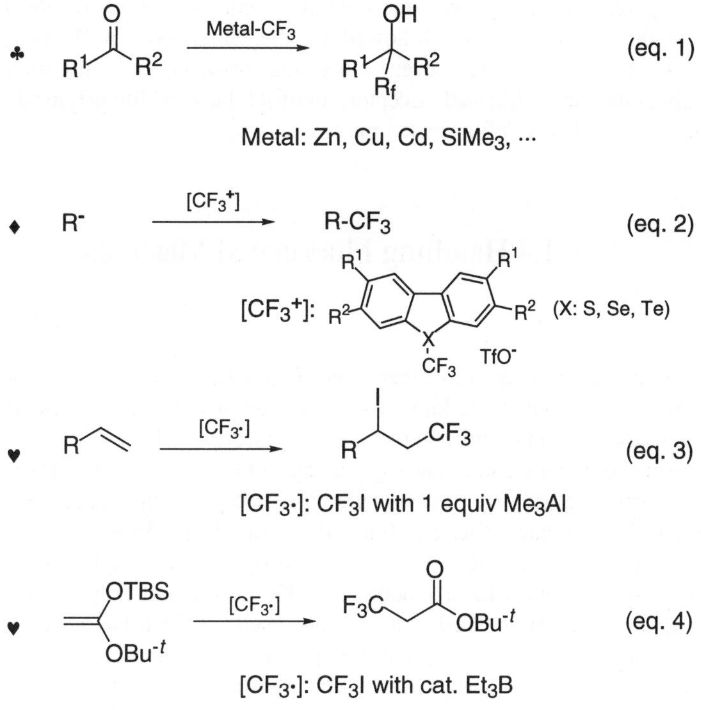
Metal-CF3
(eq. 1)
Metal: Zn, Cu, Cd, SiMe3, ···
[CF3+]
R-CF3
(eq. 2)
[CF3+]:
(X: S, Se, Te)
TfO-
[CF3·]
(eq. 3)
[CF3·]: CF3I with 1 equiv Me3Al
OTBS
OBu-t
[CF3·]
(eq. 4)
[CF3·]: CF3I with cat. Et3B

Fig. 1.19 Representative pathways to trifluomethylated molecules *via* direct introduction pathways.

esters[98] or phosphates,[99] and the latter was also reported to undergo alkylation,[100] acylation,[101] and Michael reactions.[102] Recently, it was disclosed that 1,1-difluoroallyl bromide reacted with aldehydes in the presence of indium metal in an α-regioselective manner.[103]

The instability of CF_3-metal species has already been pointed out despite the versatility of nonfluorinated organometallic reagents like R-Li or R-MgX. However, zinc,[104] copper,[104d,105] cadmium,[104d,106] and trimethylsilyl[107,108] are representative exceptions and the desired trifluoromethylated materials have been obtained in good yields *via* these species (eq. 1 in Fig. 1.19).[109]

Umemoto's group studied the development of reagents for the introduction of per- or polyfluoroalkyl moieties[110] as their formal cationic forms and found effective electrophilic trifluoromethylating agents whose potency can be controlled by changing the chalcogen atoms (sulfur, selenium, or tellurium) and/or the substituents on the aromatic rings (eq. 2).[111] These reagents can induce chirality by the addition of optically active borepin, and moderate enantiomeric excess values were recorded (45% ee).[112] Recently, some reports have been published on the incorporation of a trifluoromethyl group *via* the radical species generated by an equimolar amount of trimethylaluminum[113] or a catalytic amount of triethylborane (eq. 3 and 4).[114] It is because of the radical type reaction mode employed that the product 3,3,3-trifluoropropionate safely survives in the last instance, while under basic conditions, very facile decomposition would occur by the abstraction of an acidic hydrogen α to the carbonyl group, followed by the elimination of fluoride, as already discussed in section 1.2.1. However, since the resultant 3,3-difluorinated acrylate possesses high ability as a Michael acceptor, twofold base addition-fluoride elimination sequences furnish completely defluorinated materials.

1.4 Handling Fluorinated Materials

One must be extremely cautious when using fluoroacetic acid or its derivatives such as the corresponding esters, which are known to cause convulsions and ventricular fibrillation as described in the section on mimic and block effects (section 1.1.1).

The fluorine-containing compounds appearing in the following chapters can be placed and reacted in normal glassware unless otherwise noted. Often-employed fluorinating reagents like DAST or Ishikawa's reagent (the mixture of **35** and **36** shown in page 13) may be handled in a similar manner, but it is appropriate to employ polyethylene or polypropylene vessels when a large amount of 70% pyridinium poly(hydrogen fluoride) (pyridine-$(HF)_n$) or other HF-related materials are used. DAST has an inherent thermal stability problem and was previously reported to possibly decompose in a violent manner at temperatures above 90 °C.[115] In the case of 70% pyridine-$(HF)_n$, it is known to be stable up to 50 °C,[78] but because it easily eliminates highly toxic and corrosive hydrogen fluoride gas and is highly hygroscopic leading to the drastic reduction of fluoride nucleophilicity, all the handling procedures should be done quickly to minimize the contact of the reagent with moisture.

When one uses these toxic or reactive materials, all operations should be conducted under a well-ventilated hood and rubber gloves should be worn. Addition of these materials should be carried out using a syringe. Strong acidity of carboxylic or sulfonic acids with fluorines also requires special care.

If fluorinated substances are kept for a long time in glassware such as flasks or NMR sample tubes, the glass surface can be damaged by *in situ* generated fluoride ions by partial (or total) decomposition of the contained materials.

For compounds possessing only a small number of fluorines, the corresponding nonfluorinated prototype offers approximate boiling point information, while introduction of this atom weakens the intermolecular interaction to lower the boiling point and increase volatility.

1.5 References and Notes

1.   Banks, R. E.; Sharp, D. W. A.; Tatlow, J. C. *Fluorine: The First Hundred Years (1886-1986)*; Elsevier: New York, 1986.

2.   As representative monographs, see a) Liebman, J. F.; Greenberg, A.; Dolbier, Jr., W. R. *Fluorine-containing Molecules — Structure, Reactivity, Synthesis, and Applications*; VCH: New York, 1988.   b) Welch, J. T. *Selective Fluorination in Organic and Bioorganic Chemistry*; ACS: New York, 1991.   c) Welch, J. T.; Eswarakrishnan, S. *Fluorine in Bioorganic Chemistry*; John Wiley & Sons: New York, 1991.   d) Olah, G. A.; Chambers, R. D.; Surya Prakash, G. K. *Synthetic Fluorine Chemistry*; John Wiley & Sons: New York, 1992.   e) Banks, R. E.; Smart, B. E.; Tatlow, J. C. *Organofluorine Chemistry   Principles and Commercial Applications*; Plenum Press: New York, 1994.   f) Hudlicky, M.; Pavlath, A. E. *Chemistry of Organic Fluorine Compounds II: A Critical Review*; ACS: New York, 1995.   g) Kukhar', V. P.; Soloshonok, V. A. *Fluorine-containing Amino Acids — Synthesis and Properties —*; John Wiley & Sons: New York, 1995.   h) Chambers, R. D. *Organofluorine Chemistry — Techniques and Synthons —* (*Topics in Current Chemistry, v. 193*); Springer: New York, 1997.

3.   As representative reviews, see a) Welch, J. T. *Tetrahedron* **1987**, *43*, 3123.   b) Bravo, F.; Resnati, G. *Tetrahedron: Asym.* **1990**, *1*, 661.   c) Tsuchiya, T. *Adv. Carbohydr. Chem. Biochem.* **1990**, *48*, 91.   d) Bégué, J.-P.; Bonnet-Delpon, D. *Tetrahedron* **1991**, *47*, 3207.   e) McClinton, M. A.; McClinton, D. A. *Tetrahedron* **1992**, *48*, 6555.   f) Burton, D. J.; Yang, Z.-Y. *Tetrahedron* **1992**, *48*, 189.   g) Resnati, G. *Tetrahedron* **1993**, *49*, 9385.   h) Harper, D. B.; O'Hagan, D. *Nat. Prod. Rep.* **1994**, 123.   i) Burton, D. J.; Yang, Z.-Y.; Morken, P. A. *Tetrahedron* **1994**, *50*, 2993.

4.   See also the following special issues: a) *Enantiocontrolled Synthesis of Fluoro-Organic Compounds* (Hayashi, T.; Soloshonok, V. A.; Eds), *Tetrahedron:Asym.* **1994**, *5*, 955.   b) *Fluoroorganic Chemistry: Synthetic Challenges and Biomedicinal Rewards* (Resnati, G.; Soloshonok, V. A.; Eds), *Tetrahedron* **1996**, *52*, 1.   c) *Fluorine Chemistry* (Smart, B. E.; Ed), *Chem. Rev.* **1996**, *96*, 1555.

5.   Bondi, A. *J. Phys. Chem.* **1964**, *68*, 441.

6.   Pauling, L. *The Nature of the Chemical Bond*; Cornell University Press: New York, 1960, pp. 82.

7.   For the biological production of fluoroacetic acid, see a) Sanada, M.; Miyano, T.; Iwadare, S.; Williamson, J. M.; Arison, B. H.; Smith, J. L.; Douglas, A. W.; Liesch, J. M.; Inamine, E. *J. Antibiot.* **1986**, *39*, 259. b) Tamura, T.; Wada, M.; Esaki, N.; Soda, K. *J. Bacteriol.* **1995**, *177*, 2265. c) Hamilton, J. T. G.; Amin, M. R.; Harper, D. B.; O'Hagan, D. *Chem. Commun.* **1997**, 797. d) Nieschalk, J.; Hamilton, J. T. G.; Murphy, C. D.; Harper, D. B.; O'Hagan, D. *Chem. Commun.* **1997**, 799.

8.   Peters, R. A.; Wakelin, R. W.; Buffa, P.; Thomas, L. C. *Proc. Roy. Soc., B.* **1953**, *140*, 497.

9.   Yamazaki, T.; Haga, J.; Kitazume, T. *Bioorg. Med. Chem. Lett.* **1991**, *1*, 271.

10. Santi, D. V.; Sakai, T. T. *Biochemistry* **1971**, *10*, 3598.

11. Kaufman, S. *Biochim. Biophys. Acta* **1961**, *51*, 619.

12. Premchandran, R. H.; Ogletree, M. L.; Fried, J. *J. Org. Chem.* **1993**, *58*, 5724 and references cited therein.

13. The same strategy was employed for increasing stability of PGI$_2$: a) Bannai, K.; Toru, T.; Oba, T.; Tanaka, T.; Okamura, N.; Watanabe, K.; Hazato, A.; Kurozumi, S. *Tetrahedron* **1986**, *42*, 6735. b) Matsumura, Y.; Shimada, T.; Wang, S.-Z.; Asai, T.; Morizawa, Y.; Yasuda, A. *Bull. Chem. Soc. Jpn.* **1996**, *69*, 3523. c) Nakano, T.; Makino, M.; Morizawa, Y.; Matsumura, Y. *Angew. Chem. Int. Ed. Engl.* **1996**, *35*, 1019.

14. a) Kunimoto, S.; Komuro, K.; Nosaka, C.; Tsuchiya, T.; Fukatsu, S.; Takeuchi, T. *J. Antibiot.* **1990**, *43*, 556. b) Tsuchiya, T.; Takagi, Y. In *Anthracycline Antibiotics: New Analogues, Methods of Delivery, and Mechanisms of Action*; Priebe, W., Ed.; ACS: Washington, D. C., 1995; Chapter 6, p 100. c) Takagi, Y.; Nakai, K.; Tsuchiya, T.; Takeuchi, T. *J. Med. Chem.* **1996**, *39*, 1582.

15. The fluorine modification of aglycones has been reported. See a) Matsuda, F.; Matsumoto, T.; Ohsaki, M.; Terashima, S. *Tetrahedron Lett.* **1989**, *30*, 4259. b) Guidi, A.; Canfarini, F.; Giolitti, A.; Pasqui, F.; Pestellini, V.; Arcamone, F. *Pure Appl. Chem.* **1994**, *66*, 2319.

16. See the following as representative articles on the construction of 2-fluorinated sugars: a) Faghih, R.; Escribano, F. C.; Castillon, S.; Garcia, J.; Lukacs, G.; Olesker, A.; Thang, T.-T. *J. Org. Chem.* **1986**, *51*, 4558. b) Dessinges, A.; Escribano, F. C.; Lukacs, G.; Olesker, A.; Thang, T.-T. *J. Org. Chem.* **1987**, *52*, 1633. c) Takagi, Y.; Lim, G.-J.; Tsuchiya, T.; Umezawa, S. *J. Chem. Soc., Chem. Commun.* **1992**, 657.

17. a) Blackburn, G. M.; England, D. A.; Kolkman, F. *J. Chem. Soc., Chem. Commun.* **1981**, 1169. b) Blackburn, G. M.; Perrée, T. D.; Rashid, A. *Chem. Scr.* **1986**, *26*, 21. c) Halazy, S.; Ehrhard, A.; Danzin, C. *J. Am. Chem. Soc.* **1991**, *113*, 315. d) Smyth, M. S.; Ford, H.; Burke, Jr., T. R. *Tetrahedron Lett.* **1992**, *33*, 4137.

18. a) Adams, P. R.; Harrison, R. *Biochem. J.* **1974**, *141*, 729. b) Blackburn, G. M.; Kent, D. E. *J. Chem. Soc., Perkin Trans. 1* **1986**, 913. c) Chambers, R. D.; Jaouhari, R.; O'Hagan, D. *Tetrahedron* **1989**, *45*, 5101.

19. Blackburn, G. M.; Langston, S. P. *Tetrahedron Lett.* **1991**, *32*, 6425. See also the following: Matulic-Adamic, J.; Usman, N. *Tetrahedron Lett.* **1994**, *35*, 3227.

20. a) Wrobel, J.; Dietrich, A. *Tetrahedron Lett.* **1993**, *34*, 3543. b) Gordeev, M. F.; Patel, D. V.; Barker, P. L.; Gordon, E. M. *Tetrahedron Lett.* **1994**, *35*, 7585. c) Smyth, M.

S.; Burke, Jr., T. R. *Tetrahedron Lett.* **1994**, *35*, 551.

21. See the following works on the different targets based on the same concept: a) Phillion, D. P.; Cleary, D. G. *J. Org. Chem.* **1992**, *57*, 2763. b) Martin, S. F.; Dean, D. W.; Wagman, A. S. *Tetrahedron Lett.* **1992**, *33*, 1839.

22. For the recent articles on the construction of difluoromethylene phosphonate structures, see a) Nieschalk, J.; O' Hagan, D. *J. Chem. Soc., Chem. Commun.* **1995**, 719. b) Lequeux, T. P.; Percy, J. M. *Synlett* **1995**, 361. c) Berkowitz, D. B.; Eggen, M.; Shen, Q.-R.; Shoemaker, R. K. *J. Org. Chem.* **1996**, *61*, 4666. d) Blades, K.; Lequeux, T. P.; Percy, J. M. *Tetrahedron* **1997**, *53*, 10623. e) Yokomatsu, T.; Murano, T.; Suemune, K.; Shibuya, S. *Tetrahedron* **1997**, *53*, 815.

23. Linderman, R. J.; Jamois, E. A. *J. Fluorine Chem.* **1991**, *53*, 79.

24. Braid, M.; Isersone, H.; Lawlore, F. *J. Am. Chem. Soc.* **1954**, *76*, 4027.

25. This has been proved crystallographically. See the followings: a) Takahashi, L. H.; Rosenfield, R. E.; Meyer, Jr., E. F.; Trainor, D. A.; Stein, M. *J. Mol. Biol.* **1988**, *201*, 423. b) Takahashi, L. H.; Radhakrishnan, R.; Rosenfield, R. E.; Meyer, Jr., E. F.; Trainor, D. A. *J. Am. Chem. Soc.* **1989**, *111*, 3368. c) Brady, K.; Wei, A.-Z.; Ringe, D.; Abeles, R. H. *Biochemistry* **1990**, *29*, 7600. d) Veale, C. A.; Bernstein, P. R.; Bryant, C. B.; Ceccarelli, C.; Damewood, Jr., J. R.; Earley, R. A.; Feeney, S. W.; Gomes, B.; Kosmider, B. J.; Steelman, G. B.; Thomas, R. M.; Vacek, E. P.; Williams, J. C.; Wolanin, D. J.; Woolson, S. A. *J. Med. Chem.* **1995**, *38*, 98. e) Harel, M.; Quinn, D. M.; Nair, H. K.; Silman, I.; Sussman, J. L. *J. Am. Chem. Soc.* **1996**, *118*, 2340. f) Silva, A. M.; Cachau, R. E.; Sham, H. L.; Erickson, J. W. *J. Mol. Biol.* **1996**, *255*, 321.

26. Skiles, J. W.; Fuchs, V.; Miao, C.; Sorcek, R.; Grozinger, K. G.; Mauldin, S. C.; Vitous, J.; Mui, P. W.; Jacober, S.; Chow, G.; Matteo, M.; Skoog, M.; Weldon, S. M.; Possanza, G.; Keirns, J.; Letts, G.; Rosenthal, A. S. *J. Med. Chem.* **1992**, *35*, 641.

27. Skiles, J. W.; Miao, C.; Sorcek, R.; Jacober, S.; Mui, P. W.; Chow, G.; Weldon, S. M.; Possanza, G.; Skoog, M.; Keirns, J.; Letts, G.; Rosenthal, A. S. *J. Med. Chem.* **1992**, *35*, 4795.

28. See the following for similar types of inhibitors by different groups: a) Brown, F. J.; Andisik, D. W.; Bernstein, P. R.; Bryant, C. B.; Ceccarelli, C.; Damewood, Jr., J. R.; Edwards, P. D.; Earley, R. A.; Feeney, S. W.; Green, R. C.; Gomes, B.; Kosmider, B. J.; Krell, R. D.; Shaw, A.; Steelman, G. B.; Thomas, R. M.; Vacek, E. P.; Veale, C. A.; Tuthill, P. A.; Warner, P.; Williams, J. C.; Wolanin, D. J.; Woolson, S. A. *J. Med. Chem.* **1994**, *37*, 1259. b) Veale, C. A.; Damewood, Jr., J. R.; Steelman, G. B.; Bryant, C. B.; Gomes, B.; Williams, J. C. *J. Med. Chem.* **1995**, *38*, 86.

29. a) Angelastro, M. R.; Baugh, L. E.; Bey, P.; Burkhart, J. P.; Chen, T.-M.; Durham, S. L.; Hare, C. M.; Huber, E. W.; Janusz, M. J.; Koehl, J. R.; Marquart, A. L.; Mehdi, S.; Peet, N. P. *J. Med. Chem.* **1994**, *37*, 4538. b) Burkhart, J. P.; Koehl, J. R.; Mehdi, S.; Durham, S. L.; Janusz, M. J.; Huber, E. W.; Angelastro, M. R.; Sunder, S.; Metz, W. A.; Shum, P. W.; Chen, T.-M.; Bey, P.; Cregge, R. J.; Peet, N. P. *J. Med. Chem.* **1995**, *38*, 223.

30. a) Juaristi, E.; Cuevas, G. *Tetrahedron* **1992**, *48*, 5019. b) Graczyk, P. P.; Mikolajczyk, M. *Top. Stereochem.* **1994**, *21*, 159.

31. Klaboe, P.; Nielsen, J. R. *J. Chem. Phys.* **1960**, *33*, 1764.

32. Hirano, T.; Nonoyama, S.; Miyajima, T.; Kurita, Y.; Kawamura, T.; Sato, H. *J. Chem.*

Soc., Chem. Commun. **1986**, 606.

33.   a) Friesen, D.; Hedberg, K. *J. Am. Chem. Soc.* **1980**, *102*, 3987.   b) Fernholt, L.;
         Kveseth, K. *Acta Chem. Scand. A* **1980**, *34*, 163.

34.   a) Dixon, D. A.; Smart, B. E. *J. Phys. Chem.* **1988**, *92*, 2729.   b) Dixon, D. A.;
         Matsuzawa, N.; Walker, S. C. *J. Phys. Chem.* **1992**, *96*, 10740.   c) Wiberg, K. B. *Acc.
         Chem. Res.* **1996**, *29*, 229.   d) Berry, R. J.; Ehlers, C. J.; Burgess, Jr., D. R.; Zachariah,
         M. R.; Nyden, M. R.; Schwartz, M. *J. Mol. Struct.* **1998**, *422*, 89.

35.   a) Wolfe, S. *Acc. Chem. Res.* **1972**, *5*, 102.   b) Juaristi, E. *J. Chem. Educ.* **1979**, *56*, 438.

36.   a) Chitale, S.; Jose, C. I. *J. Chem. Soc., Faraday Trans. 1* **1986**, *82*, 663 (IR).   b) Huang,
         J.-F.; Hedberg, K. *J. Am. Chem. Soc.* **1989**, *111*, 6909 (Microwave analysis).

37.   a) Anhede, B.; Bergman, N.-Å.; Kresge, A. J. *Can. J. Chem.* **1986**, *64*, 1173.   b) Dixon,
         D. A.; Smart, B. E. *J. Phys. Chem.* **1991**, *95*, 1609.   c) Buemi, G. *J. Chem. Soc. Faraday
         Trans.* **1994**, *90*, 1211.   d) Bakke, J. M.; Bjerkeseth, L. H.; Rønnow, T. E. C. L.;
         Steinmsvoll, K. *J. Mol. Struct.* **1994**, *321*, 205.

38.   Marstokk, K.-M.; Møllendal, H. *Acta Chem. Scand. A* **1980**, *34*, 765.

39.   a) George, P.; Bock, C. W.; Trachtman, M. *J. Mol. Struct. (Theochem)* **1987**, *152*, 35.
         b) Fernández, B.; Mosquera, R. A.; Ríos, M. A.; Vázquez, S. A. *Tetrahedron Comp.
         Method.* **1989**, *2*, 85.   c) Howard, J. A. K.; Hoy, V. J.; O'Hagan, D.; Smith, G. T.
         Tetrahedron **1996**, *52*, 12613.   d) Kovács, A.; Hargittai, I. *J. Phys. Chem. A* **1998**, *102*,
         3415.

40.   However, there have been some reports on crystallographically observed OH···F
         hydrogen bonding examples.   See the followings: a) Withers, S. G.; Street, I. P.; Rettig,
         S. J. *Can. J. Chem.* **1986**, *64*, 232.   b) Shimoni, L.; Carrell, H. L.; Glusker, J. P.;
         Coombs, M. M. *J. Am. Chem. Soc.* **1994**, *116*, 8162.   c) Bravo, P.; Capelli, S.; Meille, S.
         V.; Kukhar, V. P.; Soloshonok, V. A. *Tetrahedron: Asym.* **1994**, *5*, 2009.   d) Zapata,
         A.; Bernet, B.; Vasella, A. *Helv. Chim. Acta* **1996**, *79*, 1169.   e) Blake, A. J.; Hill, S. J.;
         Hubberstey, P.; Li, W.-S. *J. Chem. Soc., Dalton Trans.* **1997**, 913.   f) Barlow, S. J.; Hill,
         S. J.; Hocking, J. E.; Hubberstey, P.; Li, W.-S. *J. Chem. Soc., Dalton Trans.* **1997**, 4701.
         g) Wiechert, D.; Mootz, D.; Dahlems, T. *J. Am. Chem. Soc.* **1997**, *119*, 12665.   h)
         Ashton, P. R.; Fyfe, M. C. T.; Glink, P. T.; Menzer, S.; Stoddart, J. F.; White, A. J. P.;
         Williams, D. J. *J. Am. Chem. Soc.* **1997**, *119*, 12514.   i) Pham, M.; Gdaniec, M.;
         Polonski, T. *J. Org. Chem.* **1998**, *63*, 3731.

41.   Warshel, A.; Papazyan, A.; Kollman, P. A. *Science* **1995**, *269*, 102.

42.   Yamazaki, T.; Kitazume, T. *J. Synth. Org. Chem. Jpn.* **1996**, *54*, 665.

43.   Taft, Jr., R. W. *Steric Effects in Organic Chemistry*; Newman, M. S., Ed.: John Wiley &
         Sons: New York, 1956, pp. 556.

44.   MacPhee, J. A.; Panaye, A.; Dubois, J.-E. *Tetrahedron* **1978**, *34*, 3553.

45.   de Riggi, I.; Virgili, A.; de Moragas, M.; Jaime, C. *J. Org. Chem.* **1995**, *60*, 27.

46.   Bott, G.; Field, L. D.; Sternhell, S. *J. Am. Chem. Soc.* **1980**, *102*, 5618.

47.   Wolf, C.; König, W. A.; Roussel, C. *Liebigs Ann.* **1995**, 781.   See also: Weseloh, G.;
         Wolf, C.; König, W. A. *Chirality* **1996**, *8*, 441.

48.   Lloyd, R. V. In *Chemical Kinetics of Small Organic Radicals*; Alfassi, Z. B., Ed.; CRC:
         Boca Raton, FL, 1988; Vol. 1, pp 1-24.

49.   See the following representative example: Deardon, D. V.; Hudgens, J. W.; Johnson, R.
         D., III; Tsai, B. P.; Kafafi, S. A. *J. Phys. Chem.* **1992**, *96*, 585.

50.   Griller, D.; Ingold, K. U.; Krusic, P. J.; Smart, B. E.; Wonchoba, E. R. *J. Phys. Chem.* **1982**, *86*, 1376.

51.   Dolbier, Jr., W. R. *Chem. Rev.* **1996**, *96*, 1557 and references cited therein.

52.   Allen, A. D.; Tidwell, T. T. In *Advances in Carbocation Chemistry*; Creary, X., Ed.: JAI Press: Greenwich, 1989; Vol. 1, pp. 1.

53.   Dixon, D. A.; Fukunaga, T.; Smart, B. E. *J. Am. Chem. Soc.* **1986**, *108*, 4027.

54.   Wiberg, K. B.; Rablen, P. R. *J. Am. Chem. Soc.* **1993**, *115*, 614.

55.   Martell, J. M.; Boyd, R. J.; Shi, Z. *J. Phys. Chem.* **1993**, *97*, 7208.

56.   For the interaction of fluorine with metals, see the following reviews: a) Plenio, H. *Chem. Rev.* **1997**, *97*, 3363. b) Murphy, E. F.; Murugavel, R.; Roesky, H. W. *Chem. Rev.* **1997**, *97*, 3425. c) Dorn, H.; Murphy, E. F.; Shah, S. A. A.; Roesky, H. W. *J. Fluorine Chem.* **1997**, *86*, 121.

57.   See the following review: Brahms, D. L. S.; Dailey, W. P. *Chem. Rev.* **1996**, *96*, 1585.

58.   a) Pierce, O. R.; McBee, E. T.; Judd, G. F. *J. Am. Chem. Soc.* **1954**, *76*, 474. b) McBee, E. T.; Roberts, C. W.; Curtis, S. G. *J. Am. Chem. Soc.* **1955**, *77*, 6387. c) Johncock, P. *J. Organomet. Chem.* **1969**, *19*, 257. See also the following recent examples: d) Gassman, P. G.; O' Reilly, N. J. *J. Org. Chem.* **1987**, *52*, 2481. e) Uno, H.; Shiraishi, Y.; Shimokawa, K.; Suzuki, H. *Chem. Lett.* **1987**, 1153.

59.   In the case of Grignard reagents, the corresponding adduct was isolated, but in low yield: Haszeldine, R. N. *J. Chem. Soc.* **1954**, 1273.

60.   Rachon, J.; Goedken, V.; Walborsky, H. M. *J. Am. Chem. Soc.* **1986**, *108*, 7435.

61.   a) Ichikawa, J.; Hamada, S.; Sonoda, T.; Kobayashi, H. *Tetrahedron Lett.* **1992**, *33*, 337. b) Patel, S. T.; Percy, J. M.; Wilkes, R. D. *Tetrahedron* **1995**, *51*, 9201. c) Purrington, S. T.; Thomas, H. N. *J. Fluorine Chem.* **1998**, *90*, 47.

62.   Drakesmith, F. G.; Stewart, O. J.; Tarrant, P. *J. Org. Chem.* **1967**, *33*, 280.

63.   Hong, F.; Tang, X.-Q.; Hu, C.-M. *J. Chem. Soc., Chem. Commun.* **1994**, 289.

64.   a) Yamazaki, T.; Mizutani, K.; Kitazume, T. *Tetrahedron: Asym.* **1993**, *4*, 1059. b) Yamazaki, T.; Mizutani, K.; Kitazume, T. *J. Org. Chem.* **1995**, *60*, 6046.

65.   Takaoka, A.; Iwakiri, H.; Ishikawa, N. *Bull. Chem. Soc. Jpn.* **1979**, *52*, 3377. Its prototype, Yarovenko' s reagent, which is less stable than Ishikawa' s reagent, has been used in a similar manner: Yarovenko, N. N.; Raksha, M. A. *J. Gen. Chem., USSR* **1959**, *29*, 2125.

66.   Kanagasabapathy, V.; Sawyer, J. F.; Tidwell, T. T. *J. Org. Chem.* **1985**, *50*, 503. See also the following review: Richard, J. P. *Tetrahedron* **1995**, *51*, 1535.

67.   Drabicky, M. J.; Myhre, P. C.; Reich, C. J.; Schmittou, E. R. *J. Org. Chem.* **1976**, *41*, 1472.

68.   a) Hine, J.; Ghirardelli, R. G. *J. Org. Chem.* **1958**, *23*, 1550. b) McBee, E. T.; Battershell, R. D.; Braendlin, H. P. *J. Am. Chem. Soc.* **1962**, *84*, 3157. c) Bordwell, F. G.; Brannen, Jr., W. T. *J. Am. Chem. Soc.* **1964**, *86*, 4645.

69.   Hagiwara, T.; Tanaka, K.; Fuchikami, T. *Tetrahedron Lett.* **1996**, *37*, 8187.

70.   a) Lin, J.-T.; Yamazaki, T.; Kitazume, T. *J. Org. Chem.* **1987**, *52*, 3211. b) Kitazume, T.; Yamazaki, T. Effect of the Fluorine Atom on Stereocontrolled Synthesis In *Selective Fluorination in Organic and Bioorganic Chemistry (ACS Symposium Series No. 456)*; Welch, J. T., Ed.: ACS: New York, 1991, pp 175. c) Fujisawa, T.; Sugimoto, T.; Shimizu, M. *Tetrahedron: Asym.* **1994**, *5*, 1095. d) Petschen, I.; Malo, E. A.; Bosch, M.

P.; Guerrero, A. *Tetrahedron: Asym.* **1996**, *7*, 2135.

71. a) Mikami, K.; Yajima, T.; Terada, M.; Uchimaru, T. *Tetrahedon Lett.* **1993**, *34*, 7591.
     b) Ramachandran, P. V.; Teodorovic, A. V.; Gong, B.-Q.; Brown, H. C. *Tetrahedron:
     Asym.* **1994**, *5*, 1075. c) Ramachandran, P. V.; Gong, B.-Q.; Brown, H. C. *J. Org.
     Chem.* **1995**, *60*, 41.

72. As far as we know, there have been very few examples of the S_N2 type reactions of
     2,2,2-trifluoroethylated materials reported: a) Tsushima, T.; Kawada, K.; Ishihara, S.;
     Uchida, N.; Shiratori, O.; Higaki, J.; Hirata, M. *Tetrahedron* **1988**, *44*, 5375. b)
     Canney, D. J.; Lu, H.-F.; McKeon, A. C.; Yoon, K.-W.; Xu, K.; Holland, K. D.;
     Rothman, S. M.; Ferrendelli, J. A.; Covey, D. F. *Bioorg. Med. Chem.* **1998**, *6*, 43. c)
     Katagiri, T.; Irie, M.; Uneyama, K. 74th Japan Chemical Society Meeting **1998**,
     3D525. Dr. Fuchikami, one of the authors of an article already shown in ref. 69,
     kindly informed us that they have succeeded in construction of a new carbon-carbon
     bond by reaction of sulfonates with sodium malonate.

73. a) Pegolotti, J. A.; Young, W. G. *J. Am. Chem. Soc.* **1961**, *83*, 3258. b) Pegolotti, J. A.;
     Young, W. G. *J. Am. Chem. Soc.* **1961**, *83*, 3251.

74. Hanzawa, Y.; Ishizawa, S.; Kobayashi, Y. *Chem. Pharm. Bull.* **1988**, *36*, 4209. See also,
     Hanzawa, Y.; Ishizawa, S.; Kobayashi, Y.; Taguchi, T. *Chem. Pharm. Bull.* **1990**, *38*,
     1104.

75. Yamazaki, T.; Umetani, H.; Kitazume, T. *Tetrahedron Lett.* **1997**, *38*, 6705.

76. a) Lipshutz, B.; James, B. *Tetrahedron Lett.* **1993**, *34*, 6689. b) Bertz, S. H.; Miao, G.-
     B.; Rossiter, B. E.; Snyder, J. P. *J. Am. Chem. Soc.* **1995**, *117*, 11023. c) Snyder, J. P. *J.
     Am. Chem. Soc.* **1995**, *117*, 11025. d) Eriksson, M.; Johansson, A.; Nilsson, M.; Olsson,
     T. *J. Am. Chem. Soc.* **1996**, *118*, 10904. e) Lipshutz, B.; Aue, D. H.; James, B.
     Tetrahedron Lett. **1996**, *37*, 8471.

77. a) Sharts, C. M.; Shepard, W. A. *Org. React.* **1974**, *21*, 125. b) Kanemoto, S.; Shimizu,
     M.; Yoshioka, H. *J. Chem. Soc., Chem. Commun.* **1989**, 690. c) Wilkinson, J. A. *Chem.
     Rev.* **1992**, *92*, 505. d) Oliver, J. E.; Waters, R. M.; Lusby, W. R. *Synthesis* **1994**, 273.

78. Olah, G. A.; Li, X.-Y. Fluorination with Onium Poly(hydrogen fluoride): The Taming
     of Anhydrous Hydrogen Fluoride for Synthesis. In *Synthetic Fluorine Chemistry*;
     Olah, G. A.; Chambers, R. D.; Surya Prakash, G. K., Eds.; John Wiley & Sons: New
     York, 1992, pp. 163. See also the following: Kanie, K.; Tanaka, Y.; Shimizu, M.;
     Kuroboshi, M.; Hiyama, T. *Chem. Commun.* **1997**, 309.

79. a) Hudlicky, M. *Org. React.* **1988**, *35*, 513. b) Kornilov, A. M.; Sorochinsky, A. E.;
     Kukhar, V. P. *Tetrahedron: Asym.* **1994**, *5*, 1015. c) Niihata, S.; Ebata, T.; Kawakami,
     H.; Matsushita, H. *Bull. Chem. Soc. Jpn.* **1995**, *68*, 1509. d) Matsumura, Y.; Wang, S.-
     Z.; Asai, T.; Shimada, T.; Morizawa, Y.; Yasuda, A. *Synlett* **1995**, 260. e) Gree, D. M.;
     Kermarrec, C. J. M.; Martelli, J. T.; Gree, R. L. *J. Org. Chem.* **1996**, *61*, 1918. f) Sato,
     Y.; Ueyama, K.; Maruyama, T.; Richman, D. D. *Nucleosides Nucleotides* **1996**, *15*, 109.
     g) Yoshimura, Y.; Kitano, K.; Satoh, H.; Watanabe, M.; Miura, S.; Sakata, S.; Sasaki, T.;
     Matsuda, A. *J. Org. Chem.* **1996**, *61*, 822.

80. For other fluorinating agents, see Munyemana, F.; Frisque-Hesbain, A.; Devos, A.;
     Ghosez, L. *Tetrahedron Lett.* **1989**, *30*, 3077.

81. a) Gerstenberger, M. R. C.; Haas, A. *Angew. Chem. Int. Ed. Engl.* **1981**, *20*, 647. b)
     Allvernhe, G.; Laurent, A.; Haufe, G. *Synthesis* **1987**, 562. c) Camps, F.; Chamorro,

E.; Gasol, V.; Guerrero, A. *J. Org. Chem.* **1989**, *54*, 4294. d) Chehidi, I.; Chaabouni, M. M.; Baklouti, A. *Tetrahedron Lett.* **1989**, *30*, 3167. e) Suga, H.; Hamatani, T.; Guggisberg, Y.; Schlosser, M. *Tetrahedron* **1990**, *46*, 4255. f) Kuroboshi, M.; Hiyama, T. *Tetrahedron Lett.* **1991**, *32*, 1215. g) Michel, D.; Schlosser, M. *Synthesis* **1996**, 1007.

82. a) Ayi, A. I.; Remli, M.; Condom, R.; Guedj, R. *J. Fluorine Chem.* **1981**, *17*, 565. b) Muehlbacher, M.; Poulter, C. D. *J. Org. Chem.* **1988**, *53*, 1026. c) Sutherland, J. K.; Watkins, W. J.; Bailey, J. P.; Chapman, A. K.; Davis, G. M. *J. Chem. Soc., Chem. Commun.* **1989**, 1386. d) Takano, S.; Yanase, M.; Ogasawara, K. *Chem. Lett.* **1989**, 1689. e) Suga, H.; Hamatani, T.; Schlosser, M. *Tetrahedron* **1990**, *46*, 4247. f) Landini, D.; Penso, M. *Tetrahedron Lett.* **1990**, *31*, 7209. g) Mori, K.; Nakayama, T.; Sakuma, M. *Bioorg. Med. Chem.* **1996**, *4*, 401.

83. Filler, R. *Isr. J. Chem.* **1978**, *71*, 17.

84. a) Watanabe, S.; Fujita, T.; Usui, Y.; Kitazume, T. *J. Fluorine Chem.* **1986**, *31*, 247. b) Shiuey, S.-J.; Partridge, J. J.; Uskokovic, M. R. *J. Org. Chem.* **1988**, *53*, 1040. c) Fritz-Langhals, E. *Tetrahedron: Asym.* **1994**, *5*, 981.

85. a) Venkiah, J.; Cornille, F.; Deshayes, C.; Doutheau, A. *J. Fluorine Chem.* **1990**, *49*, 183. b) Matheu, I.; Echarri, R.; Castillón, S. *Tetrahedron Lett.* **1993**, *34*, 2361. c) Chen, S.-H.; Huang, S.; Farina, V. *Tetrahedron Lett.* **1994**, *35*, 41.

86. For a recent review, see Lal, G. S.; Pez, G. P.; Syvret, R. G. *Chem. Rev.* **1996**, *96*, 1737. See also the following recent works by various research groups: a) Umemoto, T.; Tomita, K. *Tetrahedron Lett.* **1986**, *27*, 3271. b) Singh, S.; DesMarteau, D. D.; Zuberi, S. S.; Witz, M.; Huang, H.-N. *J. Am. Chem. Soc.* **1987**, *109*, 7149. c) Differding, E.; Lang, R. W. *Helv. Chim. Acta* **1989**, *72*, 1248. d) Umemoto, T.; Fukami, S.; Tomizawa, G.; Harasawa, K.; Kawada, K.; Tomita, K. *J. Am. Chem. Soc.* **1990**, *112*, 8563. e) Davis, F. A.; Han, W. *Tetrahedron Lett.* **1991**, *32*, 1631. f) Differding, E.; Wehrli, M. *Tetrahedron Lett.* **1991**, *32*, 3819. g) Differding, E.; Rüegg, G. M. *Tetrahedron Lett.* **1991**, *32*, 3815. h) Differding, E.; Rüegg, G. M.; Lang, R. W. *Tetrahedron Lett.* **1991**, *32*, 1779. h) DesMarteau, D. D.; Xu, Z.-Q.; Witz, M. *J. Org. Chem.* **1992**, *57*, 629. i) Ihara, M.; Taniguchi, N.; Kai, T.; Satoh, K.; Fukumoto, K. *J. Chem. Soc. Perkin Trans. 1* **1992**, 221. j) Davis, F. A.; Reddy, R. E. *Tetrahedron: Asym.* **1994**, *5*, 955. k) Shimizu, I.; Ishii, H. *Tetrahedron* **1994**, *50*, 487. l) Ihara, M.; Kawabuchi, T.; Tokunaga, Y.; Fukumoto, K. *Tetrahedron: Asym.* **1994**, *5*, 1041. m) Davis, F.; Han, W.; Murphy, C. K. *J. Org. Chem.* **1995**, *60*, 4730. n) Umemoto, T.; Tomizawa, G. *J. Org. Chem.* **1995**, *60*, 6563. o) Yin, W.-W.; DesMarteau, D. D.; Gotoh, Y. *Tetrahedron* **1996**, *52*, 15. p) Umemoto, T.; Nagayoshi, M.; Adachi, K.; Tomizawa, G. *J. Org. Chem.* **1998**, *63*, 3379.

87. Patrick, T. B.; Mortezania, R. *J. Org. Chem.* **1988**, *53*, 5153.

88. a) Liotta, C. L.; Harris, P. H. *J. Am. Chem. Soc.* **1974**, *96*, 2250. b) Rozen, S. *Chem. Rev.* **1996**, *96*, 1717.

89. a) Laurent, E.; Tardivel, R.; Thiebault, H. *Tetrahedron Lett.* **1983**, *24*, 903. b) Laurent, E.; Marquet, B.; Tardivel, R.; Thiebault, H. *Bull. Soc. Chim. Fr.* **1986**, *6*, 955.

90. a) Differding, E.; Lang, R. W. *Tetrahedron Lett.* **1988**, *29*, 6087. b) Hann, G. L.; Sampson, P. *J. Chem. Soc., Chem. Commun.* **1989**, 1650.

91. For recent examples, see a) Graham, S. M.; Prestwich, G. D. *J. Org. Chem.* **1994**, *59*, 2956. b) Pu, Y.-M.; Torok, D. S.; Ziffer, H.; Pan, Z.-Q.; Meshnick, S. R. *J. Med. Chem.*

1995, *38*, 4120. c) Parisi, M. F.; Gattuso, G.; Notti, A.; Raymo, F. M.; Able, R. H. *J. Org. Chem.* **1995**, *60*, 5174. d) Rich, R. H.; Bartlett, P. A. *J. Org. Chem.* **1996**, *61*, 3916.

92. Kuroboshi, M.; Hiyama, T. *Synlett* **1991**, 909.

93. a) Xu, Z.-Q.; DesMarteau, D. D.; Gotoh, Y. *J. Chem. Soc., Chem. Commun.* **1991**, 179.
b) Banks, R. E.; Lawrence, J. J.; Popplewell, A. L. *J. Chem. Soc., Chem. Commun.* **1994**, 343.

94. Wang, C.-L. J. *Org. React.* **1985**, *34*, 319.

95. a) Matthews, D. P.; Whitten, J. P.; McCrathy, J. R. *Tetrahedron Lett.* **1986**, *27*, 4861. b) Zupan, M.; Bregar, Z. *Tetrahedron Lett.* **1990**, *31*, 3357. c) Kuroboshi, M.; Hiyama, T. *Chem. Lett.* **1992**, 827. d) Kuroboshi, M.; Suzuki, K.; Hiyama, T. *Tetrahedron Lett.* **1992**, *33*, 4173. e) Kuroboshi, M.; Hiyama, T. *Tetrahedron Lett.* **1992**, *33*, 4177. f) Furuta, S.; Hiyama, T. *Synlett* **1996**, 1199.

96. Hartgraves, G. A.; Burton, D. J. *J. Fluorine Chem.* **1988**, *39*, 425.

97. a) Rico, I.; Cantacuzene, D.; Wakselman, C. *J. Chem. Soc. Perkin Trans. 1* **1982**, 1063.
b) Konno, T.; Kitazume, T. *Chem. Commun.* **1996**, 2227.

98. For recent articles, see a) Braun, M.; Vonderhagen, A.; Waldmüller, D. *Liebigs Ann.* **1995**, 1447, and references cited therein. b) Andrés, J. M.; Martínez, M. A.; Pedrosa, R.; Pérez-Encabo, A. *Synthesis* **1996**, 1070. c) Yoshida, M.; Suzuki, D.; Iyoda, M. *Synth. Commun.* **1996**, *26*, 2523.

99. a) Obayashi, M.; Kondo, K. *Tetrahedron Lett.* **1982**, *23*, 2327. b) Martin, S. F.; Dean, D. W.; Wagman, A. S. *Tetrahedron Lett.* **1992**, *33*, 1839.

100. Berkowitz, D.; Sloss, D. G. *J. Org. Chem.* **1995**, *60*, 7047.

101. Blades, K.; Lequeux, T. P.; Percy, J. M. *Tetrahedron* **1997**, *53*, 10623.

102. a) Lequeux, T. P.; Percy, J. M. *Synlett* **1995**, 361. b) Blades, K.; Lapôtre, D.; Percy, J. M. *Tetrahedron Lett.* **1997**, *38*, 5895.

103. Kirihara, M.; Takuwa, T.; Takizawa, S.; Momose, T. *Tetrahedron Lett.* **1997**, *38*, 2853.

104. a) O' Reilly, N. J.; Maruta, M.; Ishikawa, N. **1984**, 517. b) Kitazume, T.; Ishikawa, N. *J. Am. Chem. Soc.* **1985**, *109*, 5186. c) Francèse, C.; Tordeux, M.; Wakselman, C. *J. Chem. Soc., Chem. Commun.* **1987**, 642. d) Burton, D. J. Organometallics in Synthetic Organofluorine Chemistry. In *Synthetic Fluorine Chemistry*; Olah, G. A.; Chambers, R. D.; Surya Prakash, G. K., Eds.; John Wiley & Sons: New York, 1992, pp. 205.

105. a) Kobayashi, Y.; Yamamoto, K.; Kumadaki, I. *Tetrahedron Lett.* **1979**, 4071. b) Matsui, K.; Tobita, E.; Ando, M.; Kondo, K. *Chem. Lett.* **1981**, 1719. c) Chen, Q.-Y.; Wu, S.-W. *J. Chem. Soc., Chem. Commun.* **1989**, 705. d) Su, D.-B.; Duan, J.-X.; Chen, Q.-Y. *Tetrahedron Lett.* **1991**, *32*, 7689. For a recent example, see Mawson, S. D.; Weavers, R. T. *Tetrahedron Lett.* **1993**, *34*, 3139.

106. Bouillon, J.-P.; Maliverney, C.; Merényi, R.; Viehe, H. G. *J. Chem. Soc. Perkin Trans. 1* **1991**, 2147.

107. a) Ruppert, I.; Schlich, K.; Volbach, W. *Tetrahedron Lett.* **1984**, *25*, 2195. b) Kurishnamurti, R.; Bellew, D. R.; Prakash, G. K. S. *J. Org. Chem.* **1991**, *56*, 984. c) Prakash, G. K. S. Nucleophilic Perfluoroalkylation of Organic Compounds Using Perfluoroalkyltrialkylsilane. In *Synthetic Fluorine Chemistry*; Olah, G. A.; Chambers, R. D.; Surya Prakash, G. K., Eds.; John Wiley & Sons: New York, 1992, pp. 227. For recent examples, see a) Iseki, K.; Nagai, T.; Kobayashi, Y. *Tetrahedron Lett.* **1994**, *35*,

3137. b) Wang, Z.-Q.; Lu, S.-F.; Chao, L.; Yang, C.-J. *Bioorg. Med. Chem. Lett.* **1995**, *5*, 1899. c) Munier, P.; Giudicelli, M.-B.; Picq, D.; Anker, D. *J. Carbohydr. Chem.* **1996**, *15*, 739. d) Takagi, Y.; Nakai, K.; Tsuchiya, T.; Takeuchi, T. *J. Med. Chem.* **1996**, *39*, 1582.

108.   For the introduction of other fluorinated groups, see the followings: a) Urata, H.; Fuchikami, T. *Tetrahedron Lett.* **1991**, *32*, 91. b) Hagiwara, T.; Fuchikami, T. *Synlett* **1995**, 717. c) Yudin, A. K.; Prakash, G. K. S.; Deffieux, D.; Bradley, M.; Bau, R.; Olah, G. A. *J. Am. Chem. Soc.* **1997**, *119*, 1572.

109.   See also the following related articles: a) Yokoyama, Y.; Mochida, K. *Synlett* **1996**, 1191. b) Prakash, G. K. S.; Yudin, A. K.; Deffieux, D.; Olah, G. A. *Synlett* **1996**, 151.

110.   Umemoto, T. *Chem. Rev.* **1996**, *96*, 1757.

111.   a) Umemoto, T.; Ishihara, S. *Tetrahedron Lett.* **1990**, *31*, 3579. b) Umemoto, T.; Ishihara, S. *J. Am. Chem. Soc.* **1993**, *115*, 2156.

112.   Umemoto, T.; Adachi, K. *J. Org. Chem.* **1994**, *59*, 5692.

113.   Maruoka, K.; Sano, H.; Fukutani, Y.; Yamamoto, H. *Chem. Lett.* **1985**, 1689.

114.   a) Miura, K.; Taniguchi, M.; Nozaki, K.; Oshima, K.; Utimoto, K. *Tetrahedron Lett.* **1990**, *31*, 6391. b) Miura, K.; Takeyama, Y.; Oshima, K.; Utimoto, K. *Bull. Chem. Soc. Jpn.* **1991**, *64*, 1542. c) Iseki, K.; Nagai, T.; Kobayashi, Y. *Tetrahedron Lett.* **1993**, *34*, 2169. d) Yamanaka, H.; Takekawa, T.; Morita, K.; Ishihara, T.; Gupton, J. T. *Tetrahedron Lett.* **1996**, *37*, 1829.

115.   Middleton, W. J.; Bingham, E. M. *J. Org. Chem.* **1980**, *45*, 2883.

2 Preparation of Monofluorinated Materials

2.1 Fluorination

2.1.1 Nucleophilic Fluorination (Substitution)

Ethyl 4-fluorobut-2-ynoate

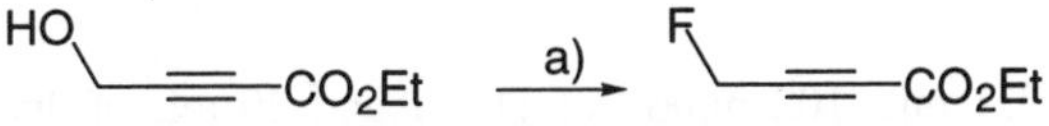

a) DAST / CH$_2$Cl$_2$

Synthetic method

A solution containing 5.20 g (32.3 mmol) of DAST in 10 mL of CH$_2$Cl$_2$ was cooled to -78 ℃ before the addition of 4.88 g (38.1 mmol) of ethyl 4-hydroxybut-2-ynoate in 5 mL of CH$_2$Cl$_2$. The resulting mixture was allowed to stir for 45 min at -78 ℃ and 200 min at room temperature before the addition of 10 mL of H$_2$O. The layers were separated, and the aqueous layer was extracted with CH$_2$Cl$_2$. The combined organic fractions were then washed with H$_2$O and dried over anhydrous MgSO$_4$. The solvent was removed at reduced pressure, and the residue was flash distilled to yield 2.49 g of ethyl 4-fluorobut-2-ynoate in 59% yield.

Spectral data

bp. 28.5-29.0 ℃/0.86 mmHg; ^{1}H NMR (CDCl$_3$): δ 1.33 (3 H, t, *J*=7.1 Hz), 4.25 (2 H, q, *J*=7.1 Hz), 5.08 (2 H, d, *J*= 46.9 Hz); ^{19}F NMR (CDCl$_3$): δ -226.6 (t); IR (CCl$_4$): v 2990, 2950, 2910, 2880, 1725, 1480, 1470, 1455, 1380, 1265, 1090, 1030 cm^{-1}.

Caution

DAST was previously reported to possibly decompose in a violent manner at temperatures above 90 ℃. See Middleton, W. J.; Bingham, E. M. *J. Org. Chem.* **1980**, *45*, 2883.

Note

Hammond and his co-workers recently reported that fluorination of an α-phosphonylated propargylic alcohol proceeded in an S$_N$2 fashion, while the S$_N$2' reaction with DAST occurred for the corresponding allylic alcohol. See Sanders, T. C.; Hammond, G. B. *J. Org. Chem.* **1993**, *58*, 5598 and Benayoud, F.; Hammond, G. B. *Chem. Commun.* **1996**, 1447 for the construction of the corresponding difluorinated phosphonate.

On the other hand, DAST seems to fluorinate simple allylic alcohols in an S$_N$1 manner. See for example, Middleton, W. J. *J. Org. Chem.* **1975**, *40*, 574. For a review on DAST, Hudlicky, M. *Org. React.* **1988**, *35*, 513.

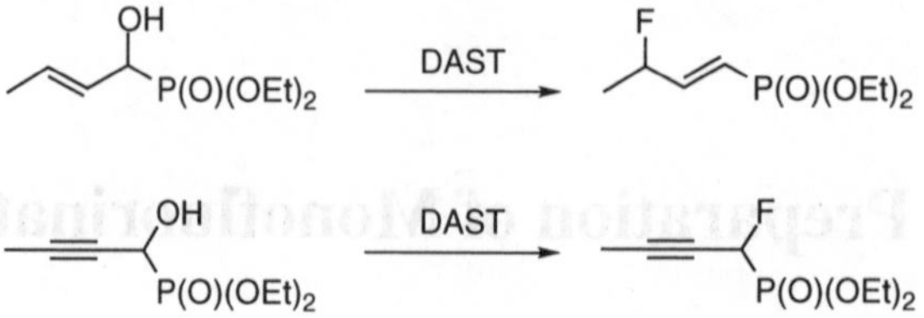

Poulter, C. D., *et al. J. Org. Chem.* **1981**, *46*, 1532.

(*R*)-2-Fluorobutan-4-olide

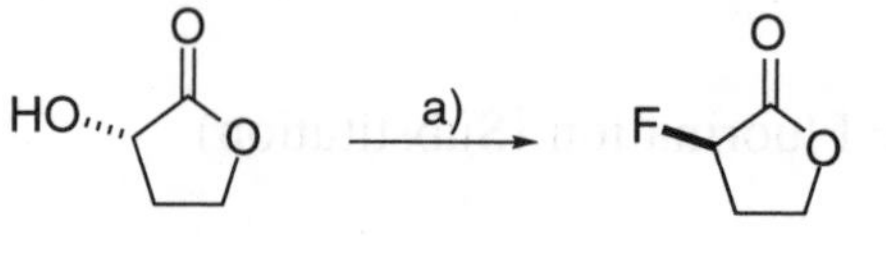

a) DAST / CH$_2$Cl$_2$

Synthetic method

To a mixture of 13.4 mL (102 mmol) of DAST and 20 mL of dry CH$_2$Cl$_2$ at -70 °C was added dropwise a solution of 3.50 g (34.3 mmol) of (*S*)-2-hydroxybutan-4-olide in 30 mL of dry CH$_2$Cl$_2$.   The mixture was stirred at -70 °C for 1 h, at 0 °C for 0.5 h, and at room temperature for 1 h.   The mixture was poured into a stirring mixture of sat. NaHCO$_3$ aq. and ice chips.   The mixture was extracted with CH$_2$Cl$_2$.   The extract was washed with H$_2$O and dried over MgSO$_4$, filtered, and evaporated under reduced pressure.   The oil was purified by column chromatography on silica gel (Et$_2$O:CH$_2$Cl$_2$=3:7→7:3) to afford 2.70 g (76%) of (*R*)-2-fluorobutan-4-olide as a volatile oil.

Spectral data

[α]$_D^{25}$ +50.3 (*c* 0.95, CHCl$_3$); ^{1}H NMR (CDCl$_3$): δ 2.20-2.84 (2 H, m), 4.26-4.64 (2 H, m), 5.18 (1 H, dt, *J*=52, 8 Hz); IR (CDCl$_3$): ν 1798 cm^{-1}.

Note

Although an elevated temperature is required for the conversion of a carboxyl group into an acyl fluoride by DAST, fluorination of 2-hydroxy-2-phenylpropionic acid furnished the desired fluorinated acid only in 21% yield, and the major product was found to be acetophenone *via* decarboxylation.   Fluorination proceeded in good (71%) yield when the corresponding ethyl ester was employed.   See Schlosser, M., *et al. Tetrahedron* **1996**, *52*, 8257.

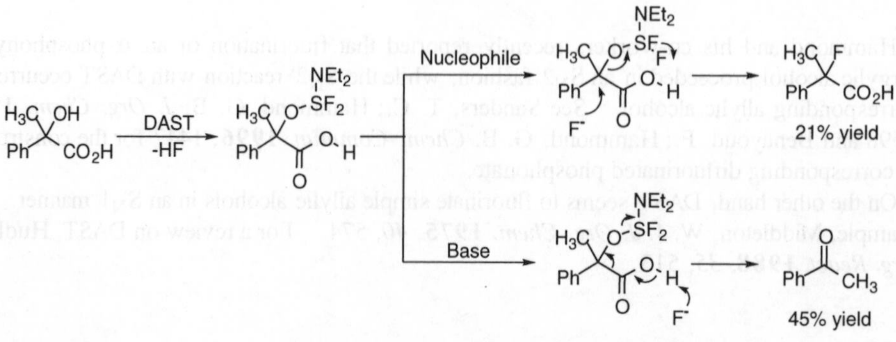

See also the following articles on DAST fluorination for systems in which neighboring groups possibly participate: i) Walba, D. M., *et al. J. Am. Chem. Soc.* **1988**, *110*, 8686, ii) Chen, S.-H., *et al. Tetrahedron Lett.* **1994**, *35*, 41, iii) Young, D. W., *et al. Tetrahedron Lett.* **1994**, *35*, 5485, iv) Shellhammer, D. F., *et al. J. Chem. Soc., Perkin Trans.* 2 **1996**, 973, v) Gree, R. L., *et al. J. Org. Chem.* **1996**, *61*, 1918, vi) Charvillon, F. B.; Amouroux, R. *Tetrahedron Lett.* **1996**, *37*, 5103, vii) Haigh, D., *et al. J. Chem. Soc., Perkin Trans.* 1 **1996**, 2895, viii) Momose, T., *et al. Chem. Commun.* **1996**, 1103, ix) Momose, T., *et al. Chem. Commun.* **1997**, 599.

Enantioselective fluorination with chiral aminosulfur trifluoride was also reported. See Hann, G. L.; Sampson, P. *J. Chem. Soc., Chem. Commun.* **1989**, 1650.

Shiuey, S.-J., *et al. J. Org. Chem.* **1988**, *53*, 1040.

2-Fluorobutanoic acid

a) NaNO$_2$ / pyridine-(HF)$_n$,

Synthetic method

To 2.1 g (20 mmol) of 2-aminobutanoic acid dissolved in 50 mL of 70% pyridine-(HF)$_n$ (70 mL) was slowly added, with good stirring, 2.1 g of NaNO$_2$ (30 mmol). After being stirred at room temperature for 4 h, the reaction mixture was quenched and extracted with Et$_2$O. The Et$_2$O layer was again washed with ice H$_2$O and dried over anhydrous Na$_2$SO$_4$. Et$_2$O evaporation gave a crude product, from which 2-fluorobutanoic acid (1.79 g, 80% yield) was obtained upon distillation.

Spectral data

bp. 90-91 °C/12 mmHg; ^{13}C NMR (CDCl$_3$): δ 8.0, 24.5 (d, *J*=21.2 Hz), 88.7 (d, *J*=184.2 Hz), 175.1 (d, *J*=24.0 Hz).

α-Amino acid	Yield (%)	bp. (°C/mmHg) [mp. (°C)]	α-Amino acid	Yield (%)	bp. (°C/mmHg) [mp. (°C)]
glycine	38	163/760	alanine	96	65-66/13
valine	84	[38-39]	leucine	88	95-96/10
isoleucine	75	97-98/10	phenylalanine	98	[73-75]
tyrosine	47	57/0.5	serine	80	[95-95.5]
threonine	54		aspartic acid	52	[141-143]
glutamic acid	28	[105-107]			

(Reproduced with permission from Olah, G. A., *et al. J. Org. Chem.* **1979**, *44*, 3872)

Note

When optically active amino acids were employed in this reaction, the original stereochemical information was retained due to the participation of a carboxyl moiety in an S$_N$2 manner for the elimination of the resulting N$_2$, followed by the attack of fluoride from the opposite side of the carboxylate. However, when phenylalanine was used as a substrate, 3-fluoro-2-phenylpropionic acid was the only isolated product possible because of the intramolecular assistance of the phenyl group, followed by the fluoride attack. See Faustini, F., *et al. Tetrahedron Lett.* **1981**, *22*, 4533. See also the following related works: i) Keck, R.; Retey, J. *Helv. Chim. Acta* **1980**, *63*, 769, ii) Leeper, F. J.; Rock, M. *J. Fluorine Chem.* **1991**, *51*, 381.

For the synthesis of 2-fluorocarboxylic acids or their derivatives, see the following examples:
i) Watanabe, S., et al. J. Fluorine Chem. **1986**, 31, 247, ii) Focella, A., et al. Synth. Commun.
1991, 21, 2165, iii) Stelzer, U.; Effenberger, F. Tetrahedron: Asym. **1993**, 4, 161, iv) Khrimian,
A. P., et al. Tetrahedron: Asym. **1996**, 7, 37.   The following deals with the formation of the
same structure via asymmetric hydrogenation: Saburi, M., et al. Tetrahedron Lett. **1992**, 33, 7877.

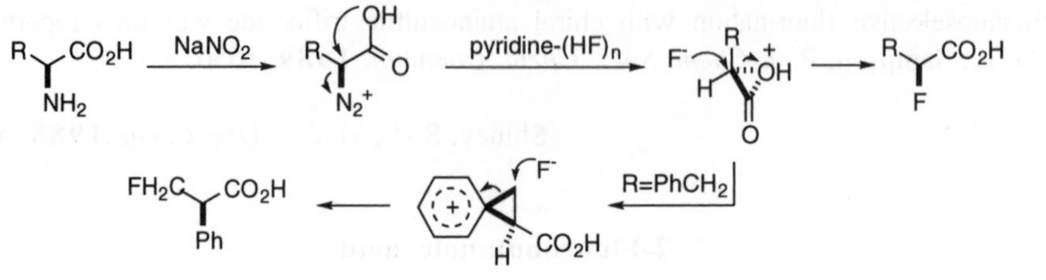

Olah, G. A., et al. J. Org. Chem. **1979**, 44, 3872.
Olah, G. A., et al. Helv. Chim. Acta **1981**, 64, 2528 (spectral data).

4-Cyano-(1-fluoroethyl)benzene

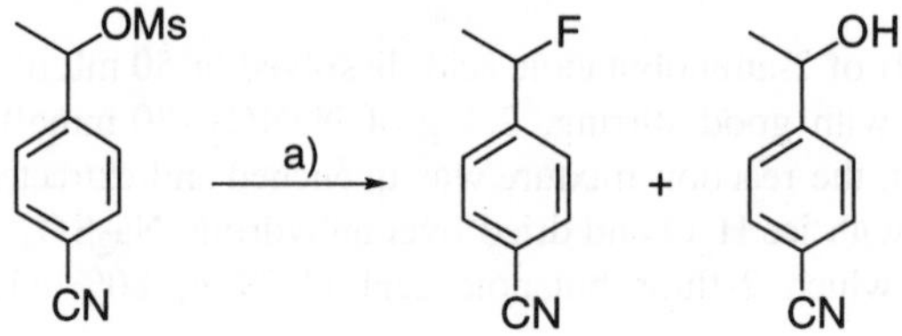

a) CsF / N-methylformamide

Synthetic method
        Cesium fluoride was added to freshly distilled N-methylformamide at 60 °C under argon
where the crystalline methanesulfonate (the corresponding alcohol was proved to be 96.2% ee) was
added without solvent (CsF: 4 equiv, a 2-3 M solution).   For work-up the mixture was diluted
with H$_2$O and extracted with Et$_2$O or methyl tert-butyl ether.   The extract was washed with H$_2$O,
dried (MgSO$_4$) and evaporated to dryness.   The ratio of fluorinated and hydrolyzed products was
determined by the integration of the benzylic protons by ^{1}H NMR.   Pure fluorinated material was
obtained by fractional distillation in 61% yield.
Spectral data
        $[\alpha]_D^{20}$ +27.1 (c 1.00, CH$_2$Cl$_2$), 96.0% ee; bp. 85-86 °C/1.7 mmHg; ^{1}H NMR (CDCl$_3$): δ
1.53 (3 H, dd, J=23, 8 Hz), 5.68 (1 H, dq, J=48, 8 Hz), 7.45-7.70 (4 H, m).

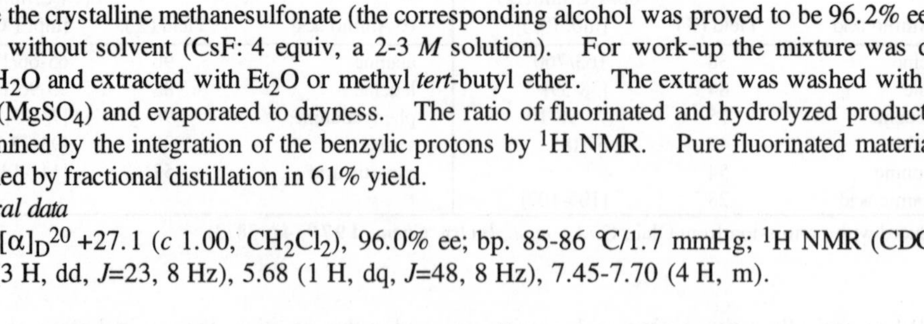

X: NO$_2$ (96.4% ee)   75 (91.4% ee) : 25
X: CO$_2$Et (83.0% ee)   46 (73.0% ee) : 54

98.7% ee   83% yield
                                96% ee

Note

For the representative fluorination of chiral materials reported by Fritz-Langhals *et al.*, see the examples on the previous page.

Fritz-Langhals, E. *Tetrahedron: Asym.* **1994**, *5*, 981.

(*E*)-1-Fluoro-6,10-dimethylundeca-5,9-dien-2-one

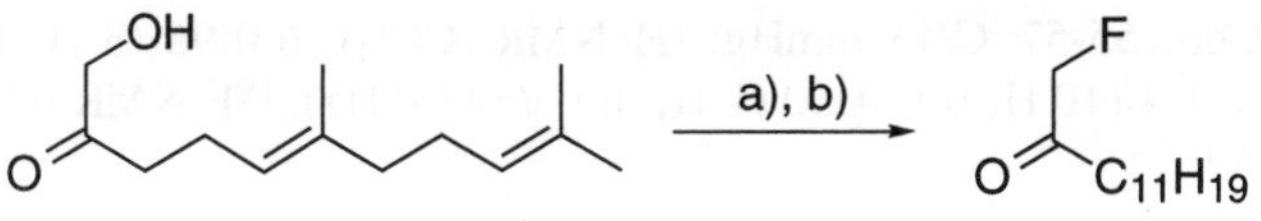

a) (CF$_3$SO$_2$)$_2$O, 2,6-lutidine / CH$_2$Cl$_2$, b) TBAF / THF

Synthetic method

(*E*)-1-Hydroxy-6,10-dimethylundeca-5,9-dien-2-one (0.95 g, 4.56 mmol) was dissolved in CH$_2$Cl$_2$ (20 mL) under nitrogen.   2,6-Lutidine (1.3 mL, 10.85 mmol) was added, and the solution was cooled to 0 °C.   Triflic anhydride (0.92 mL, 5.47 mmol) was added, and the solution was stirred at 0 °C for 45 min.   After the solvent was removed under reduced pressure, the residue was diluted with AcOEt, successively washed with 10% CuSO$_4$, KHCO$_3$, and brine, dried and concentrated to give a red oil.

This oil was immediately dissolved in dry THF under nitrogen.   TBAF (9.0 mL, 9.0 mmol) was added, and the solution was stirred for 1.5 h.   The solution was concentrated, and the residue was chromatographed on silica gel, eluting with hexane:AcOEt=15:1 to give 0.583 g of a brown oil, which was further purified by silica gel chromatography with hexane:AcOEt=20:1 to give 445 mg of (*E*)-1-fluoro-6,10-dimethylundeca-5,9-dien-2-one in 46% yield.

Spectral data

^{1}H NMR (CDCl$_3$): δ 1.59 (3 H, s), 1.62 (3 H, s), 1.67 (3 H, s), 1.95-2.10 (4 H, m), 2.25-2.35 (2 H, m), 2.55 (2 H, td, *J*=7, 3 Hz), 4.78 (2 H, d, *J*=48 Hz), 5.00-5.10 (2 H, m); ^{13}C NMR (CDCl$_3$): δ 15.9, 17.6, 21.4, 25.6, 26.5, 38.3, 39.6, 84.9 (d, *J*=184 Hz), 121.9, 124.0, 131.3, 136.8, 206.5 (d, *J*=20 Hz); ^{19}F NMR (CDCl$_3$): δ -111.6 (t, *J*=47 Hz) from C$_6$F$_6$; IR (neat): ν 2968, 2922, 2858, 1728, 1437 cm^{-1}.

Dolence, J. M.; Poulter, C. D. *Tetrahedron* **1996**, *52*, 119.

(*R*)-2-Fluorooctane

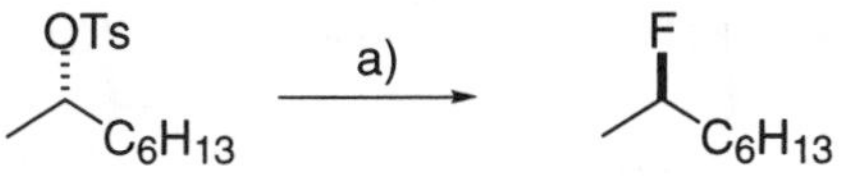

a) KF / triethylene glycol

Synthetic method

Into a dry two-necked 50 mL flask containing a Teflon-coated stirrer bar was placed 7.25 g (125 mmol) of anhydrous KF.   One neck was capped with a rubber septum, the other was connected to a cold trap, and the apparatus was flushed with nitrogen.   Anhydrous triethylene glycol (25 mL) and 7.10 g (25 mmol) of (S)-2-octyl p-toluenesulfonate (99.4% ee) were added by a syringe.   The rubber septum was replaced with a glass stopper and the flask was heated to 110 °C with vigorous stirring under reduced pressure (4 mmHg).   The volatile materials were allowed to distill from the reaction mixture and collected in a cold trap (-50 °C) over a period of 3 h. Analysis of the crude distillate by GC indicated a 52% yield of 2-fluorooctane accompanied by a 27% yield of octene(s).   The crude distillate was treated with a slight excess of bromine in CS_2, washed with $Na_2S_2O_3$, dried ($MgSO_4$), and distilled to afford 1.45 g (44%) of (R)-2-fluorooctane in an almost completely stereochemical inversion.

Spectral data

[α]$_D^{20}$ -9.99; bp. 55-57 °C/43 mmHg; ^{1}H NMR (CCl_4): δ 0.96 (3 H, t), 1.26 (3 H, dd, J=23.0, 7.0 Hz), ca. 1.4 (10 H, m), 4.50 (1 H, dm, J=48.0 Hz); ^{19}F NMR ($CDCl_3$): δ -165 (m); IR (CCl_4): ν 870 cm^{-1}.

Note

Wakselman and his co-workers repeated this reaction afterwards with some other fluorinating reagents to investigate the reaction mechanism and found that 2-fluorooctane obtained possessed 90.7% ee.   The above ^{19}F NMR data are taken from their article.   See Wakselman, C., et al. *J. Org. Chem.* **1979**, 44, 3406.

Recently, Matsumura and his co-workers reported that the selectivity between fluorination and elimination as well as the optical purity of the fluorinated products was highly dependent on the combination of the fluorinating reagents used and the leaving group at the hydroxy group.   See Matsumura, Y., et al. *Tetrahedron* **1995**, 51, 8771.

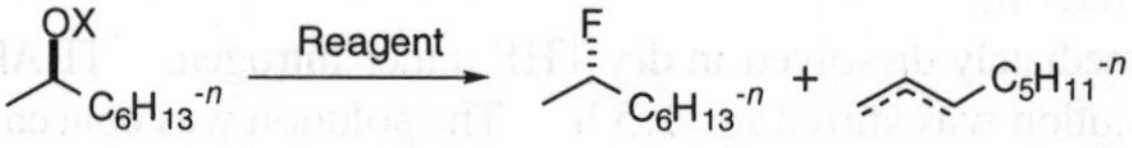

X	Reagent	Selectivity F-Octane:Octenes			Optical purity of F-Octane (% ee)
H	DAST	48	:	52	97
TMS	DAST	89	:	11	98
Ts[a]	KF	50	:	50	91
Ts[a]	TBAF	63	:	37	79

[a] See the article by Wakselman, C., et al. indicated above.

San Filippo, Jr., J.; Romano, L. J. *J. Org. Chem.* **1975**, 40, 1514.

(S)-5-(Fluoromethyl)pyrrolidin-2-one

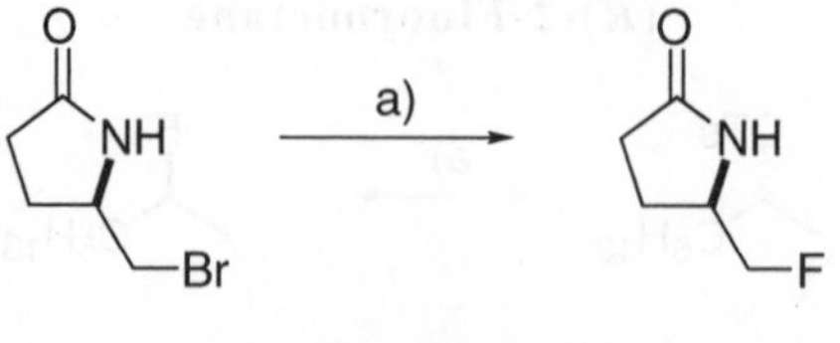

a) AgF / MeCN

Synthetic method

A solution of 1.0 g (5.62 mmol) of (S)-5-(bromomethyl)pyrrolidin-2-one in 10 mL of dry MeCN was added dropwise over 10 min to a suspension of 1.6 g (12.69 mmol) of AgF in 20 mL of dry MeCN protected from light with aluminum foil.   The reaction mixture was stirred at room temperature for 9.5 h then filtered through a pad of Celite.   The brown filtrate was concentrated *in vacuo* to give a brown oil which was triturated with CHCl$_3$; the solid formed was removed by filtration through Celite.   The light yellow filtrate was concentrated to a light orange oil which was chromatographed on silica gel (AcOEt→AcOEt:acetone=1:1) to give 560 mg of (*S*)-5-(fluoro-methyl)pyrrolidin-2-one in 85% yield as a colorless oil.   A sample of this oil was Kügelrohr distilled to give a colorless oil that solidified into a white solid upon standing.

Spectral data

bp. 73-82 °C/0.1 mmHg; mp. 24-26 °C; ^{1}H NMR (CDCl$_3$): δ 2.3 (4 H, m), 3.9 (2 H, m), 4.3 (1 H, dm, *J*=46 Hz), 7.4 (1 H, br s); IR (film): ν 3220, 1690, 1280, 1015, 970 cm^{-1}.

Note

See also the following article: Silverman, R. B.; Nanavati, S. M. *J. Med. Chem.* **1990**, *33*, 931.

Silverman, R. B.; Levy, M. A. *J. Org. Chem.* **1980**, *45*, 815.

2-Fluoropropionic acid,
S-Phenyl 2-fluoropropanethioate

a) KF / sulforane, b) 10% H$_2$SO$_4$,
c) PhSH, Et$_3$N, Me$_2$NP(O)Cl$_2$ / DME

Synthetic method

2-Fluoropropionic acid

In a three-necked flask, equipped with a magnetic stirrer, a thermometer, and a still head for distillation, were placed methyl 2-bromopropionate (41.75 g, 250 mmol), spray-dried KF (21.46 g, 370 mmol), and sulforane (75 mL).   The mixture was heated with stirring at 130 °C for 2.5 h, followed by distillation under a pressure of 50-100 mmHg at 130-150 °C to collect crude methyl 2-fluoropropionate (23.02 g).   The crude ester was mixed with 10% H$_2$SO$_4$ (500 mL) and the mixture was refluxed for 1 h.   After being cooled to room temperature, this mixture was saturated with NaCl then subjected to extraction with Et$_2$O (10x50 mL).   The ethereal extracts were washed with brine, dried over MgSO$_4$, and filtered.   The filtrate was concentrated to leave a residual oil, which was distilled under reduced pressure to give pure 2-fluoropropionic acid (11.96 g) in 52% overall yield.

Spectral data

bp. 86.0-87.0 °C/35 mmHg; ^{1}H NMR (CDCl$_3$): δ 1.63 (3 H, dd, *J*=23.0, 7.3 Hz), 5.01 (1 H, dq, *J*=48.0, 7.3 Hz), 11.57 (1 H, s); ^{19}F NMR (CDCl$_3$): δ -106.0 (dq, *J*=48.0, 23.0 Hz) from CF$_3$CO$_2$H; IR (film): ν 3680-2783, 3003, 2950, 1740, 1470, 1457, 1380, 1240, 1125, 1100, 1049, 826 cm^{-1}.

Synthetic method

S-Phenyl 2-fluoropropanethioate

To a stirred solution of 2-fluoropropionic acid (1.85 g, 20 mmol) in 1,2-dimethoxyethane (110 mL) were added successively triethylamine (4.85 g, 48 mmol), *N*,*N*-dimethylphosphoramidic dichloride (3.89 g, 24 mmol), and benzenethiol (2.64 g, 24 mmol) at such a rate that the reaction temperature did not rise above 10 ℃.   After stirring at ambient temperature for 2.5 h, the mixture was poured into a cold 5% HCl solution and was extracted with $CHCl_3$ (4x70 mL).   The organic extracts were dried over Na_2SO_4, followed by filtration and concentration.   The residue was chromatographed on a silica gel column using hexane and benzene as eluents to afford S-phenyl 2-fluoropropanethioate in 84% yield.

Spectral data

mp. 44.2-45.0 ℃; ^{1}H NMR ($CDCl_3$): δ 1.60 (3 H, dd, J=23.6, 6.8 Hz), 5.03 (1 H, dq, J=47.8, 6.8 Hz), 7.32 (5 H, s); ^{19}F NMR ($CDCl_3$): δ -114.5 (dq, J=47.8, 23.6 Hz); IR (film): ν 3057, 2937, 1704, 1584, 1482, 1444, 1373, 1319, 1148, 1086, 1068, 1020, 976, 870, 743, 706, 686 cm^{-1}.

Note

For a recent preparation of the corresponding esters, see Welch, J. T., *et al. J. Org. Chem.* **1991**, *56*, 353.

Ishihara, T., *et al. Tetrahedron* **1996**, *52*, 255.

Methyl 9-fluoro-10-hydroxydecanoate, Methyl 10-fluoro-9-hydroxydecanoate

a) Pyridine-$(HF)_n$ / CH_2Cl_2, b) KHF_2, 18-crown-6 / DMF

Synthetic method

Methyl 9-fluoro-10-hydroxydecanoate

In a dry 100-mL polypropylene flask filled with argon, 3 mL of 70% pyridinium poly-(hydrogen fluoride) (100 mmol) was cooled to 0 ℃.   Then 10 mmol of methyl 9,10-epoxy-decanoate in 6 mL of dry CH_2Cl_2 was carefully added.   After the temperature of the solution reached room temperature, the solution was additionally stirred for 6 h.   Subsequently, it was poured into 50 mL of ice-cooled 2 N ammonia and neutralized with conc ammonia.   The organic layer was separated and the aqueous layer was extracted three times with small portions of CH_2Cl_2.   Then NaCl was added and the aqueous layer was again extracted twice with CH_2Cl_2.   The combined organic layers were dried with Na_2SO_4.   Removal of CH_2Cl_2 gave methyl 9-fluoro-10-hydroxydecanoate together with about 8% of its regioisomer (10-fluoro-9-hydroxy) as a yellowish, viscose oil.   The isomer was separated by flash chromatography over silica gel (cyclohexane: AcOEt=3:1) to furnish 1.58 g of methyl 9-fluoro-10-hydroxydecanoate in 72% yield.

Spectral data

mp. 4.5 ℃; ^{1}H NMR ($CDCl_3$): δ 1.25-1.5 (10 H, br s), 1.6 (2 H, m), 2.25 (1 H, br s), 2.28 (2 H, t, J=7.3 Hz), 3.64 (3 H, s), 3.65 (2 H, m), 4.53 (1 H, m, J=50.1 Hz); ^{13}C NMR ($CDCl_3$): δ 24.8 (2C), 28.8-29.1 (3C), 30.8 (d, J=20.4 Hz), 33.9, 51.4, 64.9 (d, J=22.9 Hz), 94.6 (d, J=167.8 Hz), 174.2; ^{19}F NMR ($CDCl_3$): δ -188.9 (m).

Synthetic method

Methyl 10-fluoro-9-hydroxydecanoate

To an argon-covered solution of 10.6 g (40 mmol) of 18-crown-6 and 7.8 g (100 mmol) of KHF_2 in 120 mL of refluxing dry DMF, 25 mmol of methyl 9,10-epoxydecanoate in 40 mL of dry DMF was added. The solution was refluxed for an additional 36 h. After cooling, it was poured into 600 mL of iced H_2O and the mixture was extracted five times with a very small amount of CCl_4. The combined organic layers were repeatedly washed with H_2O and dried with Na_2SO_4. Evaporation of the solvent gave methyl 10-fluoro-9-hydroxydecanoate together with about 18% of its regioisomer (9-fluoro-10-hydroxy) as a yellowish, viscose oil. The isomer was separated by flash chromatography over silica gel (cyclohexane:AcOEt=3:1) to furnish 2.7 g of methyl 10-fluoro-9-hydroxydecanoate in 49% yield.

Spectral data

mp. 4 °C; ^1H NMR ($CDCl_3$): δ 1.28-1.4 (8 H, br s), 1.48 (2 H, m), 1.62 (2 H, m), 2.16 (1 H, s), 2.30 (2 H, t, J=7.5 Hz), 3.66 (3 H, s), 3.78-3.93 (1 H, m), 4.26 (1 H, ddd, J=48.2, 9.4, 6.8 Hz), 4.41 (1 H, ddd, J=47.1, 9.4, 3.0 Hz); ^{13}C NMR ($CDCl_3$): δ 25.1, 25.4, 29.2-29.5 (3C), 32.1 (d, J=6.4 Hz), 34.3, 51.6, 70.6 (d, J=17.8 Hz), 87.2 (d, J=167.9 Hz), 174.5; ^{19}F NMR ($CDCl_3$): δ -228.4 (td, J=47.6, 18.1 Hz).

Note

The use of less acidic $Et_3N \cdot 3HF$ instead of 70% pyridine-$(HF)_n$ gave rise to lower regio-selectivity and chemical yields. See Sattler, A.; Haufe, G. *J. Fluorine Chem.* **1994**, *69*, 185.

See the following fluorohydrin formation *via* the epoxide ring opening: i) Poulter, C. D., *et al. J. Org. Chem.* **1988**, *53*, 1026, ii) Vogel, P., *et al. J. Chem. Soc., Chem. Commun.* **1990**, 1070, iii) Lundt, I., *et al. Tetrahedron* **1993**, *49*, 7295.

Sattler, A.; Haufe, G. *Liebigs Ann. Chem.* **1994**, 921.

(1*R*, 2*S*)-1-Fluoro-1-deoxyephedrine hydrochloride

a) KF-CaF$_2$, Kryptofix [2.2.2] / MeCN,   b) 20% H$_2$SO$_4$ aq. / Et$_2$O,   c) HCl / Et$_2$O

Synthetic method

A solution of cyclic sulfamidate, prepared from (1*S*,2*S*)-pseudoephedrine and thionyl chloride, followed by oxidation with ruthenium chloride, (227 mg, 1 mmol) in dry MeCN (10 mL) was treated with KF (320 mg of a mixture of KF:CaF$_2$=1:4, 1.1 mmol) followed by 376 mg (1 mmol) of 4,7,13,16,21,24-hexaoxa-1,10-diazabicyclo[8.8.8]hexacosane (Kryptofix [2.2.2]) and heated with stirring at 80 °C for 1 h. The reaction mixture was filtered hot and the residue rinsed with boiling MeCN (10 mL). The filtrate was concentrated, and the residue was treated with Et$_2$O (5 mL) and 20% H$_2$SO$_4$ aq. (3 mL) and stirred at room temperature for 18 h. The reaction mixture was basified (pH 8) with solid NaHCO$_3$, the Et$_2$O layer removed, and the aqueous layer extracted further with AcOEt (2x25 mL). The combined organic extracts were dried over anhydrous Na$_2$SO$_4$ and concentrated, and the residue was flash chromatographed on silica gel (AcOEt:MeOH=7:3) to give (1*R*,2*S*)-1-fluoro-1-deoxyephedrine as a pale yellow oil. A portion was converted to the corresponding hydrochloride salt by treatment with ethereal HCl and recrystallization from $CH_3OH:Et_2O$ in 54% yield.

Spectral data

$[\alpha]_D$ -21.7 (*c* 1.1, MeOH); mp. 180-182 °C (dec.; MeOH:Et_2O); ^1H NMR (CD_3OD): δ 1.17 (3 H, d, *J*=6.95 Hz), 2.84 (3 H, s), 3.64-3.74 (1 H, m), 6.05 (1 H, dd, *J*=48.0, 2.1 Hz), 7.38-7.49 (5 H, m); ^{13}C NMR (CD_3OD): δ 9.57, 31.68, 60.18, 60.42, 91.78, 93.75, 125.95, 126.04, 129.93, 130.07; ^{19}F NMR (CD_3OD): δ -191.2 (dd, *J*=47.9, 26.9 Hz).

Note

A similar type of fluorination by way of the corresponding cyclic sulfates has been reported. See Gao, Y.; Sharpless, K. B. *J. Am. Chem. Soc.* **1988**, *110*, 7538 and Charvillon, F. B.; Amouroux, R. *Tetrahedron Lett.* **1996**, *37*, 5103.

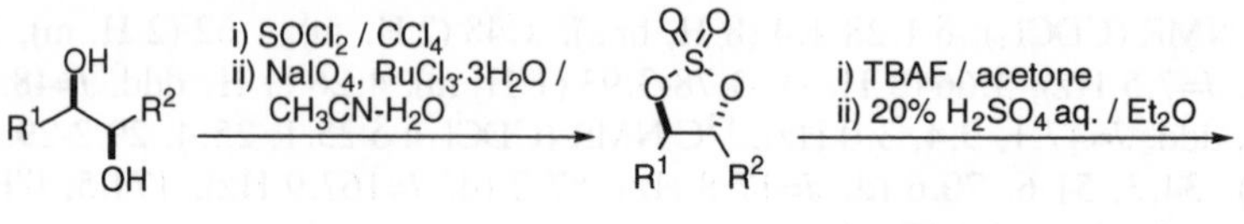

$R^1=R^2=CO_2Pr^{-i}$: 81-84% total yield
$R^1=n\text{-}C_{15}H_{31}$, $R^2=CO_2Me$: 54-60% total yield

van Dort, M. E., *et al. J. Med. Chem.* **1995**, *38*, 810.

Fluoromethyl benzoate

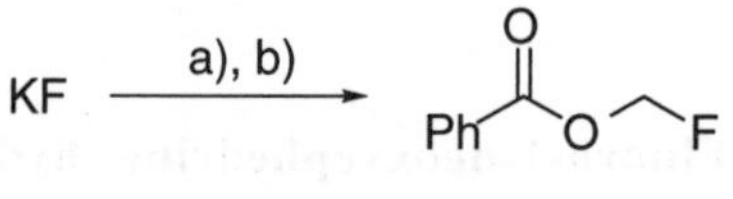

a) Formaldehyde / THF, b) PhC(O)F, 18-crown-6

Synthetic method

A stream of gaseous formaldehyde (from 3.75 g, 125 mmol of paraformaldehyde) and nitrogen was bubbled over 4 h into a vigorously stirred suspension of KF (5.81 g, 100 mmol) in anhydrous THF (50 mL) containing benzoyl fluoride (50 mmol) and 18-crown-6 (2.0 g, 7.5 mmol). After centrifugation, the supernatant solution was decanted and evaporated; the residue was distilled to give fluoromethyl benzoate in 42% yield.

Spectral data

bp. 82-83 °C/10 mmHg; mp. 10-12 °C; ^1H NMR ($CDCl_3$): δ 5.98 (2 H, d, *J*=50.5 Hz), 7.5 (2 H, m), 7.7 (1 H, m), 8.1 (2 H, m); ^{19}F NMR ($CDCl_3$): δ -94.0 (t, *J*=50 Hz).

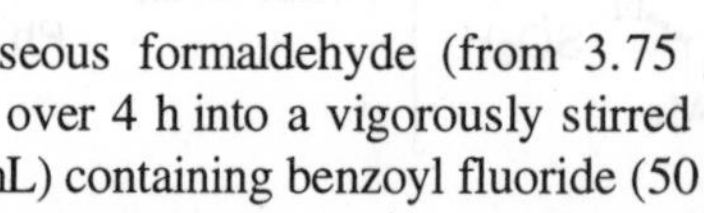

R	Yield (%)
(furyl)	52
$c\text{-}C_6H_{11}$	50
$n\text{-}C_6H_{13}$	49

Schlosser, M.; Limat, D. *Tetrahedron* **1995**, *51*, 5807.

2-Fluoro-2-methylpropionic acid

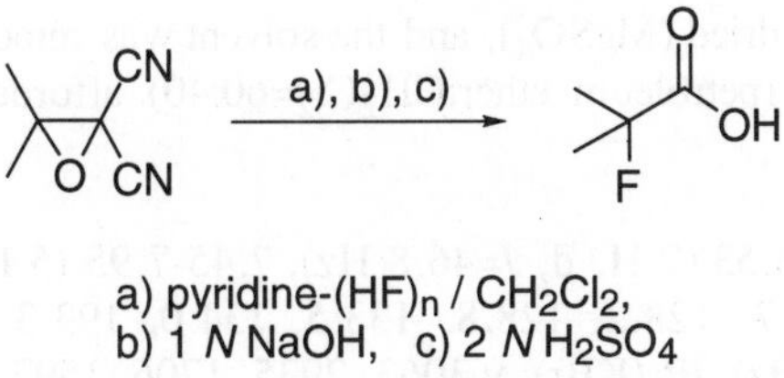

a) pyridine-(HF)$_n$ / CH$_2$Cl$_2$,
b) 1 N NaOH,   c) 2 N H$_2$SO$_4$

Synthetic method

70% Pyridine-(HF)$_n$ (10 mL), CH$_2$Cl$_2$ (10 mL) and 1,1-dicyano-2-methyl-1,2-epoxy-propane (10 mmol) was stirred at 25 °C for 72 h in a polyethylenic flask.   The mixture was shaken with cold H$_2$O (50 mL) and extracted with Et$_2$O.   The Et$_2$O layer and 1 N NaOH aq. was stirred for 3 h, then treated with 2 N H$_2$SO$_4$ aq. until acidic pH was attained.   The organic layer was dried over anhydrous MgSO$_4$, the solvent was evaporated and the residue was purified by chromatography on silica gel.   The yield was 60%.

Spectral data

bp. 90 °C/15 mmHg; ^{1}H NMR : δ 1.63 (6 H, d, J=22.4 Hz), 6.00 (1 H, br s); ^{19}F NMR : δ -148.6 (hept., J=22.4 Hz); IR (CHCl$_3$): ν 3500, 3400, 1720 cm^{-1}.

Note

Slight modification of this method allowed the direct formation of 2-fluoroesters or amides in a range of 60 to 70% yields.   Construction of 3-fluoropyruvates was also possible by a similar reaction starting with 1-cyano-1,2-epoxyesters.   See Baklouti, A., *et al. Synth. Commun.* **1996**, *26*, 237.   Oliver and his co-workers reported the reaction of 1,1,1-trichloroalkan-2-ols with TBAF and CsF in THF, furnishing 2-fluorocarboxylic acids by way of 1,1-dichloro-1,2-epoxides.   See Oliver, J. E., *et al. Synthesis* **1994**, 273, and *Tetrahedron: Asym.* **1996**, *7*, 37.

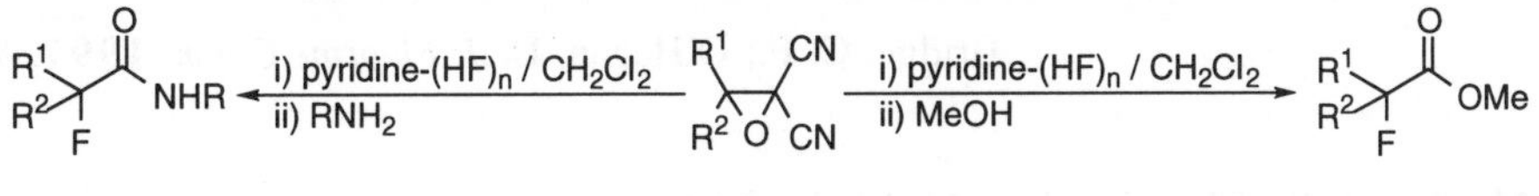

Baklouti, A., *et al. Synth. Commun.* **1996**, *26*, 1155.

2-Fluoroacetophenone,
Methyl 4-fluoro-5-oxo-5-phenylpentanoate

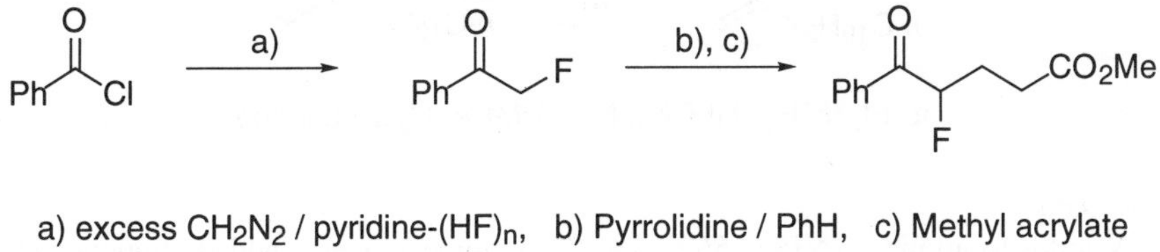

a) excess CH$_2$N$_2$ / pyridine-(HF)$_n$,   b) Pyrrolidine / PhH,   c) Methyl acrylate

Synthetic method

2-Fluoroacetophenone

To a solution of 70% pyridinium poly(hydrogen fluoride) (30 mL) at -15 °C was slowly added an ethereal solution of diazoacetophenone which was prepared *in situ* from benzoyl chloride (5.03 g, 35.8 mmol) and an excess of diazomethane.   The solution was allowed to warm to room temperature and stirred for 4 h.   The product was isolated by extraction into pentane (300 mL), and hydrogen fluoride was removed by treatment of the extract with anhydrous KF until neutral. The combined extracts were dried ($MgSO_4$), and the solvent was removed under reduced pressure. Purification over silica gel (petroleum ether:CH_2Cl_2=60:40) afforded 2.20 g of 2-fluoroaceto-phenone in 45% yield.

Spectral data

^{1}H NMR ($CDCl_3$): δ 5.53 (2 H, d, J=46.8 Hz), 7.45-7.95 (5 H, m); ^{13}C NMR ($CDCl_3$): δ 83.37 (d, J=182 Hz), 127.7, 128.2, 128.8, 133.5, 134.0, 193.3 (d, J=15.3 Hz); ^{19}F NMR ($CDCl_3$): δ -231.8 (t, J=46 Hz); IR (KBr): ν 3063, 2935, 1706, 1597 cm^{-1}.

Synthetic method

Methyl 4-fluoro-5-oxo-5-phenylpentanoate

Pyrrolidine (1.5 mmol) was added to a solution of 2-fluoroacetophenone (1 mmol) in benzene over 4 Å molecular sieves (5 g).   The solution was heated under reflux and the progress of the reaction was monitored by ^{19}F NMR.   On consumption of the ketone, methyl acrylate (10 mmol) was added and the solution was heated under reflux for 12 h. H_2O (10 mL) was added and the solution was heated for a further 1 h.   After cooling, the reaction mixture was filtered through Celite and the product was extracted into Et_2O.   The ethereal extracts were combined and washed successively with dilute HCl (25 mL), $NaHCO_3$ (25 mL) and H_2O (25 mL), dried ($MgSO_4$) and the solvent was removed under reduced pressure.   Purification over silica gel (petroleum ether: CH_2Cl_2=60:40) afforded 137 mg of methyl 4-fluoro-5-oxo-5-phenylpentanoate in 60% yield as a clear oil.

Spectral data

^{1}H NMR ($CDCl_3$): δ 2.25 (2 H, m), 2.58 (2 H, m), 3.69 (3 H, s), 5.75 (1 H, ddd, J=49.2, 9.2, 3.2 Hz), 7.45-8.01 (5 H, m); ^{13}C NMR ($CDCl_3$): δ 27.6 (d, J=21.4 Hz), 28.6 (d, J=3.4 Hz), 51.8, 91.8 (d, J=182.8 Hz), 128.8, 128.8, 128.8, 133.9, 173.0, 195.7 (d, J=18.7 Hz); ^{19}F NMR ($CDCl_3$): δ -193.02 (ddd, J=50.8, 19.2 Hz); IR (KBr): ν 2953, 1737, 1700, 1597 cm^{-1}.

Bridge, C. F.; O'Hagan, D. *J. Fluorine Chem.* **1997**, *82*, 21.

2.1.2 Nucleophilic Fluorination (Addition)

2-Fluoro-1-iodododecane

a) $Et_2NCF_2CHFCF_3$, NIS, HMPA, H_2O / toluene

Synthetic method

To *N*-iodosuccinimide (NIS, 562 mg, 2.5 mmol) were added toluene (1.5 mL), $Et_2NCF_2CHFCF_3$ (0.87 mL, 2.5 mmol), HMPA (0.35 mL, 2.0 mmol), and H_2O (45 μL, 2.5 mmol) successively at -30 °C, and the mixture was stirred at that temperature for 15 min.   A solution of 1-dodecene (169 mg, 1.0 mmol) in toluene (1.5 mL) was added and the reaction mixture was allowed to stand at room temperature for 24 h.   The usual work-up followed by purification by silica gel column chromatography gave 2-fluoro-1-iodododecane (245 mg, 78%) as a colorless oil.

Spectral data

^{1}H NMR (CCl$_4$): δ 0.68-2.20 (21 H, m), 3.22 (2 H, dd, J=6.0, 18.0 Hz), 4.43 (1 H, dm, J=48.0 Hz); ^{19}F NMR (CCl$_4$): δ -173.0 (m) from CF$_3$CO$_2$H; IR (neat): v 2940, 2860, 1470, 1420, 1380, 1195, 995, 800, 730, 640 cm^{-1}.

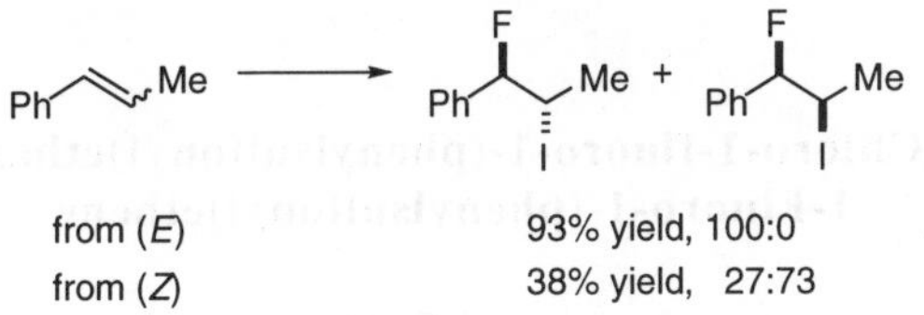

from (*E*)   93% yield, 100:0

from (*Z*)   38% yield, 27:73

Fujisawa, T., *et al. Bull. Chem. Soc. Jpn.* **1991**, 64, 2596.

5-Bromo-4-fluoropent-1-yl acetate

a) NBS, Et$_3$N·3HF / CH$_2$Cl$_2$

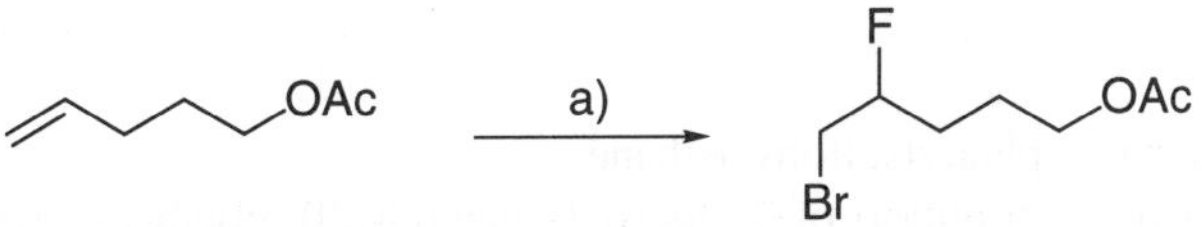

Synthetic method

N-Bromosuccinimide (NBS, 41 g, 0.23 mol) and, in the course of 15 min, Et$_3$N·3HF (37 mL, 0.23 mol) were added to a solution of pent-4-en-1-yl acetate (27 mL, 0.20 mol) in CH$_2$Cl$_2$ (400 mL).   After stirring for 10 h, the orange solution was filtered, washed with brine (2x50 mL), dried (Na$_2$SO$_4$) and evaporated.   Distillation afforded 32 g of pure 5-bromo-4-fluoropent-1-yl acetate in 71% yield.

Spectral data

bp. 99-100 °C/4 mmHg; ^{1}H NMR (CDCl$_3$): δ 1.80 (4 H, m), 2.07 (3 H, s), 3.49 (2 H, dd, J=19.1, 5.3 Hz), 4.12 (2 H, m), 4.68 (1 H, dquint, J=48.0, 5.5 Hz); ^{19}F NMR (CDCl$_3$): δ -179.2 (ddq, J=47.8, 28.8, 9.5 Hz).

Note

When dimethyl(methylthio)sulfonium tetrafluoroborate (Me$_3$S$^+$·BF$_4^-$) was employed instead of NBS, a methylthio group could be introduced at the same site as bromine.   See Haufe, G., *et al. J. Org. Chem.* **1992**, 57, 714.

Schlosser's group also reported the highly stereoselective formation of monofluoroolefins *via* the above bromofluorination followed by the elimination of HBr.   See Schlosser, M., *et al.*

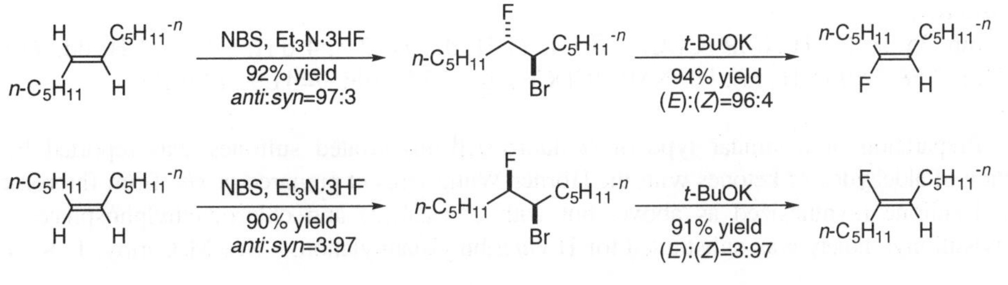

Tetrahedron **1990**, *46*, 4255.

Michel, D.; Schlosser, M. *Synthesis* **1996**, 1007.

2.1.3 Nucleophilic Fluorination (Miscellaneous)

2-Chloro-1-fluoro-1-(phenylsulfonyl)ethane,
1-Fluoro-1-(phenylsulfonyl)ethene

a) DAST, cat. $SbCl_3$ / CH_2Cl_2, b) *m*CPBA / CH_2Cl_2, c) DBU / CH_2Cl_2

Synthetic method

2-Chloro-1-fluoro-1-(phenylsulfonyl)ethane

A mixture of 3.0 g (16 mmol) of 2-chloro-1-(phenylsulfinyl)ethane, 100 mg (0.4 mmol) of antimony(III) chloride, and CH_2Cl_2 (100 mL) was treated with 4.2 mL (32 mmol) of DAST at room temperature.   The progress of the reaction was followed by GLC, and after 1 h, the light yellow solution was washed with $NaHCO_3$ aq., dried (K_2CO_3), and filtered.   The solution containing the 2-fluorosulfide was treated with 8.6 g (40 mmol) of 80% *m*CPBA and stirred at room temperature for 6 h.   The reaction was filtered, and the filtrate was washed with $NaHSO_3$ aq. and $NaHCO_3$ aq., dried ($MgSO_4$), and concentrated *in vacuo*.   Purification by flash chromatography (hexane:AcOEt=6:1) gave 2.3 g (64% yield) of 2-chloro-1-fluoro-1-(phenylsulfonyl)ethane.
Spectral data

mp. 74-76 °C; ^{1}H NMR ($CDCl_3$): δ 3.78 (1 H, ddd, *J*=13.7, 12.9, 9.5 Hz), 4.16 (1 H, ddd, *J*=32.9, 12.9, 2.2 Hz), 5.27 (1 H, ddd, *J*=48.3, 9.5, 2.3 Hz), 7.61-7.98 (5 H, m); ^{19}F NMR ($CDCl_3$): δ -180.68 (ddd, *J*=47.9, 33.4, 14.1 Hz).

Synthetic method

1-Fluoro-1-(phenylsulfonyl)ethylene

To a mixture of 2-chloro-1-fluoro-1-(phenylsulfonyl)ethane (20.9 g, 93.9 mmol) and CH_2Cl_2 (200 mL) was slowly added 15.2 g (100 mmol) of DBU.   After stirring for 2 h at room temperature, GLC showed the disappearance of the starting material.   The reaction mixture was washed with 1 *N* HCl, dried ($MgSO_4$), and concentrated.   The resulting oil was dried under high vacuum for several hours and slowly crystallized, providing 15.1 g (86% yield) of 1-fluoro-1-(phenylsulfonyl)ethylene as light tan crystals.
Spectral data

mp. 35-38 °C; ^{1}H NMR ($CDCl_3$): δ 5.43 (1 H, dd, *J*=12.5, 4.6 Hz), 5.88 (1 H, dd, *J*=41.8, 4.6 Hz), 7.58-7.99 (5 H, m); ^{19}F NMR ($CDCl_3$): δ -115.52 (dd, *J*=41.9, 12.6 Hz).
Note

Preparation of a similar type of α-fluoro-α,β-unsaturated sulfones was reported by the reaction of aldehydes or ketones with the Hörner-Wittig reagent prepared *in situ* from fluoromethyl phenyl sulfone (synthesized as above, but without catalyst) and chlorodiethylphosphate.   The phenylsulfonyl moiety was substituted for H *via* tributylstannylation.   See McCarthy, J. R., *et al.*

Tetrahedron **1996**, *52*, 45.

i) DAST
ii) *m*CPBA   i) LHMDS
ii) ClP(O)(OEt)$_2$
PhS(O)CH$_3$ ————————→ PhSO$_2$CH$_2$F ————————→
iii) R^{1}R^{2}C=O

R^1 SO$_2$Ph   i) *n*-Bu$_3$SnH   R^1 H
  ═   ii) NaOMe   ═
R^2 F   ————————→   R^2 F

Matthew, D. P.; McCarthy, J. R. *J. Org. Chem*. **1990**, *55*, 2973.

2.1.4 Electrophilic Fluorination

(4R,5S)-3-[(R)-3-(Benzyloxy)-2-fluoropropionyl]-4-methyl-5-phenyloxazolidin-2-one

BnO————————(structure)———————— a), b) ————————→ BnO————————(structure)
              Me   Ph   Me   Ph

a) NaHMDS / THF,   b) (PhSO$_2$)$_2$NF / THF

Synthetic method

In an oven-dried 250 mL, one-neck, round-bottomed flask equipped with a rubber septum, an argon inlet and a magnetic stir bar was placed 4.02 g (11.85 mmol) of (4R,5S)-3-[3-(benzyloxy)propionyl]-4-methyl-5-phenyloxazolidin-2-one in THF (90 mL). The reaction mixture was cooled to -78 °C and 11.85 mL (11.85 mmol, 1.0 *M* in THF) of NaHMDS (sodium hexamethyldisilazide) was added *via* a syringe. After stirring the reaction mixture at -78 °C for 0.5 h, it was cannulated to 4.85 g (15.4 mmol) of *N*-fluorobenzenesulfonimide in THF (28 mL) precooled to -78 °C. The solution was stirred at this temperature for 0.5 h, quenched with sat. NH$_4$Cl (5 mL) and diluted with AcOEt (20 mL). After warming to room temperature, 2 mL of sat. KI aq. was added and the resulting I$_2$ solution was treated with a sat. Na$_2$S$_2$O$_3$ solution until the iodine color disappeared. The solution was filtered through Celite and the organic phase was washed with H$_2$O (2x20 mL), brine (2x20 mL), dried over MgSO$_4$ and concentrated. The product was purified by flash chromatography (hexane:AcOEt=82:18) to afford 2.83 g of (4R,5S)-3-[(R)-3-(benzyloxy)-2-fluoropropionyl]-4-methyl-5-phenyloxazolidin-2-one in 68% yield with >99% diastereomeric excess.

Spectral data

[α]$_D^{20}$ +27.9 (*c* 0.98, CHCl$_3$); ^{1}H NMR (CDCl$_3$): δ 0.96 (3 H, d, *J*=6.6 Hz), 3.9-4.13 (2 H, m), 4.54-4.75 (3 H, m), 5.56 (1 H, d, *J*=6.96 Hz), 6.11-6.22 (1 H, dm, *J*=48.7 Hz), 7.26-7.44 (10 H, m); ^{13}C NMR (CDCl$_3$): δ 14.9, 56.0, 69.6 (d, *J*=22.4 Hz), 74.3, 80.7, 90.0 (d, *J*=183.1 Hz), 126.2, 128.6, 129.1, 129.5, 129.7, 133.2, 138.2, 153.4, 167.5 (d, *J*=22.4 Hz); ^{19}F NMR (CDCl$_3$): δ -195.5 (m); IR (neat): ν 1781, 1718 cm^{-1}.

Note

See the following reports dealing with the same type of reactions: i) Fukumoto, K., *et al. J.*

Chem. Soc. Perkin Trans. 1 **1992**, 221, ii) Staunton, J., *et al. Tetrahedron Lett.* **1996**, *37*, 3515, iii) Hoffman, R. V., *et al. Tetrahedron Lett.* **1998**, *39*, 4195.   In the second procedure, the diastereoselective electrophilic fluorination followed by the methylation of the resultant intermediate was performed for acylated oxazolidinone, while, in the usual case, the second reaction does not occur because of the difficulty in formation of the requisite enolate by the severe steric requirement.

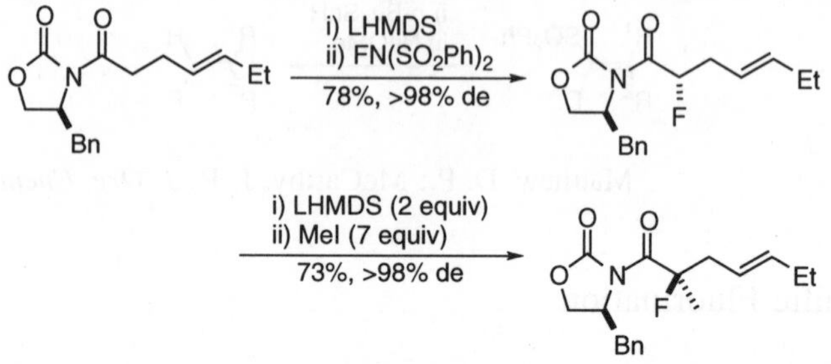

Enders and his coworkers recently reported the regio- as well as diastereoselective construction of α-fluorinated ketones.   See Enders, D., *et al. Angew. Chem. Int. Ed. Engl.* **1997**, *36*, 2362.

Davis, F. A., *et al. J. Org. Chem.* **1997**, *62*, 7546.

Ethyl 2-fluoro-1-oxo-2-cyclopentanecarboxylate

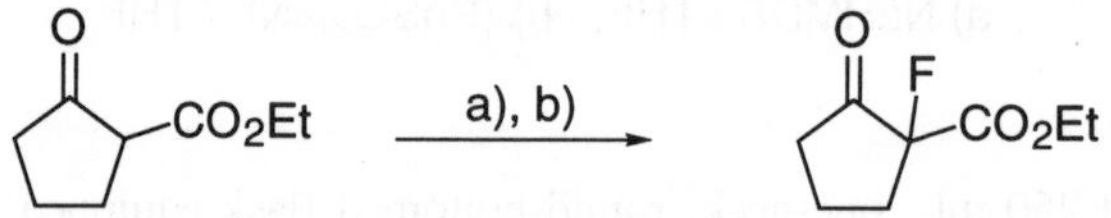

a) NaH / THF,   b) *N*-fluoro-4,6-dimethylpyridinium-2-sulfonate

Synthetic method

Under an argon atmosphere, 3.6 mmol of *N*-fluoro-4,6-dimethylpyridinium-2-sulfonate was added in several portions to a THF solution at room temperature of the sodium salt, which was prepared *in situ* treating 3 mmol of ethyl 1-oxo-2-cyclopentanecarboxylate with 3 mmol of 60% NaH in oil in 24 mL of THF at 0 °C.   After 1 h, the reaction mixture was poured into dilute HCl and extracted with Et$_2$O.   The extract was washed with NaHCO$_3$ aq. and then with H$_2$O, dried with anhydrous MgSO$_4$, filtered, and evaporated. ^{19}F NMR of the resulting residue using fluorobenzene as an internal standard showed that the fluorinated product was formed in 83% yield.
Spectral data

^{1}H NMR (CDCl$_3$): δ 1.34 (3 H, t, *J*=7.1 Hz), 2.05-2.60 (6 H, m), 4.28 (2 H, q, *J*=7.1 Hz); ^{19}F NMR (CDCl$_3$): δ -164 (t, *J*=20 Hz); IR (neat): ν 1780, 1725 cm^{-1}.
Note

The above fluorinating reagent also enables the electrophilic fluorination of aromatic compounds, enol silyl ethers, and others.   Substituents on the pyridine ring are believed to control the reactivity of the fluorinating reagents; thus, the electron-withdrawing character increases the fluorination ability but the electron-donating group lowers the reactivity.   See also the following work reported previously, Umemoto, T., *et al. J. Am. Chem. Soc.* **1990**, *112*, 8563.

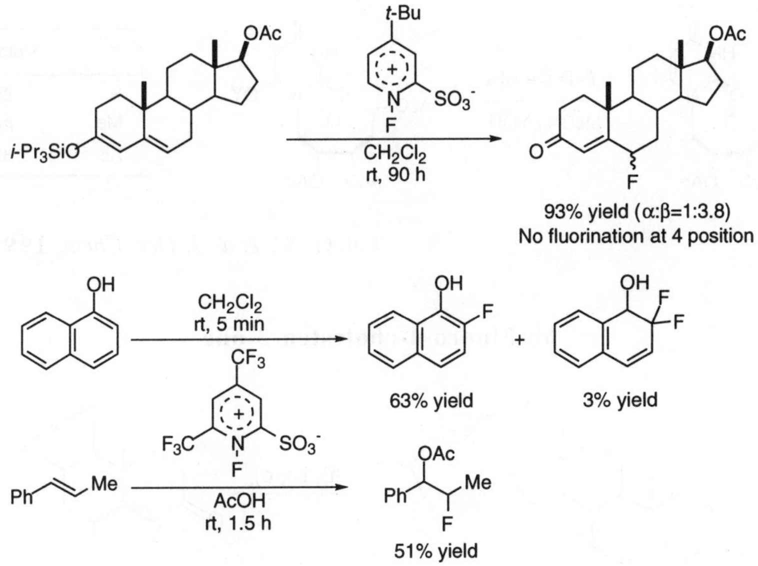

93% yield (α:β=1:3.8)
No fluorination at 4 position

63% yield   3% yield

51% yield

Umemoto, T.; Tomizawa, G. *J. Org. Chem.* **1995**, *60*, 6563.

trans-5-Fluoro-6-hydroxy-5,6-dihydrouracil

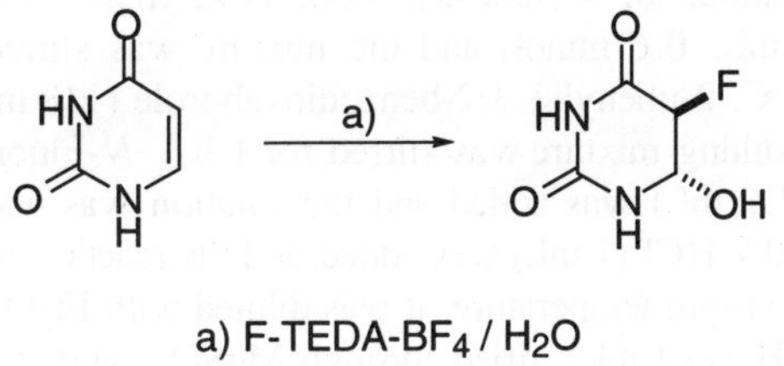

a) F-TEDA-BF$_4$ / H$_2$O

Synthetic method

A suspension of uracil (896 mg, 8 mmol) in H$_2$O (25 mL) was treated with 1-(chloromethyl)-4-fluoro-1,4-diazabicyclo[2.2.2]octane bis(tetrafluoroborate) (F-TEDA-BF$_4$ or SELECTFLUOR; 2.83 g, 8 mmol) and heated under nitrogen for 4 h at 90 °C.   On cooling, a solution of sodium tetraphenylborate (6.08 g, 17.79 mmol) in H$_2$O (25 mL) was added, and the resulting precipitate was filtered.   The filtrate was evaporated *in vacuo*, and the residue was sublimed at 210-220 °C (0.5 mm) to obtain 835 mg of the product in 82% yield as an 8:1 *trans*:*cis* mixture.

Spectral data

^{1}H NMR (DMSO-d_6): δ 7.75 (1 H, d), 10.7 (s, br), 11.5 (1 H, s, br); ^{19}F NMR (DMSO-d_6): δ -170 (d).

Note

For the recent reviews, see i) Pez, G. P., *et al. Chem. Rev.* **1996**, *96*, 1737, ii) Banks, R. E. *J. Fluorine Chem.* **1998**, *87*, 1.

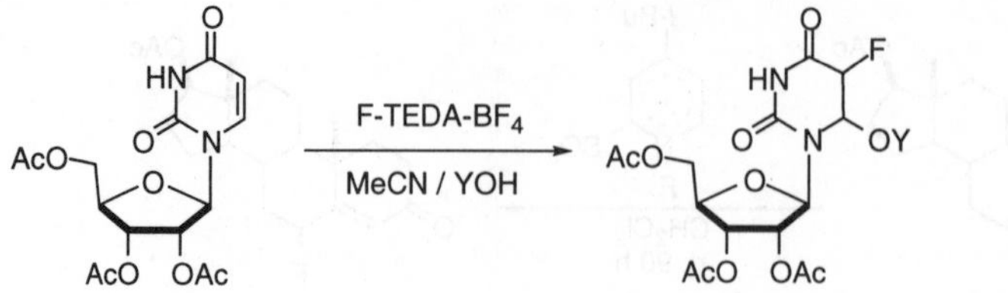

Y	Yield (%)
H	80
Me	85
Ac	82

Lal, G. S., *et al.* *J. Org. Chem.* **1995**, *60*, 7340.

6β-Fluoro-4-cholesten-3-one

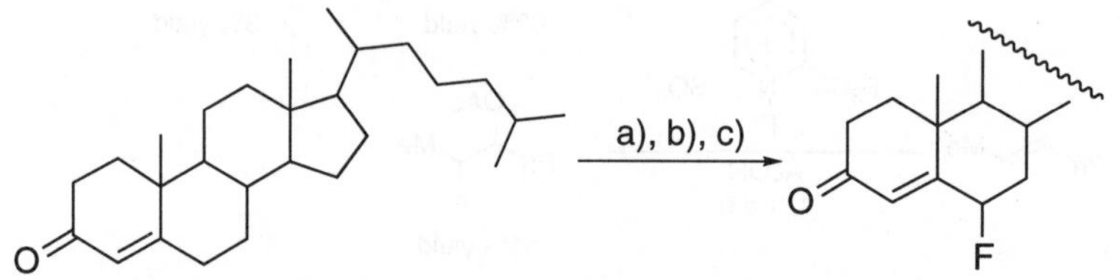

a) KH, HMPA / THF,   b) 2-phenyl-1,3,2-benzodioxaborole,   c) (PhSO₂)₂NF

Synthetic method

To a suspension of KH (69 mg, 0.6 mmol, washed with pentane to remove the oil) in THF (0.6 mL) was added a solution of 4-cholesten-3-one (192 mg, 0.5 mmol) in THF (0.5 mL), followed by HMPA (0.1 mL, 0.6 mmol) and the mixture was stirred at 22 °C for 1 h.   The mixture was cooled to -78 °C, 2-phenyl-1,3,2-benzodioxaborole (118 mg, 0.6 mmol) in THF (0.6 mL) was added and the resulting mixture was stirred for 1 h.   N-Fluorobenzenesulfonimide (193 mg, 0.6 mmol) in THF (1.2 mL) was added and the solution was warmed to room temperature over 2 h.   A solution of 10% HCl (1 mL) was added and the reaction was heated at 60 °C for 1 h. After cooling the reaction to room temperature, it was diluted with Et₂O (2 mL), washed with 10% HCl (2x2 mL), 10% NaOH (2x2 mL), dried through MgSO₄, and evaporated to afford 131 mg (62% yield) of a 1:7.8 mixture of 6α-/6β-fluoro-4-cholesten-3-one.

Spectral data

^{1}H NMR: δ 4.97 (1 H, dt, J=49.2, 2.6 Hz); ^{19}F NMR: δ -165.7 (dt, J=48.2, 10.6 Hz).

6α isomer

^{1}H NMR: δ 4.97 (1 H, dddd, J=48.0, 12.1, 6.2, 1.8 Hz); ^{19}F NMR: δ -183.4 (ddd, J=48.2, 9.2, 5.3 Hz).

Poss, A. J.; Shia, G. A. *Tetrahedron Lett.* **1995**, *36*, 4721.

2.2 Carbon-Carbon Bond-forming Reactions

2.2.1 Aldol Type Reactions

1-Fluoro-3-nitro-4-phenylbutan-2-ol,
3-Amino-1-fluoro-4-phenylbutan-2-ol

a) (COCl)$_2$, DMSO / CH$_2$Cl$_2$, b) Et$_3$N, c) PhCH$_2$CH$_2$NO$_2$, d) H$_2$, Raney Ni / EtOH

Synthetic method
1-Fluoro-3-nitro-4-phenylbutan-2-ol
To a cooled (-78 °C) solution of oxalyl chloride (2 M) in CH$_2$Cl$_2$ (23.2 mmol in 11.60 mL) was added slowly dimethyl sulfoxide (3.65 g, 3.32 mL, 46.7 mmol).   The reaction mixture was stirred for 15 min.   A solution of 2-fluoroethanol (1.16 g, 18.1 mmol) in CH$_2$Cl$_2$ (10 mL) was slowly introduced into the reaction flask.   After another 15 min of stirring, the reaction mixture was diluted with anhydrous CH$_2$Cl$_2$ (180 mL), and triethylamine (9.20 g, 12.63 mL, 90 mmol) was added to it.   Stirring was continued for another 2 h by which time the temperature changed to room temperature.   A solution of 1-nitro-2-phenylethane (2.74 g, 18.1 mmol) in anhydrous CH$_2$Cl$_2$ (10 mL) was added to the reaction mixture, and stirring was continued overnight.   The mixture was then washed with H$_2$O (2x20 mL), 4% HCl (3x20 mL), H$_2$O (2x20 mL), sat. NaHCO$_3$ (2x20 mL), and brine (2x20 mL).   Drying (Na$_2$SO$_4$) of the organic phase and solvent evaporation gave a crude material which was purified by silica gel flash chromatography (hexane: AcOEt=3:1) to give 1-fluoro-3-nitro-4-phenylbutan-2-ol as a mixture of *threo* and *erythro* diastereomers in a combined yield of 78%.
Spectral data
    Isomer I
    mp. 71-73 °C; Rf 0.46 (hexane:AcOEt=7:3); ^{1}H NMR (CDCl$_3$): δ 2.70 (1 H, d), 3.25-3.45 (2 H, m), 4.30-4.50 (2 H, m), 4.60 (1 H, m), 4.90 (1 H, m), 7.10-7.40 (5 H, m).
    Isomer II
    Rf 0.42 (hexane:AcOEt=7:3); ^{1}H NMR (CDCl$_3$): δ 2.90 (1 H, d), 3.30-3.40 (2 H, m), 4.20 (1 H, m), 4.50 (1 H, m), 4.65 (1 H, m), 4.90 (1 H, m), 7.15-7.40 (5 H, m).

Synthetic method
    3-Amino-1-fluoro-4-phenylbutan-2-ol
    A mixture of intermediate 1-fluoro-3-nitro-4-phenylbutan-2-ol (Isomer I, 0.48 g, 2.25 mmol), absolute EtOH (20 mL), and Raney nickel (50% slurry in H$_2$O, catalytic) was hydrogenated (60 psi) in a Parr apparatus for 5 h.   Filtration through a Celite pad and solvent evaporation gave 0.41 g of 3-amino-1-fluoro-4-phenylbutan-2-ol.   Similar treatment of 1-fluoro-3-nitro-4-phenylbutan-2-ol (Isomer II, 0.80 g, 3.75 mmol) gave 0.51 g of 3-amino-1-fluoro-4-phenylbutan-2-ol.
Spectral data
    Isomer I
    mp. 64-67 °C; ^{1}H NMR (CDCl$_3$): δ 1.70-2.20 (3 H, m), 2.50 (1 H, q), 2.95 (1 H, dd), 3.20

(1 H, m), 3.80 (1 H, m), 4.50 (1 H, d), 4.70 (1 H, d), 7.30 (5 H, m).
  Isomer II
  mp. 67-70 °C; ^{1}H NMR (CDCl$_3$): δ 1.65-2.20 (3 H, m), 2.55 (1 H, q), 2.95 (1 H, dd), 3.10
(1 H, m), 3.60 (1 H, m), 4.55 (1 H, d), 4.70 (1 H, d), 7.20 (5 H, m).
Caution
  2-Fluoroethanol, like fluoroacetic acid and its derivatives, is extremely toxic (LD$_{50}$ 10 mg/kg
for mice, in the case of fluoroacetic acid, LD$_{50}$ 0.6 mg/kg), and it should be handled with extreme
caution in an efficient fume hood.
Note
  The aminoalcohols obtained were further transformed into the target monofluorinated ketones
as the inhibitor of human calpain I.   *In situ* preparation of fluoroacetaldehyde and its use, see i)
Yoshioka, K., *et al. J. Org. Chem.* **1984**, *49*, 1427, ii) Teutsch, G.; Bonnet, A. *Tetrahedron Lett.*
1984, *25*, 1561.
  For a recent report on the corresponding trifluorinated materials, see Ogilvie, W., *et al. J.
Med. Chem.* **1997**, *40*, 4113.

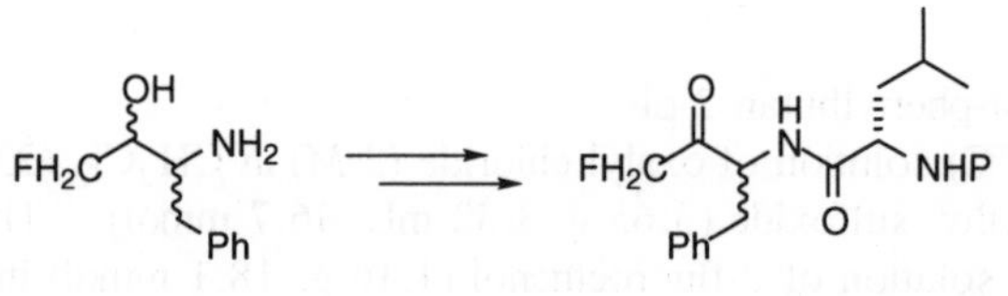

Chatterjee, S., *et al. J. Med. Chem.* **1997**, *40*, 3820.

(3R^*, 4R^*)-3-Fluoro-3-methyl-1,4-diphenylazetidin-2-one

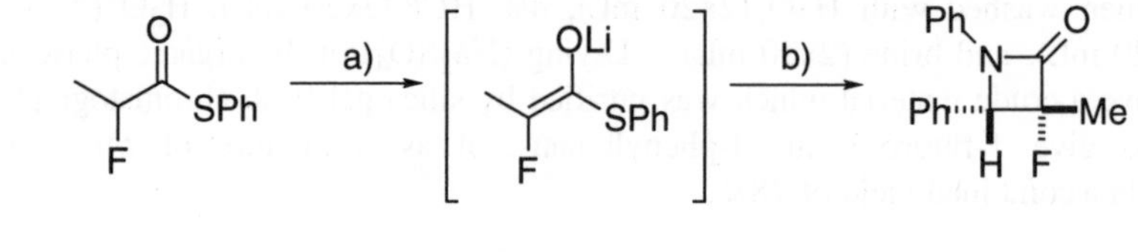

a) LDA / THF,   b) PhCH=NPh / THF

Synthetic method
  To a THF solution of LDA (2.2 mmol) was gradually added a solution of *S*-phenyl 2-fluoro-
propanethioate (0.368 g, 2.0 mmol) in THF (1 mL) at -78 °C under argon.   After stirring for 15
min at the same temperature, a solution of *N*-benzylideneaniline (0.543 g, 3.0 mmol) in THF (1
mL) was added dropwise to the reaction mixture.   The whole was stirred for 4 h at room
temperature and then poured into cold NH$_4$Cl aq.   The resultant mixture was extracted with Et$_2$O
(3x25 mL) and with CHCl$_3$ (2x25 mL). The combined extracts were washed with brine, dried over
NaSO$_4$ and concentrated *in vacuo*.   The residue was chromatographed on silica gel with hexane:
benzene=1:2 and benzene to furnish analytically pure (3R^*,4R^*)-3-fluoro-3-methyl-1,4-diphenyl-
azetidin-2-one (0.373 g) in 73% yield.   (3R^*,4R^*):(3R^*,4S^*)=>97:<3.
Spectral data
  mp. 154.3-154.7 °C; ^{1}H NMR (CDCl$_3$): δ 1.77 (3 H, d, *J*=22.0 Hz), 4.97 (1 H, d, *J*=3.6
Hz), 7.0-7.4 (5 H, m), 7.51 (5 H, s); ^{19}F NMR (CDCl$_3$): δ -82.5 (qd, *J*=22.0, 3.6 Hz) from
CF$_3$CO$_2$H; IR (KBr): ν 3032, 2988, 1744, 1600, 1498, 1460, 1391, 1366, 1206, 1149, 1114,
1080, 1050, 1027, 960, 898, 843, 825, 766, 750, 683 cm^{-1}.

Note

For the preparation of *S*-phenyl 2-fluoropropanethioate, see page 37 of this volume.

A highly *erythro* selective crossed aldol reaction of *S*-phenyl 2-fluoropropanethioate was also reported: Ishihara, T., *et al. Tetrahedron Lett.* **1995**, *36*, 8267. See also the following article on the construction of the same structure: Yokoyama, Y.; Mochida, K. *Synlett* **1998**, 37.

See also the following article on the aldol reaction of ethyl 2-fluoropropionate: Wildonger, K. J., *et al. Heterocycles* **1995**, *41*, 1891.

R[1]	R[2]	Yield	Isomer ratio
		(%)	*trans*:*cis* (R[1],Me)
p-MeOC$_6$H$_4$-	Ph	55	>97 : <3
p-MeC$_6$H$_4$-	Ph	58	>97 : <3
p-ClC$_6$H$_4$-	Ph	76	97 : 3
Ph-	*p*-MeOC$_6$H$_4$-	44	>97 : <3
CF$_3$	Ph	44	>97 : <3

(Reproduced with permission from Ishihara, T., *et al. Tetrahedron* **1996**, *52*, 255)

Ishihara, T., *et al. Tetrahedron* **1996**, *52*, 255.

anti-Ethyl 2-fluoro-3-hydroxy-3-phenylpropionate

a) LHMDS / THF, b) PhCHO

Synthetic method

To a stirred round-bottomed flask containing 30 mmol of LHMDS (prepared from hexamethyldisilazane and a 1.5 *M* solution of MeLi·LiBr in Et$_2$O) in dry THF (50 mL) were added dropwise at -85 °C under an argon atmosphere 1.79 g of HMPA (10 mmol) and 0.97 mL of ethyl fluoroacetate (10 mmol) as rapidly as possible while not allowing the temperature to rise above -85 °C. After 5 additional min, 5-10 mmol of benzaldehyde was added quickly. The mixture was allowed to stir for 10 additional min then quenched at -85 °C with 5 mL of sat. NH$_4$Cl aq. On warming to room temperature the mixture was diluted with 60 mL of distilled hexanes. Separation followed by washing with four 100-mL portions of H$_2$O, drying over MgSO$_4$, and evaporation of the solvent *in vacuo* yielded the crude product in 93% yield as a 2:1 *anti*:*syn* diastereomer mixture.

Spectral data

^{1}H NMR (CDCl$_3$): δ 1.00 (3 H, t, *J*=7 Hz), 3.9 (2 H, q, *J*=7 Hz), 4.2-5.2 (2 H, m), 7.2 (5 H, m); IR (neat): ν 3460, 3035, 3015, 1750, 1210, 730 cm^{-1}.

Caution

Ethyl fluoroacetate is extremely poisonous, causing convulsions, and ventricular fibrillation. It should only be handled using syringe techniques in an efficient hood.

Note
    See the following articles in relation to the present aldol condensation: Welch, J. T.; Seper, K. *Tetrahedron Lett.* **1984**, *25*, 5247 (with fluorinated ketones), Welch, J. T.; Plummer, J. S. *Synth. Commun.* **1989**, *19*, 1081 (other fluorinated esters), Welch, J. T.; Eswarakrishnan, S. *J. Org. Chem.* **1985**, *50*, 5403 (fluorinated amides).

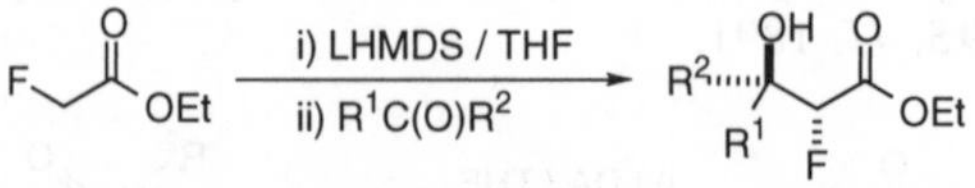

R^1	R^2	Yield (%)	Diastereoselectivity syn:anti		
t-Bu	Me	95	1	:	3.8
Et	Me	82	1	:	1
Ph	Me	96	1	:	1.6
C$_5$H$_{11}$	Me	93	1	:	1.1
C$_6$H$_{13}$	H	20	1	:	2

(Reproduced with permission from Welch, J. T., *et al. J. Org. Chem.* **1984**, *49*, 4720)

Welch, J. T., *et al. J. Org. Chem.* **1984**, *49*, 4720.

Methyl 4,6-*O*-benzylidene-2-deoxy-3-*C*-[(*S*)-(ethoxycarbonyl)fluoromethyl)]-α-D-*ribo*-hexopyranoside, Trimethyl (2*S*, 3*R*)-2-fluorocitrate

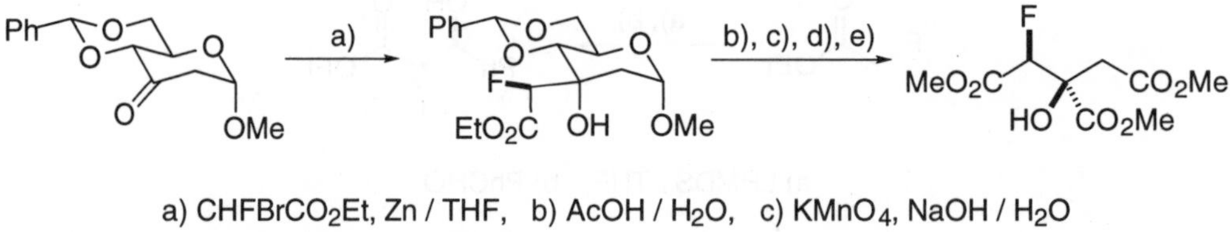

a) CHFBrCO$_2$Et, Zn / THF,   b) AcOH / H$_2$O,   c) KMnO$_4$, NaOH / H$_2$O
d) H$_3$O$^+$,   e) CH$_2$N$_2$ / Et$_2$O

Synthetic method
    Methyl 4,6-*O*-benzylidene-2-deoxy-3-*C*-[(*S*)-(ethoxycarbonyl)fluoromethyl)]-α-D-*ribo*-hexo-pyranoside
    A mixture of activated Zn dust (1.92 g, 30 mmol), methyl 4,6-*O*-benzylidene-2-deoxy-α-D-*erythro*-hexopyranoside-3-ulose (3.96 g, 15 mmol), ethyl bromofluoroacetate (3.05 g, 16.5 mmol) in THF (60 mL) was heated.   The reaction started when, or sometimes before, the boiling point was reached.   After the reaction mixture was refluxed for 10 min, it was poured into sodium dihydrogen phosphate buffer (0.1 *M*, 400 mL) and Et$_2$O (200 mL).   Filtration, extraction with Et$_2$O (2x200 mL), drying (Na$_2$SO$_4$), and concentration gave the product mixture.   Chromatography on silica gel (CH$_2$Cl$_2$:AcOEt=9:1 as an eluent) gave a practically complete separation, and the desired methyl 4,6-*O*-benzylidene-2-deoxy-3-*C*-[(*S*)-(ethoxycarbonyl)fluoromethyl)]-α-D-*ribo*-hexopyranoside and its epimer at the fluorinated carbon atom were isolated in 47% and 14% yields, respectively.
Spectral data
    [α]$_D^{25}$ +123 (*c* 1, AcOEt); mp. 131-131.5 °C (EtOH); ^{1}H NMR (CDCl$_3$): δ 1.29 (3 H, t),

2.03 (1 H, dd, J=14.9, 1.2 Hz), 2.21 (1 H, ddd, J=14.9, 3.9, 2.0 Hz), 3.41 (3 H, s), 3.52 (1 H, d, J=2.0 Hz), 4.26 (2 H, q), 3.2-4.6 (4 H, m), 4.87 (1 H, br d, J=3.9 Hz), 5.02 (1 H, d, J=47.9 Hz), 5.63 (1 H, s), 7.2-7.6 (5 H, m); ^{13}C NMR (CDCl$_3$): δ 14.08, 33.84 (d, J=1.4 Hz), 55.33, 58.87, 61.52, 69.05, 70.97 (d, J=22.0 Hz), 77.36 (d, J=1.1 Hz), 86.18 (d, J=186.8 Hz), 98.41, 101.85, 126.24, 128.09, 129.01, 137.15, 167.47 (d, J=23.8 Hz).

Minor diastereomer

[α]$_D^{25}$ +63 (c 1, AcOEt); mp. 139.5-140.5 °C (EtOH); ^{1}H NMR (CDCl$_3$): δ 1.02 (3 H, t), 2.07 (1 H, dd, J=14.6, 1.3 Hz), 2.22 (1 H, dddd, J=14.6, 4.0, 2.0, 1.2 Hz), 3.42 (3 H, s), 3.83 (1 H, m), 3.86 (2 H, q), 3.5-4.4 (4 H, m), 4.65 (1 H, d, J=47.1 Hz), 5.84 (1 H, dd, J=4.0, 1.0 Hz), 5.56 (1 H, s), 7.2-7.6 (5 H, m); ^{13}C NMR (CDCl$_3$): δ 13.64, 36.19 (d, J=3.3 Hz), 55.53, 58.65, 61.74, 69.17, 71.46 (d, J=19.5 Hz), 77.19, 88.15 (d, J=194.1 Hz), 98.21, 101.87, 126.24, 128.01, 129.09, 137.03, 167.70 (d, J=24.4 Hz).

Synthetic method

Trimethyl (2S,3R)-2-fluorocitrate

A solution of methyl 4,6-O-benzylidene-2-deoxy-3-C-[(S)-(ethoxycarbonyl)fluoromethyl)]-α-D-*ribo*-hexopyranoside (783 mg, 2.1 mmol) in a mixture of AcOH (13 mL) and H$_2$O (67 mL) was refluxed for 30 min.   After the solvents were evaporated, the residue was dissolved in 0.5 N NaOH aq. (60 mL) and a solution of KMnO$_4$ (30 mmol) in H$_2$O (155 mL) was then added.   After 15 h at 22 °C, acetone (10 mL) was added and the mixture was stirred for 5 min.   Filtration, treatment with ion exchanger (Dowex 50W-X8, H$^+$), and concentration to dryness gave a residue, which was treated with ethereal diazomethane.   Drying (MgSO$_4$), filtration, and evaporation of the solvent gave a crude product, whose purification by preparative HPLC yielded the ester as a colorless liquid in 28% yield with >99% purity.   The isomeric trimethyl fluorocitrate was obtained analogously from the above epimer in 33% yield.

Spectral data

[α]$_D^{20}$ +18.4 (c 2.0, MeOH); ^{1}H NMR (CDCl$_3$): δ 3.07 (2 H, s), 3.70 (3 H, s), 3.86 (3 H, s), 3.89 (3 H, s), 4.01 (1 H, s), 5.03 (1 H, d, J=46.9 Hz).

Trimethyl (R)-2-Fluorocitrate

[α]$_D^{20}$ +13.7 (c 2.0, MeOH); ^{1}H NMR (CDCl$_3$): δ 3.04 (2 H, s), 3.71 (3 H, s), 3.82 (3 H, s), 3.88 (3 H, s), 3.92 (1 H, s), 5.12 (1 H, d, J=47.1 Hz).

Brandänge, S., *et al. J. Am. Chem Soc.* **1981**, *103*, 4452.

2.2.2 Wittig Type Reactions

Ethyl (2Z,6E)-3-fluoromethyl-7,11-dimethyldodeca-2,6,10-trienoate

a) (EtO)$_2$P(O)CH$_2$CO$_2$Et, NaH / benzene

Synthetic method

NaH (0.124 g, 2.58 mmol) was suspended in benzene (5 mL) under nitrogen.   Triethyl phosphonoacetate (0.51 mL, 2.57 mmol) was added dropwise, and the solution was stirred for 1 h.

Monofluorinated ketone in benzene (8 mL) was added and the solution stirred overnight.   Then the solution was concentrated, and the residue was diluted with AcOEt and 1 M HCl.   The organic layer was washed with H_2O and brine, dried, and concentrated to give 0.65 g of a yellow oil.   The crude mixture was chromatographed on silica gel, eluting with hexane:AcOEt=40:1 to provide 0.465 g of the ester as a mixture of 34:66 $(E):(Z)$ isomers.   The (Z)-isomer was isolated by preparative thin layer chromatography, eluting three times with hexane:AcOEt=80:1 to afford ethyl (2Z,6E)-3-fluoromethyl-7,11-dimethyldodeca-2,6,10-trienoate in 62% yield.

Spectral data

^{1}H NMR (CDCl$_3$): δ 1.26 (3 H, t, J=8 Hz), 1.60 (6 H, s), 1.67 (3 H, s), 1.98-2.12 (4 H, m), 2.14-2.30 (2 H, m), 2.32-2.40 (2 H, m), 4.13 (2 H, q, J=7 Hz), 5.05-5.20 (2 H, m), 5.54 (2 H, d, J=48 Hz), 5.70 (1 H, s); ^{13}C NMR (CDCl$_3$): δ 14.2, 16.0, 17.6, 25.6, 26.3, 26.6, 33.6 (d, J=8 Hz), 39.6, 60.0, 82.0 (d, J=162 Hz), 115.9 (d, J=6 Hz), 122.6, 124.1, 131.3, 136.3, 158.6 (d, J=18 Hz), 165.7; ^{19}F NMR (CDCl$_3$): δ -113.8 (t, J=49 Hz) from C$_6$F$_6$; IR (neat): ν 2978, 2930, 2860, 1712, 1645, 1446, 1383 cm^{-1}.

(*E*)-isomer

^{1}H NMR (CDCl$_3$): δ 1.29 (3 H, t, J=7 Hz), 1.60 (6 H, s), 1.68 (3 H, s), 1.95-2.12 (4 H, m), 2.16-2.26 (2 H, m), 2.53-2.60 (2 H, m), 4.18 (2 H, q, J=7 Hz), 4.86 (2 H, dd, J=47, 2 Hz), 5.04-5.18 (2 H, m), 5.92 (1 H, d, J=1 Hz); ^{13}C NMR (CDCl$_3$): δ 14.5, 16.0, 17.6. 25.6, 26.6, 27.0, 29.0 (d, J=97 Hz), 39.7, 60.0, 84.1 (d, J=177 Hz), 115.1 (d, J=13 Hz), 123.0, 124.5, 131.4, 136.5, 155.8 (d, J=12 Hz), 165.9 (d, J=1 Hz); ^{19}F NMR (CDCl$_3$): δ -120.3 (t, J=49 Hz) from C$_6$F$_6$; IR (neat): ν 2962, 2924, 1718, 1660, 1446 cm^{-1}.

Note

For the preparation of the starting material, see page 35 of this volume.

Dolence, J. M.; Poulter, C. D. *Tetrahedron* **1996**, *52*, 119.

Methyl dichlorofluoroacetate,
Methyl (Z)-2-fluorohex-2-enoate

a) SbF$_3$, Br$_2$, b) butyraldehyde, Zn, cat. CuCl, Ac$_2$O, MS 4 Å / THF

Synthetic method

Methyl dichlorofluoroacetate

In a flask equipped with a distillation apparatus and mechanical stirrer were placed methyl trichloroacetate (0.15 mol) and antimony(III) fluoride (0.1 mol).   Bromine (0.16 mol) was slowly added to the well-stirred mixture under heating.   Crude products were distilled at such a rate that the boiling point of the distillate did not exceed 120 °C.   The usual treatment of the distillate followed by fractional distillation gave pure methyl dichlorofluoroacetate (bp. 117.0-118.5 °C) and methyl chlorodifluoroacetate (bp. 80-82 °C) in 65% and 10% isolated yields, respectively.

Synthetic method

Methyl (Z)-2-fluorohex-2-enoate

To a solution of activated zinc powder (0.98 g, 15 mmol), copper(I) chloride (0.15 g, 1.5 mmol) and molecular sieves 4 Å (1 g) in THF (18 mL) under an argon atmosphere, butanal (0.43 g,

6 mmol) and acetic anhydride (0.61 g, 6 mmol) were added *via* a syringe.   After this mixture was heated to 50 °C, methyl dichlorofluoroacetate (0.8 g, 5 mmol) was added dropwise and the whole mixture was stirred for 4 h at the same temperature.   The reaction mixture was diluted with Et$_2$O (*ca.* 20 mL) and filtered through a Celite bed.   The filtrate was concentrated under reduced pressure, and the oil thus obtained was chromatographed on silica gel to give the analytically pure methyl 2-fluorohex-2-enoate in 89% yield as a mixture of 93:7 (*Z*):(*E*) isomers.

Spectral data

^{1}H NMR (CDCl$_3$): δ 0.93 (3 H, t, *J*=6.8 Hz), 1.2-1.8 (2 H, m), 2.21 (2 H, dt, *J*=7.6, 7.6 Hz), 3.77 (3 H, s), 6.04 (1 H, dt, *J*=7.6, 33 Hz); ^{19}F NMR (CDCl$_3$): δ -85.7 (d, *J*=33.0 Hz); IR (film): ν 1744, 1678 cm^{-1}.

Note

For the utilization of the present procedure, see also the following work: Bold, G., *et al. Helv. Chim. Acta* **1992**, *75*, 865.

For the preparation of the same structure by the Hörner-Wittig reagent, see i) Pirrung, M. C., *et al. J. Org. Chem.* **1993**, *58*, 5683, ii) Patrick, T. B., *et al. J. Org. Chem.* **1994**, *59*, 1210, iii) Bartlett, P. A., *et al. J. Org. Chem.* **1995**, *60*, 3107, iv) Kiyota, H., *et al. Biosci. Biotech. Biochem.* **1996**, *60*, 1076, v) Rolando, C., *et al. Chem. Commun.* **1997**, 1489, iv) Kvicala, J., *et al. Synlett* **1997**, 986, vi) de Lera, A. R., *et al. J. Org. Chem.* **1997**, *62*, 310.

R	Yield (%)	(*Z*):(*E*)	R	Yield (%)	(*Z*):(*E*)
i-Pr	69	99:1	*t*-Bu	69	100:0
n-C$_6$H$_{13}$	79	93:7	(*E*)-CH$_3$CH=CH	83	93:7
(*E*)-CH$_3$CH=C(CH$_3$)	76	99:1	Ph	78	100:0
p-CH$_3$C$_6$H$_4$-	83	100:0	*p*-Cl-C$_6$H$_4$-	75	100:0
p-CH$_3$O-C$_6$H$_4$-	76	100:0			

(Reproduced with permission from Ishihara, T.; Kuroboshi, M. *Chem. Lett.* **1987**, 1145)

Ishihara, T.; Kuroboshi, M. *Chem. Lett.* **1987**, 1145.

Ethyl 2-fluoro-2-(phenylsulfinyl)acetate, Ethyl (*Z*)-2-fluorooct-2-enoate

a) *m*CPBA / CH$_2$Cl$_2$, b) NaH / DMF, c) *n*-C$_6$H$_{13}$Br, d) 95 °C

Synthetic method

Ethyl 2-fluoro-2-(phenylsulfinyl)acetate

A solution of 100 mmol of ethyl 2-fluoro-2-(phenylsulfenyl)acetate in 200 mL of CH$_2$Cl$_2$ was treated with a solution of 24 g of *m*CPBA (85%, 120 mmol) in 200 mL of CH$_2$Cl$_2$ at -60 °C and stirred overnight.   After warming to room temperature, the mixture was filtered and the filtrate was washed with NaHCO$_3$ aq. and NH$_4$Cl aq.   Drying, evaporation and flash chromatography of the residue with a mixture of hexane and AcOEt furnished ethyl 2-fluoro-2-(phenylsulfinyl)acetate in 80% yield as a diastereomer mixture.

Spectral data

bp. 112 °C/0.06 mmHg

diastereomer I

^{1}H NMR (CDCl$_3$): δ 1.11 (3 H, t, J=7 Hz), 4.04 (2 H, q, J=7 Hz), 5.71 (1 H, d, J=48 Hz), 7.5 (5 H, br s).

diastereomer II

^{1}H NMR (CDCl$_3$): δ 1.18 (3 H, t, J=7 Hz), 4.09 (2 H, q, J=7 Hz), 5.53 (1 H, d, J=46 Hz), 7.5 (5 H, br s).

Synthetic method

Ethyl (Z)-2-fluorooct-2-enoate

A suspension of 0.33 g of NaH (80% in oil, 11 mmol) in 10 mL of dry DMF was cooled to 0 °C under argon and treated with a solution of 10 mmol of ethyl 2-fluoro-2-(phenylsulfinyl)acetate in 4 mL of DMF.   After stirring at room temperature for 20 min, the mixture was cooled to 5 °C and 11 mmol of n-hexyl bromide was added in one portion.   The reaction mixture was stirred at room temperature for 45 min.   The mixture was then heated to 95 °C for 1 h, poured into ice-NH$_4$Cl aq. and extracted with a mixture of hexane and AcOEt.   Drying, evaporation and flash chromatography of the residue with hexane and then a mixture of hexane and AcOEt furnished ethyl 2-fluorooct-2-enoate in 77% yield with >95:5 (Z):(E) selectivity.

Spectral data

^{1}H NMR (CDCl$_3$): δ 0.9 (3 H), 1.33 (3 H, t), 1.27-1.37 (4 H), 1.45 (2 H), 2.23 (2 H), 4.28 (2 H, q), 6.10 (1 H, dt, J=34 Hz).

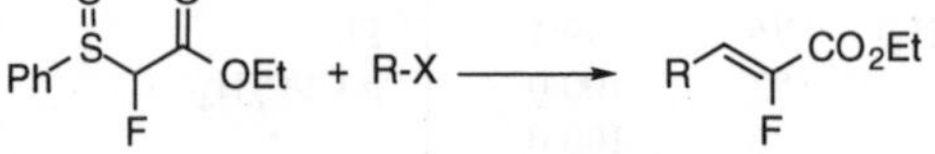

R-X	Yield (%)	R-X	Yield (%)
n-C$_6$H$_{13}$-Cl	44	n-C$_6$H$_{13}$-I	73
Et-CHMe-CH$_2$Br	64	Et-CHMe-CH$_2$OMs	43
PhCH$_2$CH$_2$-Br	76	crotyl-Cl	70
EtO$_2$C(CH$_2$)$_3$-Br	85	MeO$_2$C(CH$_2$)$_2$-Br	24
AcO(CH$_2$)$_3$-Br	47	PhthN(CH$_2$)$_3$-Br	73

(Reproduced with permission from Allmendinger, T. *Tetrahedron* **1991**, *47*, 4905)

Note

For the preparation of the starting material, see page 73 of this volume.

Ethyl 2-fluoro-2-(phenylsulfinyl)acetate also underwent Michael addition reactions with, for example, α,β-unsaturated esters.

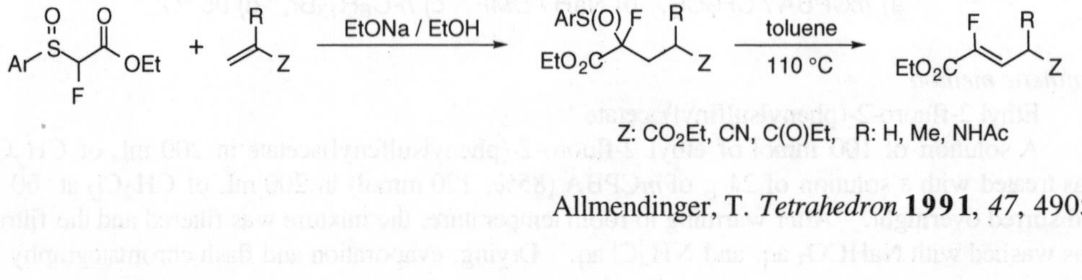

Z: CO$_2$Et, CN, C(O)Et,   R: H, Me, NHAc

Allmendinger, T. *Tetrahedron* **1991**, *47*, 4905.

2.2.3 Rearrangement and Cycloaddition

(E)-But-2-en-1-yl fluoroacetate, anti-2-Fluorohex-4-enoic acid

a) Phthaloyl chloride, b) (E)-but-2-en-1-ol, pyrdine / CH$_2$Cl$_2$
c) LDA; TMSCl / THF

Synthetic method
    (E)-But-2-en-1-yl fluoroacetate
    Sodium fluoroacetate (90% pure; 167 g, 1.67 mmol) and 375 g (1.68 mol) of phthaloyl chloride were thoroughly mixed and gently heated in a flask fitted for distillation. All the distillate of boiling point up to 90 ℃ was collected in a flask protected by a CuCl$_2$ tube, and redistilled to yield 138 g (95%) of fluoroacetyl chloride, bp. 70-71 ℃.
    To 60 mL of a 0.3 M solution of pyridine in CH$_2$Cl$_2$ at 0 ℃ was added 1.44 g of (E)-but-2-en-1-ol (20 mmol) while maintaining the temperature at 0 ℃, 1.45 mL (17 mmol, *ca.* 1.1 g/mL) of fluoroacetyl chloride was added dropwise cautiously. The reaction mixture was stirred at 0 ℃ for 1 h and at 25 ℃ for 12 h. The reaction mixture was washed with sat. CuSO$_4$ (3x10 mL), followed by H$_2$O then brine. The organic extracts were dried over MgSO$_4$, and the solvent was distilled. Distillation of the product (bp. 75-78 ℃/49 mmHg) yielded the pure (E)-but-2-en-1-yl fluoroacetate in 91% yield.
Spectral data
    ^{1}H NMR (CDCl$_3$): δ 1.60 (3 H, d, J=6.4 Hz), 4.47 (2 H, d, J=6.0 Hz), 4.68 (2 H, d, J=46.7 Hz), 5.50 (1 H, dq, J=21.4, 6.4 Hz), 5.70 (1 H, dt, J=21.4, 6.3 Hz); ^{13}C NMR (CDCl$_3$): δ 17.7, 66.0, 77.6 (d, J=181.9 Hz), 116.4, 136.6, 166.8 (d, J=21.97 Hz); ^{19}F NMR (CDCl$_3$): δ -233.21 (t, J=48.0 Hz).

Synthetic method
    anti-2-Fluorohex-4-enoic acid
    A solution of 2.8 mmol of LDA in 30 mL of THF under argon was magnetically stirred and cooled to -100 ℃, and 0.5 g (3.7 mmol) of (E)-but-2-en-1-yl fluoroacetate was added followed by the addition of 2.8 mmol of TMSCl. The solution was warmed to room temperature and heated at 40 ℃ for 2 h. The reaction was quenched with MeOH and basified with 5% NaOH; the basic layer was washed with Et$_2$O and the organic washings discarded. The extracted basic phase was acidified with conc HCl, and the acidic layer was extracted with CH$_2$Cl$_2$ several times. The combined organic layers were washed with brine (1x20 mL), dried with MgSO$_4$, and evaporated *in vacuo* to yield 0.24 g (66%) of *anti*-2-fluoro-3-methylpent-4-enoic acid.
Spectral data
    ^{1}H NMR (CDCl$_3$): δ 0.94 (3 H, d, J=7.33 Hz), 2.64 (2 H, dm, J=26.86 Hz), 4.88 (2 H, dd, J=48.34, 3.42 Hz), 5.14 (1 H, d, J=10.26 Hz), 5.34 (1 H, d, J=17.09 Hz), 5.84 (1 H, ddd, J=17.58, 10.26, 7.33 Hz), 10.1-10.6 (1 H, br s); ^{13}C NMR (CDCl$_3$): δ 13.40 (d, J=6.1 Hz), 40.16 (d, J=20 Hz), 91.23 (d, J=189.6 Hz), 116.24, 137.76, 172.18 (d, J=24.41 Hz); ^{19}F NMR (CDCl$_3$): δ -200.73 (dd, J=48.8, 28.1 Hz).

Caution

Sodium fluoroacetate and fluoroacetyl chloride are fatal poisons affecting the central nervous system, causing epileptic convulsions.   They must be handled with extreme caution in an efficient fume hood.

Note

The present [3,3]-sigmatropic rearrangement was also realized by the action of triisopropyl-silyl trifluoromethanesulfonate and triethylamine in CH_2Cl_2, furnishing the product in more than 95% yield from the above ester with 15:1 diastereoselectivity in preference to the same stereoisomer. See Araki, K.; Welch, J. T. *Tetrahedron Lett.* **1993**, *34*, 2251.

A similar structure was constructed by use of trifluorovinylsilanes.   See Tellier, F., *et al. Tetrahedron Lett.* **1998**, *39*, 5041.

Hudlicky, M. in *Chemistry of Organic Fluorine Compounds*;
Ellis Horwood: Chichester, 1976; p. 699
(for the preparation of fluoroacetyl chloride)
Welch, J. T., *et al. J. Org. Chem.* **1991**, *56*, 353.

$(3R,4S)$-4-[(S)-2,2-Dimethyl-1,3-dioxolan-4-yl)]-3-fluoro-N-(4-methoxyphenyl)azetidin-2-one

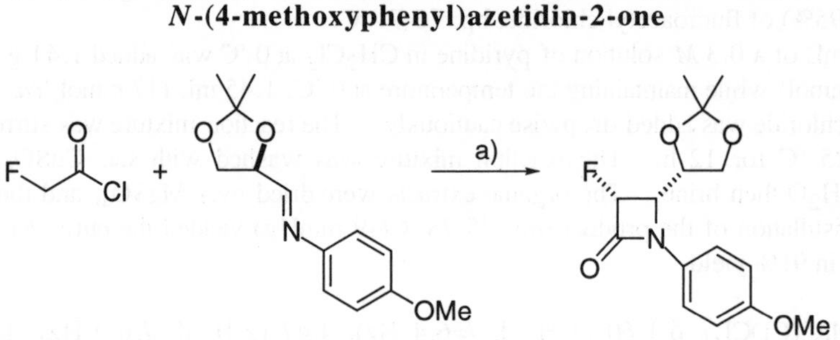

a) Et₃N / CH₂Cl₂

Synthetic method

To a CH_2Cl_2 solution (300 mL) of freshly prepared N-[(S)-2,2-dimethyl-1,3-dioxolan-4-ylidene)]-4-methoxyphenylamine (147 mmol) and triethylamine (22.2 g, 220 mmol) was added dropwise a CH_2Cl_2 solution (30 mL) of fluoroacetyl chloride (176 mmol) at room temperature with stirring.   After overnight stirring, the reaction mixture was washed with H_2O and a sat. NaCl solution and dried with $MgSO_4$.   Evaporation of the solvent gave $(3R,4S)$-4-[(S)-2,2-dimethyl-1,3-dioxolan-4-yl)]-3-fluoro-N-(4-methoxyphenyl)azetidin-2-one in 68% yield after recrystallization.   The diastereomeric excess was not less than 99%.

Spectral data

$[\alpha]_D$ +77.9 (*c* 1.5, CHCl₃); mp. 154-156 °C (EtOH); ¹H NMR (CDCl₃): δ 1.36 (3 H, s), 1.55 (3 H, s), 3.77-3.82 (1 H, m), 3.80 (3 H, s), 4.23-4.31 (2 H, m), 4.40 (1 H, m), 5.54 (1 H, dd, *J*=55.0, 5.0 Hz), 6.88 (2 H, m), 7.66 (2 H, m); ¹³C NMR (CDCl₃): δ 24.9, 26.7, 26.7, 55.5, 62.0 (d, *J*=21.3 Hz), 66.6 (d, *J*=6.5 Hz), 76.6 (d, *J*=3.0 Hz), 90.3 (d, *J*=221.3 Hz), 110.4, 114.1, 119.7, 130.5, 157.0, 160.4 (d, *J*=22.0 Hz); ¹⁹F NMR (CDCl₃): δ -202.1 (d, *J*=55 Hz); IR (KBr): ν 1749 cm⁻¹.

Caution

Fluoroacetyl chloride and sodium fluoroacetate used in preparation are fatal poisons affecting the central nervous system and causing epileptic convulsions.   Fluoroacetyl chloride was handled with extreme caution in an efficient fume hood.

Note

For the preparation of fluoroacetyl chloride, see page 57 of this volume.

Welch, J. T., *et al. J. Org. Chem.* **1993**, *58*, 2454.

2.2.4 Alkylations

2-(4-Biphenylyl)-1-chloro-1-fluoro-1-(phenylsulfinyl)ethane

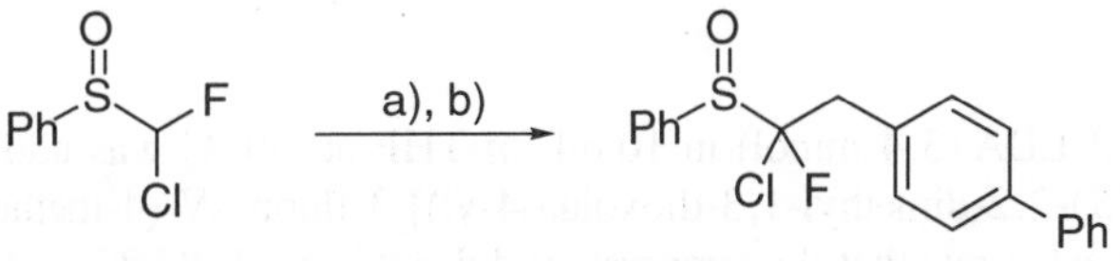

a) MeLi·LiBr / THF,   b) 4-Ph-C$_6$H$_4$-CH$_2$Br / THF

Synthetic method

An ethereal solution of MeLi·LiBr (1.2 M, 0.83 mL, 1 mmol) was slowly added to a solution of chlorofluoro(phenylsulfinyl)methane (193 mg, 1 mmol) in dry THF (10 mL) at -90 °C with stirring under an argon atmosphere over 10 min.   After the brownish yellow solution was stirred at that temperature for an additional 15 min, a THF (10 mL) solution of 4-(bromomethyl)biphenyl (248 mg, 1 mmol) was slowly added.   The resulting mixture was stirred at -90 °C for 30 min, allowed to warm to room temperature, quenched with a sat. NH$_4$Cl solution and extracted with Et$_2$O.   The ethereal extract was washed with brine, dried with Na$_2$SO$_4$ and evaporated.   The residue was chromatographed on silica gel (hexane:CH$_2$Cl$_2$) to give 2-(4-biphenylyl)-1-chloro-1-fluoro-1-(phenylsulfinyl)ethane in 57% yield (a 3:1 diastereomer mixture) as colorless crystals.

Spectral data

mp. 86-88 °C (hexane:CH$_2$Cl$_2$); IR (KBr): ν 1488, 1450, 1316, 1090, 1060 cm^{-1}.

major diastereomer

^{1}H NMR (CDCl$_3$): δ 3.66 (2 H, m), 7.3-7.9 (14 H, m); ^{13}C NMR (CDCl$_3$): δ 41.39 (d, J=21 Hz), 120.20 (d, J=305 Hz), 127.06, 127.19, 127.40 (d, J=1 Hz), 127.45, 127.78, 128.84, 130.00, 131.57, 132.73, 136.90 (d, J=3 Hz), 140.46, 140.92; ^{19}F NMR (CDCl$_3$): δ -114.69 (dd, J=23, 11 Hz).

minor diastereomer

^{1}H NMR (CDCl$_3$): δ 3.43 (1 H, dd, J=26.6, 14.8 Hz), 3.91 (1 H, dd, J=18.0, 14.8 Hz), 7.3-7.9 (14 H, m); ^{13}C NMR (CDCl$_3$; typical signals): δ 42.49 (d, J=19 Hz), 131.45, 132.57, 137.45, 140.07, 140.56; ^{19}F NMR (CDCl$_3$): δ -112.63 (dd, J=27, 18 Hz).

Note

For the preparation of chlorofluoro(phenylsulfinyl)methane, see page 72 of this volume.

See the following article on the reaction of chlorofluoro(phenylsulfinyl)methane with various aldehydes: Uno, H., *et al. Bull. Chem. Soc. Jpn.* **1992**, *65*, 218.

Uno, H., *et al. Bull. Chem. Soc. Jpn.* **1992**, *65*, 210.

(3R,4S)-4-[(S)-2,2-Dimethyl-1,3-dioxolan-4-yl)]-3-fluoro-N-(4-methoxyphenyl)-3-methylazetidin-2-one

a) LDA / THF, b) MeI

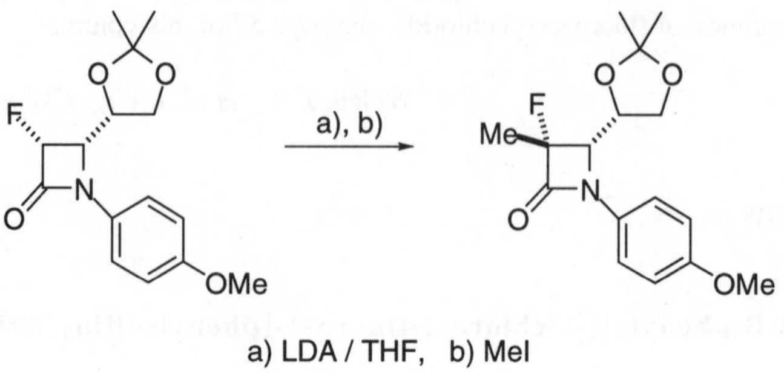

Synthetic method

To a solution of LDA (3.4 mmol) in 10 mL of THF at -90 °C was added a THF solution (7 mL) of (3R,4S)-4-[(S)-2,2-dimethyl-1,3-dioxolan-4-yl)]-3-fluoro-N-(4-methoxyphenyl)azetidin-2-one (1.76 mmol) at such a rate that the temperature did not exceed -85 °C.   After 5 min, 3.2 mmol of methyl iodide was added dropwise at -90 °C.   The reaction mixture was allowed to warm gradually to 0 °C then quenched with sat. NH_4Cl, followed by extraction with CH_2Cl_2.   The organic extracts were dried over $MgSO_4$ and concentrated *in vacuo*.   The residue was then purified by silica gel column chromatography (hexane:AcOEt) to give (3R,4S)-4-[(S)-2,2-dimethyl-1,3-dioxolan-4-yl)]-3-fluoro-N-(4-methoxyphenyl)-3-methylazetidin-2-one in 99% yield without loss of the original stereochemical integrity.

Spectral data

$[\alpha]_D$ +82.9 (c 1.5, $CHCl_3$); mp. 114-115 °C; ^1H NMR ($CDCl_3$): δ 1.35 (3 H, s), 1.55 (3 H, s), 1.71 (3 H, d, J=23.0 Hz), 3.78 (1 H, m), 3.80 (3 H, s), 4.01 (1 H, dd, J=8.3, 3.9 Hz), 4.28 (1 H, ddd, J=8.8, 6.5, 2.5 Hz), 4.37 (1 H, m), 6.88 (2 H, m), 7.68 (2 H, m); ^{13}C NMR ($CDCl_3$): δ 18.9 (d, J=26.7 Hz), 24.9, 26.7, 55.5, 66.5 (d, J=4.6 Hz), 68.1 (d, J=21.1 Hz), 76.7 (d, J=3.0 Hz), 98.8 (d, J=216.6 Hz), 110.2, 114.1, 119.8, 130.5 (d, J=4.1 Hz), 156.9, 162.6 (d, J=24.3 Hz); ^{19}F NMR ($CDCl_3$): δ -164.9 (q, J=23.9 Hz); IR (KBr): ν 1754 cm^{-1}.

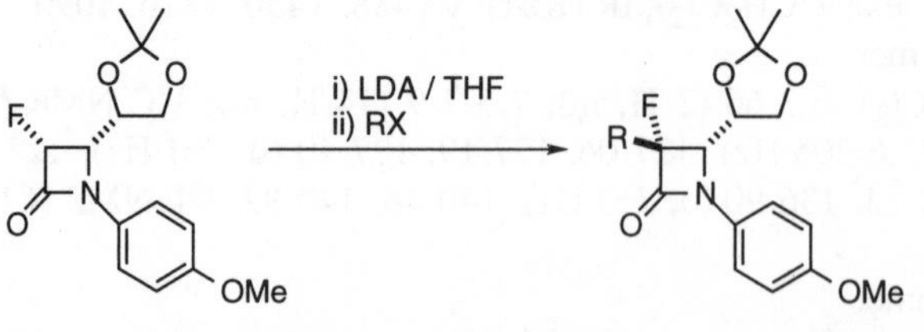

RX	Yield (%)	RX	Yield (%)
EtI	78	EtOTf	46
$C_6H_5CH_2Br$	57	n-BuI	56
i-PrI	13	CH_2=$CHCH_2Br$	80
CH≡CCH_2Br	44	$C_6H_5CH_2OCH_2Cl$	49
MeOCH$_2$CH$_2$OCH$_2$Cl	38		

(Reproduced with permission from Welch, J. T., *et al. J. Org. Chem.* **1993**, 58, 2454)

Note

For the preparation of the starting fluorinated azetidinone, see page 58 of this volume and for further application of the above products, see Welch, J. T., *et al. Bioorg. Med. Chem. Lett.* **1993**, 3, 2457.

Welch, J. T., *et al. J. Org. Chem.* **1993**, *58*, 2454.

Methyl 2-amino-2-(chlorofluoromethyl)hex-5-enoate

a) NaH / THF,   b) CHCl$_2$F,   c) 1 N HCl aq.

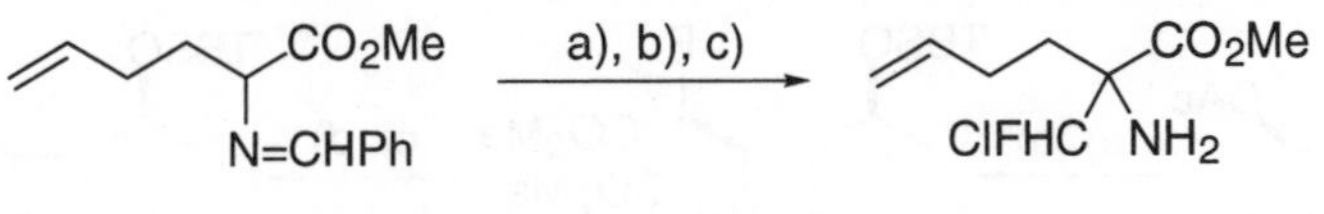

Synthetic method

A mixture of methyl 2-(benzylideneamino)hex-5-enoate (16 g, 69 mmol) and sodium hydride (45% dispersion in oil; 3.68 g, 69 mmol) in anhydrous THF (140 mL) was heated at 45 °C, under nitrogen, until completion of metalation (2 h).   The mixture was cooled to -20 °C.   Dichloro-fluoromethane (20 mL) was added through a refrigerated addition funnel.   The mixture was stirred at room temperature for 15 h.   Hydrolysis and removal of the solvent under reduced pressure left an oily residue, which was triturated and extracted with Et$_2$O (3x50 mL).   The combined organic layers were dried over anhydrous MgSO$_4$.   Filtration and removal of the solvent under reduced pressure, yielded an orange red oil.   Distillation under high vacuum yielded a 1:1 mixture (11.1 g) of methyl 2-benzylideneamino-2-(chlorofluoromethyl)hex-5-enoate and methyl 2-(benzylidene-amino)hex-5-enoate (175 °C/0.2 mmHg).

A solution of this crude mixture in 1 N HCl aq. (60 mL) was stirred at room temperature for 2 h.   The aqueous acidic solution was washed with Et$_2$O (2x50 mL) and evaporated to dryness under reduced pressure.   NaHCO$_3$ was added until saturation and the basic aqueous layer was extracted with Et$_2$O (3x50 mL).   The combined organic layers were dried over anhydrous MgSO$_4$. Filtration and removal of the solvent under reduced pressure gave a 3:2 mixture of methyl 2-amino-2-(chlorofluoromethyl)hex-5-enoate and of methyl 2-aminohex-5-enoate.   Chromatography on silica gel (cyclohexane:AcOEt=8:2) yielded the former as an oil in 14% overall yield.

Spectral data

^{1}H NMR (CDCl$_3$): δ 1.65-2.50 (6 H, s), 3.73 and 3.76 (3 H, s), 4.80-5.20 (2 H, m), 5.45-6.08 (1 H, m), 6.27 (1 H, d, *J*=50 Hz).

Note

A similar type of reaction using fluorinated carbenes has been reported.   See the following review: Brahms, D. L. S.; Dailey, W. P. *Chem. Rev.* **1996**, *96*, 1585.

Schirlin, D., *et al. J. Chem. Soc. Perkin Trans. 1* **1992**, 1053.

2.2.5 Miscellaneous

(3S,4S)-3-[(R)-1-(*tert*-Butyldimethylsilyloxy)ethyl]-4-
[bis(methoxycarbonyl)fluoromethyl]azetidin-2-one,
(3S,4S)-3-[(R)-1-(*tert*-Butyldimethylsilyloxy)ethyl]-
4-[(R)-fluoro(methoxycarbonyl)]azetidin-2-one

a) LHMDS / THF,   b) CHF(CO₂Me)₂,   c) LiOH / MeOH-H₂O,   d) heat / xylene

Synthetic method

(3S,4S)-3-[(R)-1-(*tert*-Butyldimethylsilyloxy)ethyl]-4-[bis(methoxycarbonyl)fluoromethyl]-
azetidin-2-one

To a stirred solution of dimethyl fluoromalonate (5.0 g, 33.3 mmol) in dry THF (50 mL) at
-78 °C was added 1 M LHMDS in hexane (31.8 mL, 31.8 mmol).   After being stirred for 30 min
at -78 °C, a solution of (3S,4R)-4-acetoxy-3-[(R)-1-(*tert*-butyldimethylsilyloxy)ethyl]azetidin-2-one
(8.7 g, 30.3 mmol) in dry THF (90 mL) was added.   The mixture was stirred for 3 h at that
temperature and for 9 h at room temperature.   After neutralization with 1 N HCl under ice cooling,
the mixture was concentrated *in vacuo* and the residue was taken up into AcOEt.   The extract was
washed with H₂O and brine, dried over Na₂SO₄, and evaporated.   Chromatography of the residue
on silica gel with CHCl₃:MeOH=99:1 as an eluent gave 11.0 g of (3S,4S)-3-[(R)-1-(*tert*-butyl-
dimethylsilyloxy)ethyl]-4-[bis(methoxycarbonyl)fluoromethyl]azetidin-2-one in 96% yield.

Spectral data

$[\alpha]_D^{29}$ -47.7 (c 1.1, CHCl₃); mp. 158-160 °C; ¹H NMR (CDCl₃): δ 0.07 (6 H, s), 0.88 (9
H, s), 1.06 (3 H, d, J=6.1 Hz), 3.29 (1 H, br s), 3.87 (3 H, s), 3.88 (3 H, s), 4.25-4.30 (1 H, m),
4.41(1 H, dd, J=20.8, 1.8 Hz), 5.85 (1 H, br s); IR (CHCl₃): ν 3440, 1755 cm⁻¹.

Synthetic method

(3S,4S)-3-[(R)-1-(*tert*-Butyldimethylsilyloxy)ethyl]-4-[(R)-fluoro(methoxycarbonyl)methyl]-
azetidin-2-one

To a methanol (5 mL) solution of (3S,4S)-3-[(R)-1-(*tert*-butyldimethylsilyloxy)ethyl]-4-[bis-
(methoxycarbonyl)fluoromethyl]azetidin-2-one (90 mg, 0.24 mmol) was added a solution of
LiOH·H₂O (10 mg, 0.24 mmol) in H₂O (2 mL), and the mixture was stirred for 2 h at ambient
temperature.   After concentration under reduced pressure, the residue was partitioned between
Et₂O (15 mL) and H₂O (5 mL).   The ethereal layer was washed with sat. NaHCO₃ (3 mL).   The
combined aqueous layer was acidified with 1 N HCl under ice cooling, then thoroughly extracted
with AcOEt.   The extract was washed with brine, dried (MgSO₄), and evaporated to give a 2:1
diastereomeric mixture of the acid in 82% yield.

A mixture of this acid (100 mg, 0.28 mmol) in xylene (5 mL) was heated for 3 h at 135 °C.
After evaporation of the solvent, the residue was taken up into CH₂Cl₂.   The extract was washed
with brine, dried (MgSO₄), and evaporated to afford a crude material, which was subjected to flash
chromatography on silica gel.   Elution with a mixture of hexane:AcOEt=7:3 yielded a 2:1
diastereomeric mixture of (3S,4S)-3-[(R)-1-(*tert*-butyldimethylsilyloxy)ethyl]-4-[fluoro(methoxycarbonyl)methyl]azetidin-2-one in 77% yield.

Spectral data

mp. 136-138 ℃; ^{1}H NMR (CDCl$_3$): δ 0.06 (3 H, s), 0.07 (3 H, s), 0.86 (9 H, s), 1.13 and 1.15 (3 H (2:1), each d, each *J*=6.4 Hz), 3.23-3.26 (1 H (1:2), each br s), 3.83 and 3.84 (3 H (1:2), each s), 4.00-4.15 (1 H, m), 4.20-4.30 (1 H, m), 4.91(1 H (minor isomer), dd, *J*=48.3, 5.6 Hz), 4.97 (1 H (major isomer), dd, *J*=48.3, 4.9 Hz), 5.85 and 5.90 (1 H (2:1), each br s); IR (CDCl$_3$): ν 3405, 1764 cm^{-1}.

Fukumoto, K., *et al. Heterocycles* **1996**, *42*, 437.

2-Fluoroacetophenone

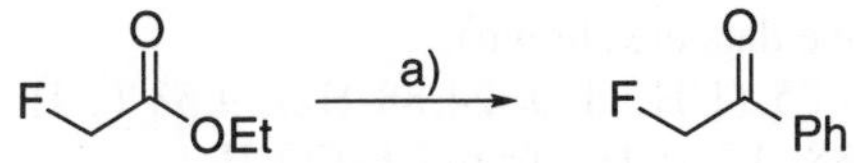

a) PhMgBr / Et$_2$O

Synthetic method

A solution of phenyl Grignard reagent (44 mmol) in Et$_2$O (80 mL) was added to a solution of ethyl fluoroacetate (4.2 g, 40 mmol) in Et$_2$O (30 mmol) at -70 ℃ under a nitrogen atmosphere. After 4 h of stirring at that temperature, the mixture was quenched with sat. NH$_4$Cl aq., and oily materials were extracted with Et$_2$O. The extract was dried over MgSO$_4$, and the solvent was removed *in vacuo*. Distillation afforded 2-fluoroacetophenone in 83% yield.

Spectral data

bp. 108-110 ℃/18 mmHg; ^{1}H NMR (CCl$_4$): δ 5.33 (2 H, d, *J*=40.5 Hz), 7.2-8.0 (5 H, m); ^{19}F NMR (CCl$_4$): δ -150 (t, *J*=40.5 Hz) from CF$_3$CO$_2$H; IR (neat): ν 1710 cm^{-1}.

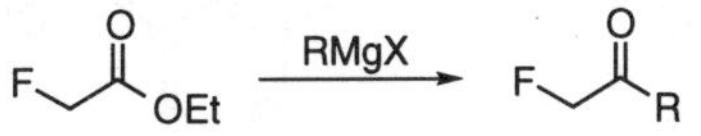

RMgX	Yield (%)	bp (°C/mmHg)
PhCH$_2$Cl	60	83-86/5
n-C$_6$H$_{13}$	57	59-60/23
n-C$_8$H$_{17}$	70	100/17

Kitazume, T., *et al. J. Fluorine Chem.* **1987**, *35*, 477.

(Z)-2-Fluoro-3-phenylprop-2-enal

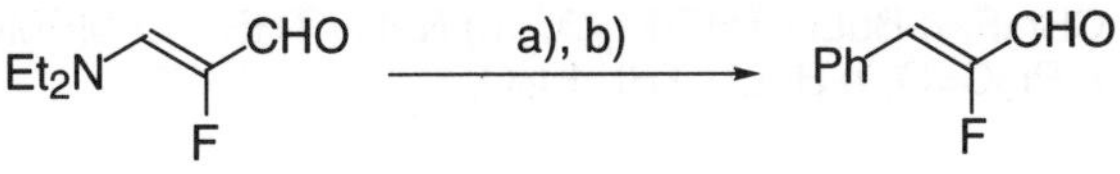

a) PhLi / THF, b) 10% HCl

Synthetic method

To a THF (2.5 mL) solution of 3-(diethylamino)-2-fluoroprop-2-enal (0.145 g, 1.0 mmol) was slowly added phenyllithium (1.17 mL of a 1.03 *M* cyclohexane-Et$_2$O solution, 1.2 mmol) at -78 ℃ under an argon atmosphere. After the mixture was stirred at -78 ℃ for 0.5 h, a mixture of THF (2.5 mL) and a 10% HCl solution (2.5 mL) was added at the same temperature. The reaction mixture was gradually warmed to room temperature, and left at this temperature for 1 h. The resultant mixture was quenched with cold H_2O (50 mL) followed by extraction with Et_2O (3x30 mL). The organic layer was washed with a sat. $NaHCO_3$ aq. solution (2x30 mL) and brine (2x30 mL), dried over anhydrous Na_2SO_4, and concentrated *in vacuo*. The residue was chromatographed on a silica gel column (hexane:AcOEt=4:1) to afford 0.139 g of analytically pure (Z)-2-fluoro-3-phenylprop-2-enal ((Z):(E)=>99:<1) in 93% yield.

Spectral data

Rf 0.43 (hexane:AcOEt=4:1); mp. 42.1-42.8 °C (hexane); ^1H NMR ($CDCl_3$): δ 6.62 (1 H, d, J= 34.20 Hz), 7.40-7.50 (3 H, m), 7.65-7.75 (2 H, m), 9.36 (1 H, d, J=16.12 Hz); ^{13}C NMR ($CDCl_3$): δ 126.89, 129.01, 130.61, 130.78 (d, J=19.85 Hz), 130.90, 154.76 (d, J=271.27 Hz), 183.98 (d, J=24.81 Hz); ^{19}F NMR ($CDCl_3$) δ -50.81 (dd, J=34.20, 16.12 Hz) from CF_3CO_2H; IR (CCl_4): ν 1648.2, 1697.5 cm^{-1}.

(E)-isomer (representative data are shown)

^1H NMR ($CDCl_3$): δ 6.85 (1 H, d, J=24.88 Hz), 9.58 (1 H, d, J=17.31 Hz); ^{19}F NMR ($CDCl_3$) δ -48.02 (dd, J=24.88, 17.31 Hz) from CF_3CO_2H.

Note

For the preparation of 3-(diethylamino)-2-fluoroprop-2-enal, see page 74 of this volume.

Funabiki, K., *et al. Chem. Lett.* **1997**, 739.

2,2-Dibromo-2-fluoro-1-(1-naphthyl)ethanol, 1,1-Dibromo-1-fluoro-2-(2-methoxyethoxy)methoxy-2-(1-naphthyl)ethane, *syn*-2-Bromo-2-fluoro-1-(2-methoxyethoxy)methoxy-1-(1-naphthyl)-3-propylhexan-3-ol

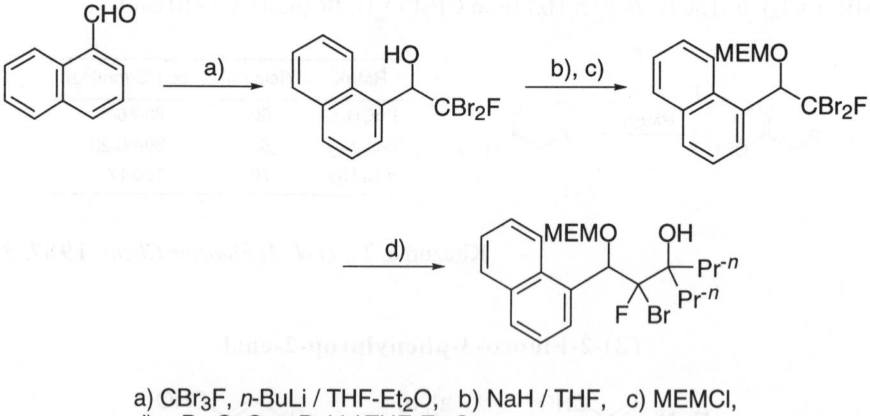

a) CBr_3F, *n*-BuLi / THF-Et_2O, b) NaH / THF, c) MEMCl,
d) *n*-Pr_2C=O, *n*-BuLi / THF-Et_2O,

Synthetic method

2,2-Dibromo-2-fluoro-1-(1-naphthyl)ethanol

To a solution of tribromofluoromethane (59 μL, 0.60 mmol) and 1-naphthaldehyde (68 μL, 0.50 mmol) in THF (3 mL)-Et_2O (1.5 mL) was added a 1.60 M hexane solution of *n*-BuLi (3.1 mL, 0.50 mmol) at -130 °C *via* a syringe over 10 min. The resulting mixture was stirred for 0.5 h at -130 °C before quenching with a sat. NH_4Cl aq. solution. The aqueous layer was extracted with Et_2O (5x20 mL). The combined extracts were dried over anhydrous Na_2SO_4 and concentrated *in vacuo*. The residue was purified by silica gel column chromatography to afford 148 mg of 2,2-

dibromo-2-fluoro-1-(1-naphthyl)ethanol in 85% yield as a yellow oil.
Spectral data
   ^{1}H NMR (CDCl$_3$): δ 3.19 (1 H, d, *J*=3.4 Hz), 6.02 (1 H, dd, *J*=8.1, 2.7, Hz), 7.42-7.66 (3 H, m), 7.85-8.13 (4 H, m); ^{13}C NMR (CDCl$_3$): δ 78.5 (d, *J*=22.6 Hz), 103.0 (d, *J*=326.3 Hz), 123.5 (d, *J*=3.1 Hz), 125.1, 125.8, 126.7, 127.0, 129.1, 130.3, 131.2, 131.6, 133.7; ^{19}F NMR (CDCl$_3$): δ -60.3 (d, *J*=8.1 Hz); IR (neat) 3425, 3050, 1510, 1395, 1350, 1260, 1230, 1205, 1170, 1095, 1080, 1030, 1010, 980, 920, 865, 815 cm^{-1}.

Synthetic method
   1,1-Dibromo-1-fluoro-2-(2-methoxyethoxy)methoxy-2-(1-naphthyl)ethane
   A solution of 2,2-dibromo-2-fluoro-(1-naphthyl)ethan-1-ol (1.01 g, 2.9 mmol) in THF (5 mL) and 2-methoxyethoxymethyl chloride (0.41 ml, 3.6 mmol) were added successively to NaH (60% in oil, 0.14 g, 3.6 mmol) in THF (5 mL) at 0 °C under an argon atmosphere.   The reaction mixture was stirred for 3 h at 0 °C and for 1 h at room temperature, then treated with a sat. NH$_4$Cl aq. solution (5 mL).   The organic layer was separated and the aqueous layer was extracted with Et$_2$O.   The combined organic phase was dried over anhydrous Na$_2$SO$_4$ and concentrated *in vacuo*.   The residue was purified by silica gel column chromatography to give 1.09 g of 1,1-dibromo-1-fluoro-2-(2-methoxyethoxy)methoxy-2-(1-naphthyl)ethane in 86% yield as a pale yellow oil.
Spectral data
   ^{1}H NMR (CDCl$_3$): δ 3.27 (3 H, s), 3.29-3.60 (3 H, m), 3.92-4.01 (1 H, m), 4.70 (1 H, d, *J*=7.0 Hz), 4.95 (1 H, d, *J*=7.0 Hz), 6.03 (1 H, d, *J*=9.4 Hz), 7.36 (3 H, m), 7.89 (3 H, m), 8.15 (1 H, d, *J*=8.0 Hz); ^{13}C NMR (CDCl$_3$): δ 58.9, 67.9, 71.4, 85.4 (d, *J*=21.5 Hz), 94.2, 99.6 (d, *J*=322.7 Hz), 123.6, 124.9, 125.8, 126.6, 127.9, 128.4, 128.9, 130.2, 132.3, 133.6; ^{19}F NMR (CDCl$_3$): δ -59.4 (d, *J*=7.8 Hz); IR (neat) 3053, 2926, 2893, 1597, 1512, 1452, 1396, 1367, 1288, 1240, 1199, 1176, 1116, 1055, 987, 931, 852, 821, 798, 775, 723 cm^{-1}.

Synthetic method
   syn-2-Bromo-2-fluoro-1-(2-methoxyethoxy)methoxy-1-(1-naphthyl)-3-propylhexan-3-ol
   A hexane solution of *n*-BuLi (1.62 *M*, 0.30 mL, 0.48 mmol) was added to a solution of 1,1-dibromo-1-fluoro-2-(2-methoxyethoxy)methoxy-2-(1-naphthyl)ethane (174 mg, 0.4 mmol) and 4-heptanone (120 μL, 0.8 mmol) in THF (3 mL) and Et$_2$O (1.5 mL) at -130 °C *via* a syringe over 10 min.   The resulting mixture was stirred for 1 h at -130 °C and warmed up to -78 °C before quenching with a sat. NH$_4$Cl aq. solution.   The aqueous layer was extracted with Et$_2$O (5x20 mL). The combined extracts were dried over anhydrous Na$_2$SO$_4$ and concentrated *in vacuo*.   The residue was purified by silica gel column chromatography to afford 105 mg of 2-bromo-2-fluoro-1-(2-methoxyethoxy)methoxy-1-(1-naphthyl)-3-propylhexan-3-ol in 55% yield as a mixture of diastereomers (*syn*:*anti*=83:17; the relative stereochemical relationship between OH and Br).
Spectral data
   IR (neat) 3497, 3053, 2963, 2874, 1597, 1512, 1456, 1396, 1294, 1234, 1199, 1174, 1045, 925, 912, 854, 798, 777, 738 cm^{-1}.
   ^{1}H NMR (CDCl$_3$): δ 0.88-1.07 (6 H, m), 1.42-1.61 (4 H, m), 1.92-2.06 (4 H, m), 3.32 (3 H, s), 3.35-3.65 (3 H, m), 3.76-3.94 (1 H, m), 4.62 (1 H, d, *J*=6.9 Hz), 4.79 (1 H, dd, *J*=6.9, 1.8 Hz), 6.27 (1 H, s), 7.41-7.59 (3 H, m), 7.84-8.18 (4 H, m); ^{13}C NMR (CDCl$_3$): δ 14.9, 15.0, 17.0 (d, *J*=3.1 Hz), 17.6 (d, *J*=2.3 Hz), 37.4, 38.6, 58.9, 69.0, 71.6, 73.2 (d, *J*=32.4 Hz), 80.3 (d, *J*=21.0 Hz), 93.6, 121.4 (d, *J*=261.7 Hz), 123.2 (d, *J*=4.2 Hz), 124.7, 125.4, 126.3, 128.1, 128.8, 129.4, 131.4, 132.9, 133.4; ^{19}F NMR (CDCl$_3$): δ -112.7 (d, *J*=7.8 Hz).
   anti-isomer (representative data are shown)
   ^{1}H NMR (CDCl$_3$): δ 3.30 (3 H, s), 4.39 (1 H, d, *J*=6.8 Hz), 4.67 (1 H, d, *J*=6.8 Hz), 6.37 (1 H, d, *J*=23.7 Hz); ^{19}F NMR (CDCl$_3$): δ -130.0 (d, *J* = 23.7 Hz).

Note

In the above case, the protective group played an important role in determining the product distribution.   For example, silyl groups led to the formation of epoxides and the *syn:anti* ratio was decreased by non-chelating methyl ether.

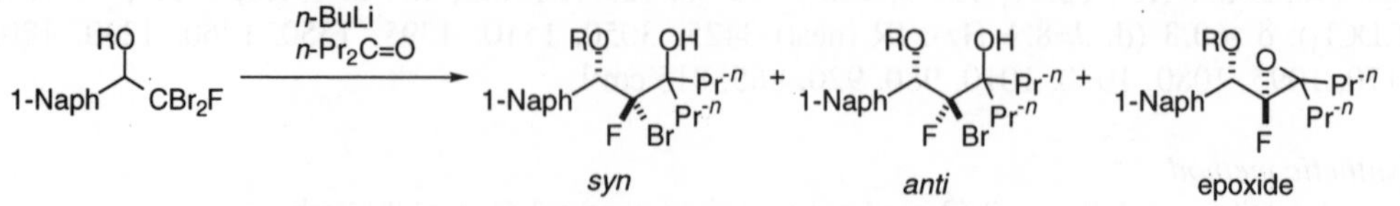

R	Yield (%)	Product ratio				
		syn	:	*anti*	:	epoxide
SiEt$_3$	38	57	:	25	:	18
SiMe$_2$Bu^{-t}	47	67	:	0	:	33
Me	49	60	:	40	:	0
MOM	55	85	:	15	:	0
MEM	55	83	:	17	:	0

(Reproduced with permission from Shimizu, M., *et al. Tetrahedron Lett.* **1996**, *37*, 7387)

Shimizu, M., *et al. Tetrahedron Lett.* **1996**, *37*, 7387.

(*R*)-3-Fluoro-1-(*p*-toluenesulfinyl)propan-2-one

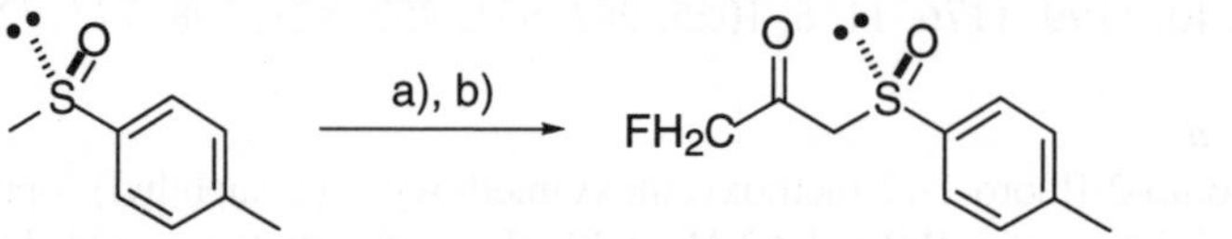

a) LDA / THF, b) CH$_2$FCO$_2$Et / THF

Synthetic method

A solution of (*R*)-*p*-tolyl methyl sulfoxide (10 mmol) in dry THF (20 mL) was added dropwise to a stirred solution of LDA (10.5 mmol), maintaining the temperature at -75 °C under an argon atmosphere.   After 3 min, a solution of ethyl fluoroacetate (15 mmol) in THF (40 mL) was added at -75 °C and stirring was continued for 5 min at that temperature.   The reaction was quenched with 60 mL of sat. NH$_4$Cl aq.   The pH was adjusted to 7 with dilute HCl, the layers were separated, and the aqueous layer was extracted with AcOEt (3x70 mL).   The organic layers were combined and dried over Na$_2$SO$_4$.   Evaporation and recrystallization from AcOEt gave (*R*)-3-fluoro-1-(*p*-toluenesulfinyl)propan-2-one in 75% yield.

Spectral data

$[\alpha]_D^{20}$ +242 (*c* 1.0, CHCl$_3$); mp. 120-122 °C; ^{1}H NMR (DMSO-d_6): δ 2.38 (3 H, s), 4.03 (1 H, dd, *J*=14.5, 2.5 Hz), 4.21 (1 H, d, *J*=1.5 Hz), 5.03 (2 H, d, *J*=46.5 Hz), 7.40-7.59 (4 H, m); ^{13}C NMR (DMSO-d_6): δ 20.85, 62.89, 85.66 (d, *J*=182 Hz), 124.15, 129.85, 140.13, 141.38, 196.99 (d, *J*=10.0 Hz); IR (nujol): ν 1725 cm^{-1}.

Note

Further manipulation of the resultant β-ketosulfoxide has also been reported: i) Bravo, P., *et al. Gazz. Chim. Ital.* **1988**, *118*, 115, ii) Bravo, P., *et al. J. Org. Chem.* **1989**, *54*, 5171, iii)

Bravo, P., *et al. J. Chem. Soc. Perkin Trans. 1* **1989**, 1201, iv) Bravo, P., *et al. J. Chem. Soc. Perkin Trans. 1* **1991**, 1315, v) Bravo, P., *et al. J. Med. Chem.* **1992**, *35*, 3102, vi) Bravo, P., *et al. Tetrahedron* **1995**, *51*, 8289.

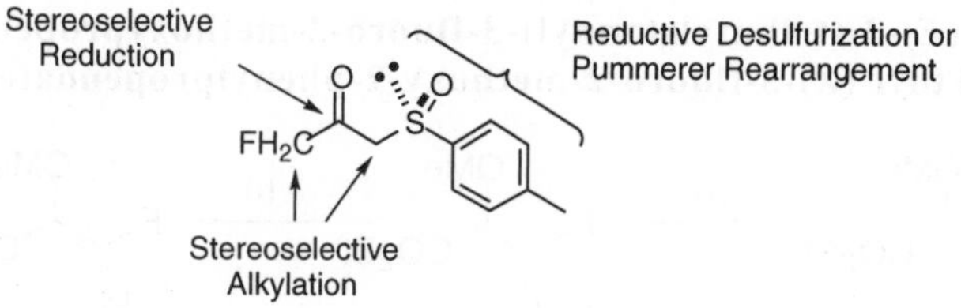

Bravo, P., *et al. Synthesis* **1986**, 579.

2-Fluoro-*N,N*-dimethylhexanamide

a) *n*-BuLi / Et$_2$O, b) *n*-BuLi

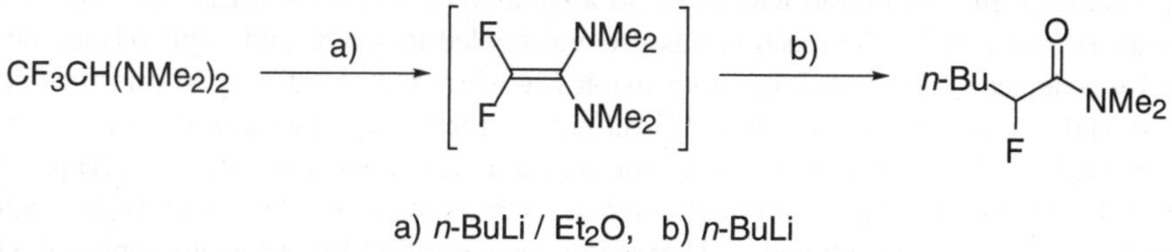

Synthetic method

To a solution of 1,1-bis(dimethylamino)trifluoroethane (1.7 g, 10 mmol) in Et$_2$O (5 mL) at -78 °C under nitrogen was added *n*-BuLi (2.5 *M*, 4.5 mL). Then the reaction mixture was allowed to warm to room temperature and stirred for 10 h. ^{19}F NMR spectrum showed that the starting material was gone and a new peak (δ at -114.1 ppm) occurred. *n*-BuLi (2.5 *M*, 6 mL) was added slowly and the reaction mixture was stirred at room temperature for 8 h. 100 mL of Et$_2$O was added to the mixture, the solution was washed with H$_2$O and dried over MgSO$_4$. After the solvent was removed, the residue was distilled at reduced pressure to give 1.37 g of 2-fluoro-*N,N*-dimethylhexanamide in 85.1% yield.

Spectral data

bp. 90-92 °C/6 mmHg; ^{1}H NMR (CDCl$_3$): δ 0.86 (3 H, t, *J*=6.9 Hz), 1.26-1.83 (6 H, m), 2.87 (3 H, d, *J*=1.0 Hz), 2.97 (3 H, d, *J*=1.7 Hz), 5.00 (1 H, ddd, *J*=49.3, 8.6, 4.4 Hz); ^{13}C NMR (CDCl$_3$): δ 13.6, 22.1, 26.7 (d, *J*=3.5 Hz), 31.5 (d, *J*=21.2 Hz), 36.1 (d, *J*=50.8 Hz), 89.6 (d, *J*=178.8 Hz), 168.7 (d, *J*=21.1 Hz); ^{19}F NMR (CDCl$_3$): δ -186.03 (ddd, *J*=51.2, 32.2, 21 Hz); IR (neat): ν 2958.4, 1659.8, 1503.3, 1467.6, 1419.8, 1380.0, 1354.4, 1261.8, 1166.0, 1120.8, 1015.8, 889.1, 781.8, 717.5 cm^{-1}.

Note

For the preparation of 2,2,2-trifluoro-1,1-bis(dimethylamino)ethane, see page 176 of this volume.

The above difluorinated intermediate was found to react with aldehydes to give 2,2-difluoro-3-hydroxyamides in good yields, while this process was limited only for aldehydes without α-hydrogen. See Xu, Y.-L.; Dolbier, Jr., W. R. *J. Org. Chem.* **1997**, *62*, 6503.

Dolbier, Jr., W. R., *et al. J. Org. Chem.* **1997**, *62*, 1576.

2.3 Removal of Fluorine from Difluorinated Materials

Ethyl (*E*)-3-(tributylstannyl)-3-fluoro-2-methoxypropenoate,
Ethyl (*Z*)-3-fluoro-2-methoxy-2-phenylpropenoate

a) (Bu₃Sn)₂CuLi / THF, b) PhI, Pd(PPh₃)₄, CuI / DMF

Synthetic method

Ethyl (*E*)-3-(tributylstannyl)-3-fluoro-2-methoxypropenoate

Bu₃SnH (2.7 mL, 10 mmol) was added to a solution of 10 mmol of LDA in THF (15 mL) at 0 °C.   After 10 min at 0 °C, the reaction mixture was cooled to -15 °C and copper(I) iodide (1.0 g, 5.0 mmol) was added.   The resulting suspension was stirred at -15 °C for 30 min then cooled to -78 °C.   A THF solution (2 mL) of ethyl 3,3-difluoro-2-methoxypropenoate (0.84 g, 5 mmol) was added over 2 min.   After 30 min at -78 °C, the reaction was quenched with sat. NH₄Cl (60 mL) and extracted with Et₂O.   The combined ethereal extracts were washed with H₂O (2x30 mL), dried over Na₂SO₄, and concentrated.   The residue was purified by chromatography on silica gel (petroleum ether:AcOEt=9:1) to afford 3.5 g of ethyl (*E*)-3-(tributylstannyl)-3-fluoro-2-methoxy-propenoate in 80% yield as a colorless oil.

Spectral data

¹H NMR (CDCl₃): δ 0.89 (9 H, t, *J*=7.1 Hz), 1.05-1.72 (18 H, m), 1.31 (3 H, t, *J*=7.1 Hz), 3.78 (3 H, s), 4.25 (2 H, q, *J*=7.1 Hz); ¹⁹F (CDCl₃): δ -31.0 (s) from CF₃CO₂H.

Synthetic method

Ethyl (*Z*)-3-fluoro-2-methoxy-3-phenylpropenoate

Ethyl (*E*)-3-(tributylstannyl)-3-fluoro-2-methoxypropenoate (0.44 g, 1.0 mmol) and iodo-benzene (0.20 g, 1.0 mmol) were dissolved in DMF (10 mL) under nitrogen at room temperature. Pd(PPh₃)₄ (0.12 g, 0.10 mmol) and purified copper(I) iodide (0.14 g, 0.75 mmol) were then added.   The mixture was stirred at room temperature and monitored by TLC for the disappearance of the starting organostannane.   The reaction mixture was diluted with Et₂O (20 mL), filtered, and washed with H₂O (2x20 mL).   The ethereal phase was stirred with 20% KF aq. for 30 min before being dried and concentrated.   The residue was purified by column chromatography on silica gel (petroleum ether:AcOEt=9:1) to afford 0.20 g of ethyl (*Z*)-3-fluoro-2-methoxy-3-phenylpropenoate in 95% yield.

Spectral data

¹H NMR (CDCl₃): δ 1.07 (3 H, t, *J*=7.0 Hz), 3.81 (3 H, s), 4.14 (2 H, q, *J*=7.0 Hz), 7.37 (5 H, s); ¹⁹F (CDCl₃): δ -23.0 (s) from CF₃CO₂H.

Note

For the preparation of starting material, see page 107 of this volume.

Shi and his co-workers employed other aromatic iodides, triflates, or vinylic triflate, to successfully construct the new carbon-carbon usually in a range of 80% to 90% yields in every instance.   The corresponding 2-benzyloxy compound was also reported to undergo addition-elimination type reactions with a variety of nucleophiles.   See Shi, G.-Q.; Cao, Z.-Y. *J. Chem. Soc., Chem. Commun.* **1995**, 1969.

Shi, G.-Q., *et al. J. Org. Chem.* **1995**, *60*, 6608.

(*E*)-1-Fluoronon-1-en-3-ol

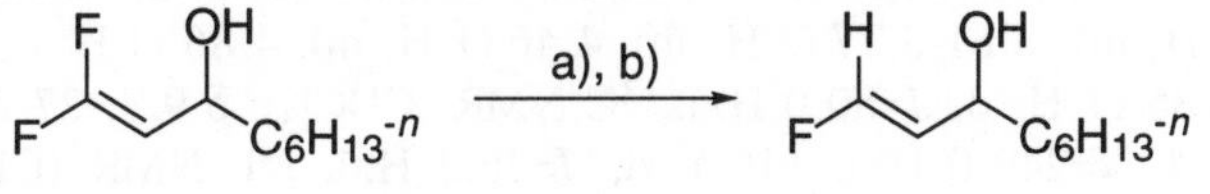

a) MeLi / Et$_2$O,   b) LiAlH$_4$

Synthetic method

A solution of methyllithium (20 mmol) was added to a stirred solution of 1,1-difluoronon-1-en-3-ol (prepared from 20 mmol of a carbonyl compound) in Et$_2$O (50 mL) at -30 °C.   The resultant mixture was stirred at 0 °C for 15 min and powdered lithium aluminum hydride (13 mmol) was added.   The temperature was increased to 20 °C and after 4 h the reaction was complete. AcOEt (5 mL) was added to the mixture followed by a 1 *N* H$_2$SO$_4$ solution and extraction with Et$_2$O.   The organic layer was washed successively with sat. NaHCO$_3$ aq. and brine, and dried over MgSO$_4$.   The solvent was evaporated (≤20 °C) when the crude 1-fluoronon-1-en-3-ol was obtained usually in 80-90% yield as a 95:5 (*E*):(*Z*) mixture.

Spectral data

^{1}H NMR (CDCl$_3$): δ 0.9 (3 H, t), 1.3 (8 H, m), 1.5 (2 H, m), 4.1 (1 H, dt, *J*=8.2, 6.6 Hz), 5.4 (1 H, ddd, *J*=18.1, 11.0, 8.2 Hz), 6.7 (1 H, dd, *J*=83.3, 11.0 Hz); ^{13}C NMR (CDCl$_3$): δ 14.1, 22.6, 25.3, 29.1, 31.8, 37.6, 68.0 (d, *J*=11.0 Hz), 115.2 (d, *J*=7.4 Hz), 150.6 (d, *J*=259.2 Hz); ^{19}F NMR (CDCl$_3$): δ -130.2 (dd, *J*=84, 18 Hz); IR (neat): ν 3340, 2920, 2850, 1670, 1460, 1100, 915 cm^{-1}.

(*Z*)-isomer (only distinct signals are shown)

^{1}H NMR (CDCl$_3$): δ 4.65 (1 H, dt, *J*=8.8, 7 Hz), 4.9 (1 H, ddd, *J*=42.3, 8.8, 4.1 Hz), 6.5 (1 H, dd, *J*=84.1, 4.1 Hz); ^{19}F NMR (CDCl$_3$): δ -128.2 (dd, *J*=84, 43 Hz).

Note

For the preparation of the starting allylic alcohol, see page 103 of this volume.

Tellier, F.; Sauvêtre, R. *J. Fluorine Chem.* **1996**, *76*, 181.

(*E*)-1-Fluoro-2-[(2-methoxyethoxy)methoxy]pent-1-en-3-ol

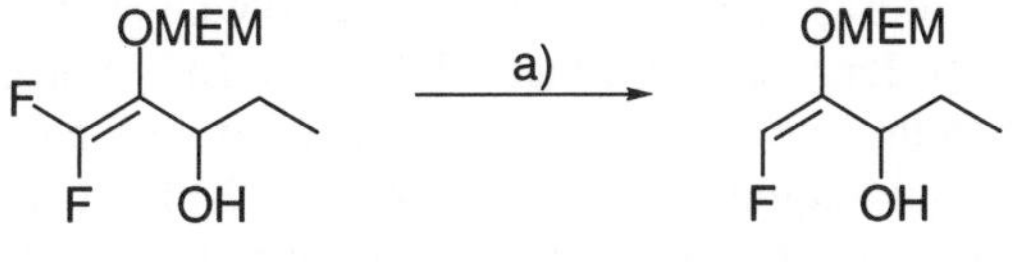

a) Red-Al / pentane

Synthetic method

1,1-Difluoro-2-[(2-methoxyethoxy)methoxy]pent-1-en-3-ol (4.0 g, 17.7 mmol) and Red-Al (a 65% solution in toluene; 13.3 mL, 68.0 mmol) were heated to reflux in pentane (40 mL) over 3 h.

The reaction mixture was cooled and poured onto an ice-H_2O mixture (80 mL). The resultant white suspension was acidified to *ca.* pH 3 with conc HCl, and the mixture was extracted with AcOEt (3x100 mL). The combined organic extracts were washed with brine (50 mL), dried (MgSO$_4$) and concentrated *in vacuo* to give a pale yellow oil. Purification by flash column chromatography, using hexane:AcOEt=3:2 as an eluent, gave 1-fluoro-2-[(2-methoxyethoxy)-methoxy]pent-1-en-3-ol in 72% yield as a 10:1 (*E*):(*Z*) mixture.

Spectral data

^{1}H NMR (CDCl$_3$): δ 0.85 (3 H, t, *J*=7.4 Hz), 1.47-1.73 (2 H, m), 2.52 (1 H, br s), 3.32 (3 H, s), 3.47-3.53 (2 H, m), 3.61-3.77 (2 H, m), 4.46 (1 H, m), 4.86 (1 H, d, *J*=6.5 Hz), 4.90 (1 H, d, *J*=6.5 Hz), 6.85 (1 H, d, *J*=80.0 Hz); ^{13}C NMR (CDCl$_3$): δ 9.7, 27.2, 58.9, 67.4, 67.7, 71.5, 95.4, 138.9 (d, *J*=249.0 Hz), 146.4 (d, *J*=26.3 Hz); ^{19}F NMR (CDCl$_3$): δ -172.8 (d, *J*=80.0 Hz); IR (film): ν 3434, 1650, 1462 cm^{-1}.

(Z)-isomer (only distinct signals are shown)

^{1}H NMR (CDCl$_3$): δ 0.84 (3 H, t, *J*=7.4 Hz), 6.40 (1 H, d, *J*=77.0 Hz); ^{13}C NMR (CDCl$_3$): δ 10.0, 27.0, 68.5, 71.6, 96.4; ^{19}F NMR (CDCl$_3$): δ -157.8 (1 F, d, *J*=77.0 Hz).

Note

For the preparation of the starting allylic alcohol, see page 110 of this volume.

See Suda, M. *Tetrahedron Lett.* **1981**,22, 1421 for the NaBH$_4$ reduction of β,β-difluoro-α,β-unsaturated carbonyl compounds. For the selective reduction of simple difluorinated olefins, see Ishikawa, N., *et al.* *Chem. Lett.* **1979**, 983.

Percy, J. M., *et al.* *Tetrahedron* **1995**, *51*, 11327.

2.4 Optically Active Materials by Bioorganic Methods

(*R*)-2-Fluoro-1-phenylethanol

a) Baker's yeast (no sucrose) / H_2O

Synthetic method

2-Fluoroacetophenone (0.155 g, 1.12 mmol) was added with stirring at 30 °C to a mixture of baker's yeast (Fleischmann or Itaiguara, 35 g) and H_2O (20 mL). Stirring was continued at that temperature for 24 h. Then the reaction mixture was saturated with NaCl and the products were extracted with CHCl$_3$ in a liquid-liquid extractor over 48 h. 2-Fluoro-1-phenylethanol (0.105 g, 0.75 mmol, 67% yield) was isolated as a colorless oil by silica gel column chromatography using CHCl$_3$ as an eluent. The optical purity was 97%.

Spectral data

[α]$_D$^{25} -51.7 (*c* 1.6, CHCl$_3$); ^{1}H NMR (CCl$_4$): δ 2.7 (1 H, br s), 4.23 (1 H, ddd, *J*=47.8, 9.2, 7.4 Hz), 4.40 (1 H, ddd, *J*=47.8, 9.2, 3.6 Hz), 4.86 (1 H, ddd, *J*=13.0, 7.4, 3.6 Hz), 7.25 (5 H, s).

Note

For the preparation of 2-fluoroacetophenone, see pages 41 and 63 of this volume.

See also the following related work: i) Guanti, G., *et al. J. Chem. Soc., Chem. Commun.* **1986**, 138, ii) Kitazume, T., *et al. J. Fluorine Chem.* **1987**, *35*, 537, iii) Vidal-Cros, A., *et al. J. Org. Chem.* **1989**, *54*, 498, iv) Moretti, I., *et al. Tetrahedron* **1989**, *45*, 7505.   For the recent review on the chiral fluoroorganic compounds *via* enzymatic methods, see Bravo, P.; Resnati, G. *Tetrahedron: Asym.* **1990**, *1*, 661.

For the resolution example by chemical methods, see i) Miyashita, O., *et al. Chem. Pharm. Bull.* **1982**, *30*, 3005, ii) Bailey, P. D., *et al. Tetrahedron Lett.* **1989**, *30*, 7457,

Rodrigues, J. A. R., *et al. Tetrahedron* **1991**, *47*, 2073.

Methyl (*S*)-9-acetoxy-10-fluorodecanoate

a) Lipase PS, Ac$_2$O / toluene

Synthetic method

4.27 mmol of methyl 10-fluoro-9-hydroxydecanoate and 0.4 mL (0.436 g, 4.27 mmol) of acetic anhydride were dissolved in 10 mL of distilled toluene.   After the addition of 31 mg of lipase PS (*Pseudomonas cepacia*, Amano Pharmaceuticals, Japan), absorbed on 0.126 g Celite 577, the resulting suspension was magnetically stirred at room temperature for approximately 24 h until 50% conversion was reached.   For reaction control, 0.2 mL of the suspension was filtered over silica gel (toluene, 5-10 mL), neutralized with 5% NaHCO$_3$ aq., dried and submitted to GC analysis.   After completion of the reaction, the whole suspension was treated in the same manner and dried with MgSO$_4$.   Chromatography on silica gel (cyclohexane:AcOEt=3:1) provided 0.536 g (2.05 mmol, 48% yield) of methyl (*S*)-9-acetoxy-10-fluorodecanoate and 0.329 g (1.49 mmol, 35% yield) of methyl (*R*)-10-fluoro-9-hydroxydecanoate.

Spectral data

[α]$_D^{20}$ +5.16 (*c* 2.5, CHCl$_3$), 94% ee; bp. 172-174 °C/15 mmHg; ^{1}H NMR (CDCl$_3$): δ 1.25-1.35 (8 H, br s), 1.6 (4 H, br s), 2.07 (3 H, s), 2.29 (2 H, t, *J*=7.6 Hz), 3.65 (3 H, s), 4.35 (1 H, ddd, *J*=47.2, 10.1, 6.7 Hz), 4.50 (1 H, ddd, *J*=47.2, 10.1, 5.2 Hz), 5.0 (1 H, m, *J*=21.5 Hz); ^{13}C NMR (CDCl$_3$): δ 20.96, 24.8-29.45 (6C), 33.97, 51.36, 72.25 (d, *J*=20.3 Hz), 83.54 (d, *J*=172.9 Hz), 170.50, 174.12; ^{19}F NMR (CDCl$_3$): δ -230.0 (dt, *J*=47.7, 22.8 Hz).

Methyl (*R*)-10-fluoro-9-hydroxydecanoate

[α]$_D^{20}$ +3.23 (*c* 2.5, CHCl$_3$), 66% ee.   All other physical and spectroscopic data were identical to that of the racemic material.

Note

For the preparation of the starting alcohol, see page 38 of this volume.

The optically resolved materials obtained were further transformed into the fluorinated analog of phoracantholide, one of the three major components of the metasternal gland secretion of the eucarypt longicorn (*Phoracantha synonyma*).

Sattler, A.; Haufe, G. *Tetrahedron: Asym.* **1995**, *6*, 2841.

2.5 Substitution Reactions

Chlorofluoro(phenylsulfenyl)methane,
Chlorofluoro(phenylsulfinyl)methane,
Chlorofluoro(phenylsulfonyl)methane

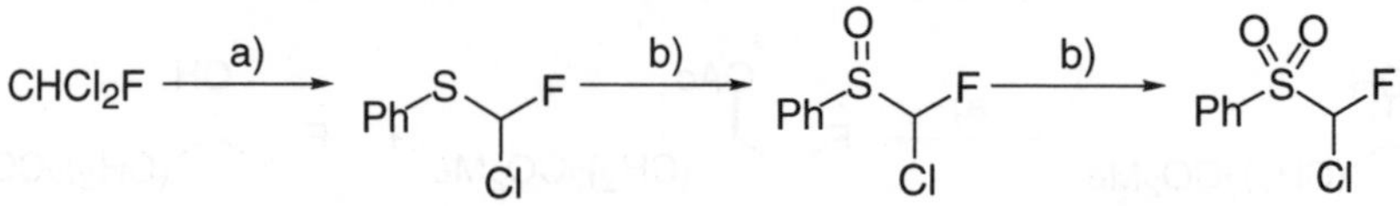

a) PhSH, NaOH, cat. BnMe$_3$NCl / PhH-H$_2$O,   b) *m*CPBA / CH$_2$Cl$_2$

Synthetic method

Chlorofluoro(phenylsulfenyl)methane

Thiophenol (2.0 mL, 20 mmol), benzyltrimethylammonium chloride (0.2 g), NaOH (5 g), benzene (20 mL), and H$_2$O (10 mL) were placed in a 100-mL stainless steel autoclave. Dichlorofluoromethane was introduced into the vessel with vigorous stirring for 10 min. During this period, the temperature rose to about 50 °C and the pressure became 3-5 kg/cm^2. After stirring for 4 h, the vessel was opened and the contents poured into H$_2$O. The mixture was extracted with Et$_2$O and the ethereal extract was washed five times with a cold 5% NaOH solution, dried over Na$_2$SO$_4$ and concentrated *in vacuo*. Distillation of the residue gave 2.47 g of chlorofluoro-(phenylsulfenyl)methane in 70% yield as a pale yellow oil.

Spectral data

bp. 37-38 °C (bath temperature)/0.2 mmHg; ^{1}H NMR (CDCl$_3$): δ 7.03 (1 H, d, *J*=55.8 Hz), 7.38 (3 H, m), 7.58 (2 H, m); IR (neat): ν 3030, 1456, 1290, 1185 cm^{-1}.

Synthetic method

Chlorofluoro(phenylsulfinyl)methane

m-Chroloperbenzoic acid (*m*CPBA; 3.61 g, 70% purity, 14.6 mmol) was added to a stirred solution of chlorofluoro(phenylsulfenyl)methane (2.030 g, 13 mmol) in CH$_2$Cl$_2$ (50 mL) at 0 °C. After disappearance of the starting material (overnight), the reaction was quenched by adding several drops of 1 *M* Na$_2$S$_2$O$_3$ aq. The solvent was evaporated and 50 mL of benzene was added to a white residual solid. The suspension was filtered and the white solid washed twice with 30 mL of benzene. The combined benzene solution was washed with 0.5 *M* NaOH aq. (five times) and brine, dried over Na$_2$SO$_4$ and evaporated. The residue was chromatographed on silica gel (hexane:CH$_2$Cl$_2$) to give 2.453 g of crude chlorofluoro(phenylsulfinyl)methane as a pale yellow oil. Distillation of the oil through a Kügelrohr apparatus gave 2.15 g of pure chlorofluoro(phenyl-sulfinyl)methane in 86% yield as a colorless oil (a 2:1 diastereomer mixture), which solidified below 0 °C.

Spectral data

bp. 65-68 ℃ (bath temperature)/0.2 mmHg; IR (neat): ν 3030, 1470, 1440, 1180, 1055 cm⁻¹.

major diastereomer

^{1}H NMR (CDCl₃): δ 6.65 (1 H, d, J=50.0 Hz), 7.57 (3 H, m), 7.74 (2 H, m); ^{13}C NMR (CDCl₃): δ 110.44 (d, J=288 Hz), 125.67 (d, J=1 Hz), 129.33, 132.79, 137.27 (d, J=5 Hz).

minor diastereomer

^{1}H NMR (CDCl₃): δ 6.62 (1 H, d, J=50.4 Hz), 7.57 (3 H, m), 7.74 (2 H, m); ^{13}C NMR (CDCl₃): δ 108.91 (d, J=287 Hz), 126.15 (d, J=1 Hz), 129.29, 132.95, 137.33 (d, J=3 Hz).

Chlorofluoro(phenylsulfonyl)methane

bp. 85-87 ℃ (bath temperature)/0.2 mmHg; ^{1}H NMR (CDCl₃): δ 6.61 (1 H, d, J=48.8 Hz), 7.65 (2 H, m), 7.80 (1 H, m), 8.01 (2 H, m); ^{13}C NMR (CDCl₃): δ 104.46 (d, J=284 Hz), 129.54, 130.72 (d, J=1 Hz), 132.02, 135.77; IR (neat): ν 3040, 1575, 1440, 1340, 1250, 1240, 1080, 1060 cm⁻¹.

Uno, H., *et al. Bull. Chem. Soc. Jpn.* **1992**, 65, 210.

(Z)-3-(Diethylamino)-2-fluoroprop-2-enal

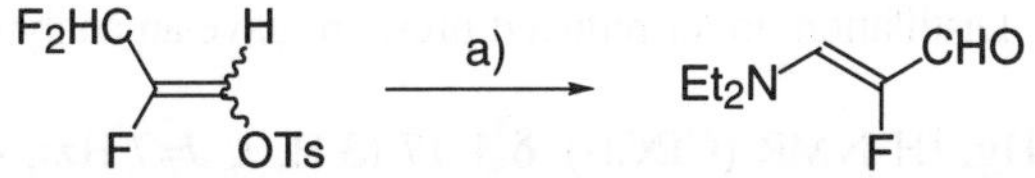

a) Et₂NH, Et₃N, cat. TBAF / MeCN

Synthetic method

To a solution of diethylamine (0.080 g, 1.1 mmol), triethylamine (0.101 g, 1.0 mmol), and a catalytic amount of tetra-*n*-butylammonium fluoride (TBAF; 0.1 mL of a 1.0 *M* THF solution, 0.1 mmol) in MeCN (1 mL) was added a solution of 2,3,3-trifluoroprop-1-en-1-yl *p*-toluenesulfonate (0.266 g, 1.0 mmol) in MeCN (2 mL) at 0 ℃. The mixture was stirred at ambient temperature for 3 h and was then quenched with brine (50 mL). The resulting mixture was extracted with CH₂Cl₂ (3x30 mL), dried over anhydrous Na₂SO₄, and concentrated *in vacuo*. The residue was chromatographed on a silica gel column (AcOEt) to give 0.144 g of analytically pure (Z)-3-(diethyl-amino)-2-fluoroprop-2-enal in 99% yield.

Spectral data

Rf 0.35 (AcOEt); ^{1}H NMR (CDCl₃): δ 1.22 (6 H, t, J=7.0 Hz), 3.33 (4 H, q, J=7.0 Hz), 6.15 (1 H, d, J=28.0 Hz), 8.34 (1 H, d, J=20.4 Hz); ^{19}F NMR (CDCl₃): δ -90.3 (dd, J=28.0, 20.4 Hz) from CF₃CO₂H; IR (neat): ν 3480, 3087, 2996, 2942, 2880, 1675, 1665, 1600, 1467, 1435, 1385, 1375, 1357, 1318, 1267, 1250, 1175, 1133, 1096, 1080, 1068, 992, 938, 868, 821, 792, 775, 763 cm⁻¹.

Note

For the preparation of the starting material, see page 204 of this volume.

The same starting compound was also employed for the introduction of aromatic groups and further reaction with NaOH in an alcohol solvent furnished the monofluorinated ketoacetals. See

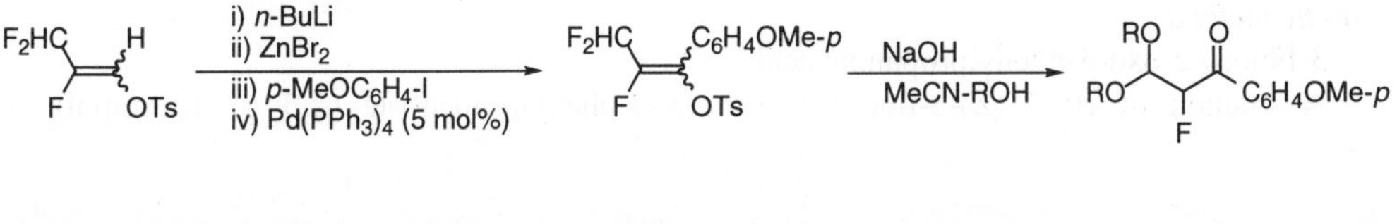

Funabiki, K., *et al. J. Chem. Soc., Perkin Trans. 1* **1997**, 2679.

Funabiki, K., *et al. Chem. Lett.* **1997**, 739.

Ethyl 2-fluoro-2-(phenylsulfenyl)acetate

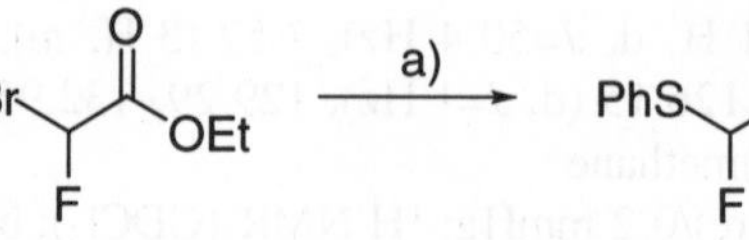

a) PhSH, Et₃N / THF

Synthetic method

To a stirred solution of ethyl bromofluoroacetate (2.22 g, 12 mmol) in THF (25 mL) were added successively Et₃N (2.5 mL, 18 mmol) and thiophenol (1.24 mL, 12 mmol) and the mixture was heated at reflux for 1 h. The precipitate was removed by filtration and the filtrate was concentrated by a rotary evaporator below 35 °C. The residue was purified by silica gel column chromatography (hexane:Et₂O=10:1) to afford ethyl 2-fluoro-2-(phenylsulfenyl)acetate as an oil (2.56 g , 95.6% yield). Distillation under reduced pressure gave an analytical sample.

Spectral data

bp. 125 °C/14 mmHg; ^{1}H NMR (CDCl₃): δ 1.17 (3 H, t, J=7 Hz), 4.17 (2 H, q, J=7 Hz), 6.10 (1 H, d, J=52 Hz), 7.2-7.7 (5 H, m); ^{19}F NMR (CDCl₃): δ -158.7 (d, J=53 Hz); IR (neat): ν 1760, 1585, 1045 cm^{-1}.

Note

Various substitution reactions at the 2 position were also discussed.

Takeuchi, Y., *et al. J. Chem. Soc. Perkin Trans. 1* **1988**, 1149.

2.6 Miscellaneous Reactions

3-Fluoro-2-oxo-3-phenylpropanoic acid,
anti-3-Fluorophenylalanine

a) TMSI / CHCl₃,   b) NH₃ aq.,   c) NaBH₄

Synthetic method

3-Fluoro-2-oxo-3-phenylpropanoic acid

A solution of ethyl (Z)-3-fluoro-2-methoxy-3-phenylpropenoate (2.3 g, 10 mmol) and

Me$_3$SiI (30 mmol) in CHCl$_3$ (20 mL) was refluxed under nitrogen for 16 h.   The reaction mixture was then concentrated *in vacuo*, and the residue was taken up in MeOH (20 mL).   After being allowed to stand at room temperature for 3 h, the methanolic solution was concentrated and the solid thus obtained was washed with cold CHCl$_3$ to afford 1.73 g of 3-fluoro-2-oxo-3-phenyl-propanoic acid in 95% yield.   This compound was found to be sensitive to oxygen and exist in solution as an enol form.

Spectral data

mp. 131-133 °C (hexane:Et$_2$O); ^{1}H NMR (acetone-d_6): δ 7.45 (3 H, m), 7.90 (2 H, m); ^{19}F (acetone-d_6): δ -63.5 (s) from CF$_3$CO$_2$H.

Synthetic method

anti-3-Fluorophenylalanine

3-Fluoro-2-oxo-3-phenylpropanoic acid (2.3 g, 10 mmol) was dissolved in 25% NH$_3$ aq. (10 mL), and the solution was maintained at 40 °C for 3 h.   The resultant brown solution was cooled to 10 °C, and NaBH$_4$ (0.51 g, 15 mmol) was added.   The reaction mixture was then evacuated with a water pump for 20 min while a stream of nitrogen was bubbled in.   After an additional 2 h at 30 °C, the reaction was again cooled in an ice bath and acidified to pH 1-2 using a 2.5 *N* HCl solution.   Precipitates coming out of the solution were dissolved in propan-2-ol.   The clear solution thus obtained was then treated with 1 liter of acidic ion exchange resin (AG 50W-X8) by sequentially using 50% propan-2-ol aq., H$_2$O, and 1 *N* NH$_3$ aq. as eluents.   Removal of the solvent from the last eluent at 25 °C under reduced pressure gave 0.36 g (40%) of 3-fluorophenyl-alanine with *anti* relative stereochemistry as a white solid.

Spectral data

mp. 161-162 °C (dec); ^{1}H NMR (D$_2$O): δ 4.33 (1 H, dd, J=16.3, 3.4 Hz), 6.17 (1 H, dd, J=44.0, 3.4 Hz), 7.44 (5 H, m); ^{19}F NMR (D$_2$O): δ -21.3 (dd, J=44.0, 16.3 Hz) from C$_6$F$_6$; IR (KBr): ν 3420-2400, 1660-1550, 1520, 1450, 1410, 1390, 1370, 1330, 1135, 1029, 708 cm^{-1}.

Note

For the preparation of the starting material, see page 68 of this volume.

The work-up method of the second experiment in the above case and spectral data were obtained from an earlier work by the Shionogi group.   See Tsushima, T., *et al. J. Org. Chem.* **1984**, *49*, 1163.

Shi, G.-Q., *et al. J. Org. Chem.* **1995**, *60*, 6608.

2-Fluoro-*N*-[(*R*)-1-phenyl-2-piperidinoethyl]acetamide, (*R*)-*N*-(2-Fluoroethyl)-1-phenyl-2-piperidinoethylamine

a) (*R*)-1-Phenyl-2-piperidinoethylamine, MeONa / EtOH,   b) BH$_3$·THF / THF

Synthetic method

N-[(*R*)-1-Phenyl-2-piperidinoethyl]-2-fluoroacetamide

Ethyl fluoroacetate (1.6 mL, 16.6 mmol) and sodium methoxide (96%, 160 mg, 2.85 mmol)

were added to a solution of (R)-1-phenyl-2-piperidinoethylamine (1.90 g, 9.30 mmol) in EtOH (40 mL), and the resulting solution was heated under reflux for 47 h. The solvent was evaporated *in vacuo*, and sat. NH_4Cl aq. (2 mL) was added to the residue. The aqueous layer was basified by the addition of sat. $NaHCO_3$ aq., and was extracted with CH_2Cl_2 (3x70 mL). The combined organic extracts were washed with brine (40 mL), dried (K_2CO_3), filtered, and concentrated *in vacuo* to give a pale yellow oil. This crude oil was treated with a 35% solution of HCl in EtOH (6 mL). The solvent was evaporated *in vacuo*, and the residue was suspended in PhH, which was evaporated *in vacuo*. This process was repeated three times to give a colorless amorphous solid. Recrystallization from Et_2O:EtOH=15 mL:3 mL gave 2.07 g of the monohydrochloride of N-[(R)-1-phenyl-2-piperidinoethyl]-2-fluoroacetamide in 74% yield as colorless prisms.

This salt (1.85 g) was dissolved in 10% NaOH aq. (20 mL) at 0 °C and then the solution was extracted with hexane (3x40 mL). The combined organic extracts were washed with brine (25 mL), dried (Na_2SO_4), and concentrated to dryness *in vacuo* to give 1.45 g of N-[(R)-1-phenyl-2-piperidinoethyl]-2-fluoroacetamide in 89% yield as a colorless oil.

Spectral data

N-[(R)-1-phenyl-2-piperidinoethyl]-2-fluoroacetamide monohydrochloride

$[\alpha]_D^{25}$ -65.8 (c 1.03, EtOH); mp. 152.5-153.0 °C; IR (KBr): ν 1690, 1550 cm^{-1}.

N-[(R)-1-phenyl-2-piperidinoethyl]-2-fluoroacetamide

$[\alpha]_D^{25}$ -73.1 (c 0.98, EtOH); ^{1}H NMR (CDCl$_3$): δ 1.1-1.7 (6 H, br s), 2.0-2.6 (6 H, br m), 4.80 (2 H, d, J=48 Hz), 4.95 (1 H, dd, J=7.4, 6.6 Hz), 7.3 (5 H, m), 7.3-7.4 (1 H, br s); IR (neat): ν 1660, 1550 cm^{-1}.

Synthetic method

(R)-N-(2-Fluoroethyl)-1-phenyl-2-piperidinoethylamine

A solution of N-[(R)-1-phenyl-2-piperidinoethyl]-2-fluoroacetamide (2.07 g, 7.83 mmol) in THF (10 mL) was added to a 1.0 M solution of borane-tetrahydrofurane complex in THF (47 mL, 47 mmol) at 0 °C, and the resulting mixture was stirred at 0 °C for 4 h. The reaction was quenched with MeOH (50 mL) and then conc HCl (50 mL) was added at the same temperature. The mixture was stirred for 38 h at room temperature. After removal of the solvent *in vacuo*, MeOH (100 mL) was added to the residue and the solvent was concentrated *in vacuo*. This process was repeated three times. The pale yellow residue was washed with Et_2O (2x40 mL) and was then dissolved in H_2O (50 mL). Solid $NaHCO_3$ (33 g) was added to this solution and the aqueous layer was extracted with Et_2O (3x60 mL). The combined organic extracts were washed with brine (30 mL), dried (K_2CO_3), filtered, and concentrated *in vacuo* to give a yellow oil (2.66 g). Purification by silica gel column chromatography (CHCl$_3$:MeOH) gave a yellow oil, which was converted to the corresponding dihydrochloride salt by the procedure described above. Recrystallization from Et_2O (17 mL) gave 1.81 g of the dihydrochloride of (R)-N-(2-fluoroethyl)-1-phenyl-2-piperidinoethyl-amine in 72% yield as colorless needles.

This salt, which was converted to the free (R)-N-(2-fluoroethyl)-1-phenyl-2-piperidinoethyl-amine by a procedure similar to the one described above, gave (R)-N-(2-fluoroethyl)-1-phenyl-2-piperidino-ethylamine as a colorless oil.

(R)-N-(2-Fluoroethyl)-1-phenyl-2-piperidinoethylamine dihydrochloride

$[\alpha]_D^{25}$ -3.4 (c 0.94, MeOH); mp. 217-219 °C.

N-[(R)-1-phenyl-2-piperidinoethyl]-2-fluoroacetamide mono-hydrochloride

$[\alpha]_D^{25}$ -79.4 (c 0.96, MeOH); ^{1}H NMR (CDCl$_3$): δ 1.0-1.8 (6 H, m), 1.8-2.9 (9 H, br m), 3.80 (1 H, dd, J=11, 4 Hz), 4.46 (2 H, ddd, J=47, 5, 5 Hz), 6.9-7.6 (5 H, m), 7.3-7.4 (1 H, br s); ^{13}C NMR (CDCl$_3$): δ 24.3, 26.0, 47.4 (d, J=19 Hz), 54.4, 59.4, 66.3, 83.3 (d, J=166 Hz), 127.0, 127.2, 128.2, 142.6.

Note

The above monofluorinated diamine was reported to be a potent reagent for the enantio-selective deprotonation of 4-substituted cyclohexanone. The corresponding 2,2-di- or 2,2,2-trifluoroethylated, or 2,2,3,3,3-pentafluoropropylated diamines were also synthesized, and very high efficiency (92% yield, 95% ee) was eventually attained by the 2,2,2-trifluoroethylated chiral diamine.

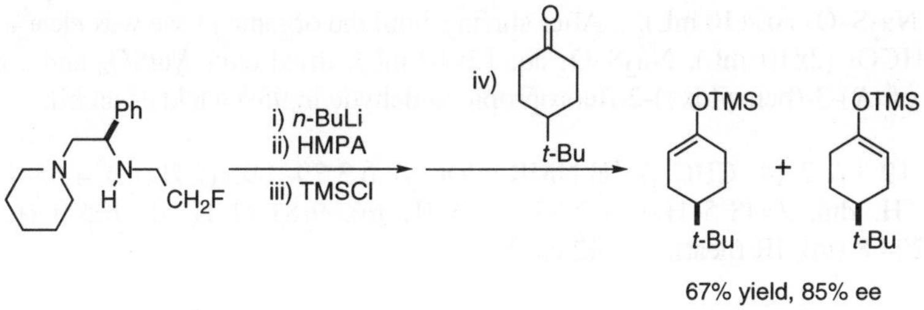

67% yield, 85% ee

Koga, K., *et al. Tetrahedron* **1997**, *53*, 13641.

(S)-3-(Benzyloxy)-2-fluoropropan-1-ol,
(R)-3-(Benzyloxy)-2-fluoropropionaldehyde

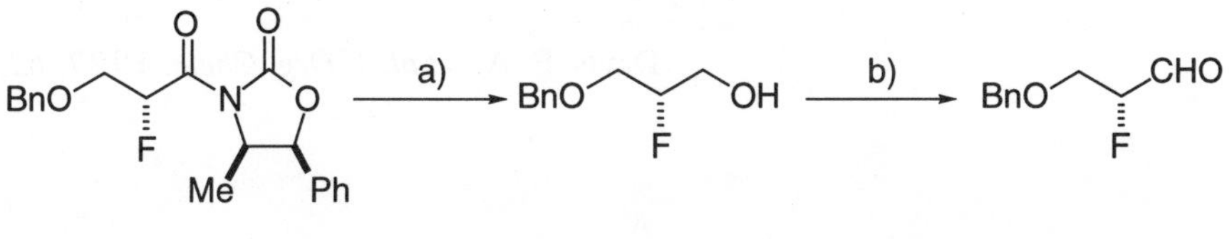

a) LiBH₄ / THF, b) Dess-Martin reagent / CH₂Cl₂

Synthetic method

(S)-3-(Benzyloxy)-2-fluoropropan-1-ol

In an oven-dried 250-mL, one-neck, round-bottomed flask equipped with a rubber septum, an argon inlet and a magnetic stir bar, was placed 0.669 g (1.87 mmol) of (4R,5S)-3-[(R)-3-(benzyloxy)-2-fluoropropionyl]-4-methyl-5-phenyloxazolidin-2-one in THF (75 mL). The reaction mixture was cooled to 0 °C, 1.12 mL (2.24 mmol, 2.0 M in THF) of lithium borohydride was added and the solution was stirred until TLC indicated the absence of starting material (2-3 h). At this time, the solution was warmed to room temperature, quenched with sat. NH₄Cl (5 mL) and diluted with AcOEt (30 mL). The organic phase was washed with brine (2x15 mL), dried over MgSO₄ and concentrated to give the product, which was purified by flash chromatography (hexane:AcOEt=75:25) to afford 0.320 g of (S)-3-(benzyloxy)-2-fluoropropan-1-ol in 93% yield as an oil and 0.268 g of (4R,5S)-4-methyl-5-phenyloxazolidin-2-one in 81% yield.

Spectral data

[α]$_D^{20}$ +5.7 (*c* 1.13, MeOH); ¹H NMR (CDCl₃): δ 2.0 (1 H, br t), 3.7 (1 H, dd, *J*=21.5, 4.66 Hz), 3.8 (1 H, dt, *J*=23.1, 4.8 Hz), 4.6 (2 H, s), 4.63-4.83 (1 H, dq, *J*=53.2 Hz), 7.30-7.40 (5 H, m); ¹³C NMR (CDCl₃): δ 62.6 (d, *J*=21.9 Hz), 68.5 (d, *J*=22.8 Hz), 73.6, 92.3 (d, *J*=172.3 Hz), 127.7, 128.4, 137.4; ¹⁹F NMR (CDCl₃): δ -196.7 (m); IR (neat): ν 3410 cm⁻¹.

Synthetic method

(*R*)-3-(Benzyloxy)-2-fluoropropionaldehyde

In an oven-dried 25-mL, one-neck, round-bottomed flask equipped with an argon inlet, a rubber septum and a magnetic stir bar, was placed 0.9 g (2.17 mmol) of the Dess-Martin periodinate in CH_2Cl_2 (10 mL) and stirred at room temperature.   (*S*)-3-(Benzyloxy)-2-fluoropropan-1-ol, 0.363 g (1.97 mmol, >97% ee) in CH_2Cl_2 (10 mL), was added *via* a cannula to the reaction mixture and after 10 min, the solution was diluted with Et_2O (20 mL), sat. $NaHCO_3$ (10 mL) and $Na_2S_2O_3$ aq. (10 mL).   After stirring until the organic phase was clear it was washed with sat. $NaHCO_3$ (2x10 mL), $Na_2S_2O_3$ aq. (2x10 mL), dried over $MgSO_4$ and concentrated to give 0.341 g of (*R*)-3-(benzyloxy)-2-fluoropropionaldehyde in 95% yield as an oil.

Spectral data

[α]$_D^{20}$ +9.3 (*c* 2.14, CHCl$_3$); ^{1}H NMR (CDCl$_3$): δ 3.79-3.95 (2 H, m), 4.55-4.64 (2 H, m), 4.84-5.00 (1 H, dm, *J*=48.5 Hz), 7.27-7.37 (5 H, m), 9.81 (1 H, d, *J*=5.9 Hz); ^{19}F NMR (CDCl$_3$): δ -204.8 (m); IR (neat): ν 1742 cm^{-1}.

Note

For the preparation of the starting material, see page 45 of this volume.

Oxidation of (*S*)-3-(benzyloxy)-2-fluoropropan-1-ol by the Swern or Moffat methods led to complete racemization and the only reagent to give satisfactory yields and ee's was the Dess-Martin periodinate reagent in CH_2Cl_2 (Dess, D. B.; Martin, J. C. *J. Am. Chem. Soc.* **1991**, *113*, 7277. Ireland, R. E.; Liu, L. *J. Org. Chem.* **1993**, *58*, 2899).   However, much lower ee's was observed on longer reaction time and instability towards silica gel was proved.   So, the final aldehyde was not subjected to further purification.

Davis, F. A., *et al. J. Org. Chem.* **1997**, *62*, 7546.

3 Preparation of Difluorinated Materials

3.1 Fluorination

3.1.1 Nucleophilic Fluorination

6,6-Difluoroundecane

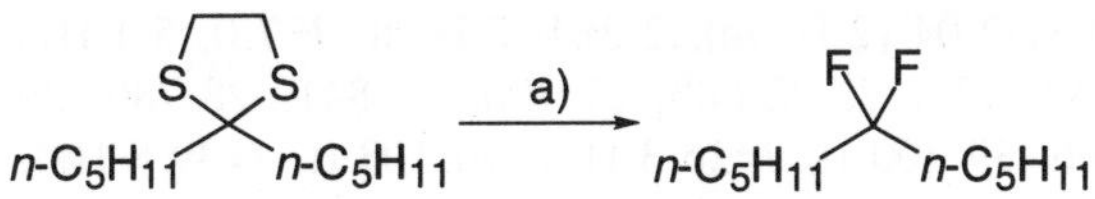

a) pyridine-(HF)$_n$, DBH / CH$_2$Cl$_2$

Synthetic method

1,3-Dibromo-5,5-dimethylhydantoin (DBH; 0.58 g, 2.03 mmol) was dissolved in 4.10 mL of dry CH$_2$Cl$_2$ and stirred under nitrogen. The mixture was cooled to -78 °C and 1.00 mL (4.40 mmol, *i.e.*, 40.6 equiv of F$^-$) of 70% pyridinium poly(hydrogen fluoride) was added *via* a polypropylene-polyethylene syringe, followed by the dropwise addition of 2,2-di-*n*-pentyl-1,3-dithiolane (0.50 g, 2.03 mmol). After 10 min, the reaction mixture, which was deep red, was diluted with hexane (30 mL) and filtered through a column (a 50-mL polypropylene-polyethylene syringe with a cotton plug) of basic alumina. Flash chromatography (hexane:Et$_2$O=49:1) yielded 312 mg (80%) of 6,6-difluoroundecane.

Spectral data

^{1}H NMR (CDCl$_3$): δ 0.90 (6 H, t), 1.26-1.50 (12 H, m), 1.67-1.85 (4 H, m); ^{13}C NMR (CDCl$_3$): δ 13.8, 22.0 (t, *J*=4.75 Hz), 22.4, 31.6, 36.3 (t, *J*=25.5 Hz), 125.4 (t, *J*=240 Hz); ^{19}F NMR (CDCl$_3$): δ -98.14 (q, *J*=16 Hz).

Note

As the promoter for the second fluorination process, nitrosonium tetrafluoroborate was reported to be comparably effective, one advantage being its availability even for compounds with electron-rich aromatic rings. Application of the above procedure to such materials has possibly led to the formation of the ring brominated products. See Olah, G. A., *et al. Tetrahedron* **1996**, *52*, 9. This type of reaction was realized from the corresponding hydrazones instead of dithioacetals. See Olah, G. A., *et al. Synlett* **1990**, 594.

For the formation of difluoromethylene compounds from dithioacetals under milder conditions, TBAH$_2$F$_3$ (preparable from TBAF and KHF$_2$. See Landini, D., *et al. Synthesis* **1988**, 953) was proved to be effective. See Kuroboshi, M.; Hiyama, T. *Synlett* **1991**, 909.

Sondej, S. C.; Katzenellenbogen, J. A. *J. Org. Chem.* **1986**, *51*, 3508.

6,6-Difluoro-2-methyloctadec-7-yne

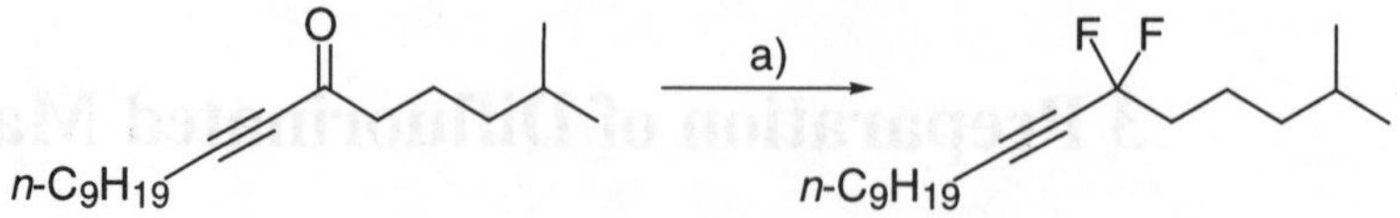

a) DAST (neat)

Synthetic method

A solution of 2-methyloctadec-7-yn-6-one (169.6 mg, 0.609 mmol) in neat DAST (0.8 mL) was stirred for 92 h at 55 °C in a sealed screw cap vial. The reaction mixture was poured into 50 mL of hexane then neutralized cautiously with solid $NaHCO_3$ and sat. $NaHCO_3$ aq. until the pH of the aqueous layer was ≥8. The hexane layer was washed with H_2O (20 mL), 1 *N* HCl (2x20 mL), H_2O (20 mL), and brine then dried. Concentration and chromatography (hexane) gave 92.3 mg (0.307 mmol, 50%) of the desired 6,6-difluoro-2-methyloctadec-7-yne as a clear oil.

Spectral data

Rf 0.51 (hexane); ^1H NMR (CDCl$_3$): δ 0.86-0.93 (9 H, m), 1.19-1.41 (16 H, m+br s), 1.49-1.63 (5 H, m), 1.87-2.04 (2 H, m), 2.264 (2 H, tt, *J*=7.0, 5.1 Hz); ^{13}C NMR (CDCl$_3$): δ 14.069, 18.351, 20.847, 22.417, 22.688, 27.790, 27.841, 28.780, 29.052, 29.325, 29.488, 29.562, 31.912, 38.206, 39.700 (t, *J*=26.3 Hz), 74.103 (t, *J*=39.8 Hz), 88.554 (t, *J*=6.4 Hz), 115.085 (t, *J*=231 Hz).

Caution

DAST was previously reported to possibly decompose in a violent manner at temperatures above 90 °C. See Middleton, W. J.; Bingham, E. M. *J. Org. Chem.* **1980**, *45*, 2883.

Note

For a review of DAST, see Hudlicky, M. *Org. React.* **1988**, *35*, 513.

Graham, S. M.; Prestwich, G. D. *J. Org. Chem.* **1994**, *59*, 2956.

(*E*)-1,1-Difluoronon-2-ene

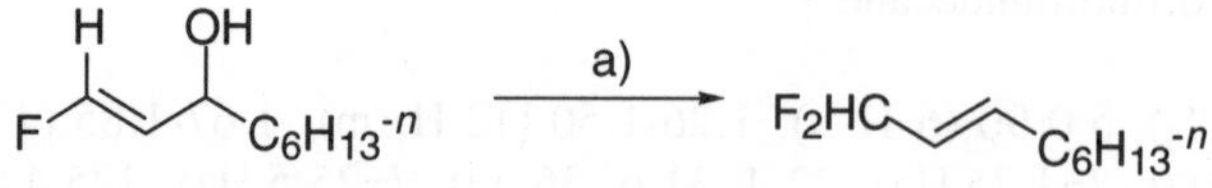

a) DAST / CH$_2$Cl$_2$

Synthetic method

DAST (2.5 mL, 20 mmol) was added over 5 min at -70 °C to a solution of the crude 1-fluoronon-1-en-3-ol (prepared from 20 mmol of a carbonyl compound) in CH$_2$Cl$_2$ (50 mL). After 15 min at -70 °C, the temperature was allowed to warm to 0 °C within 15 min. The reaction mixture was hydrolzyed by means of H_2O (30 mL) at 0 °C and extracted with Et$_2$O. The organic phase was successively washed with sat. $NaHCO_3$ aq. and brine. It was then dried over MgSO$_4$ and concentrated *in vacuo*. The crude residue was filtered through a small column packed with silica (cyclohexane). The solvent was evaporated and the residue was distilled to give (*E*)-1,1-difluoronon-2-ene as the sole product after isolation.

Spectral data

bp. 55 °C/10 mmHg; ^{1}H NMR (CDCl$_3$): δ 0.9 (3 H, t), 1.3 (6 H, m), 1.4 (2 H, m), 2.1 (2 H, m), 5.6 (1 H, dtdt, J=16.0, 7.7, 6.0, 1.6 Hz), 6.02 (1 H, td, J=56.1, 6.0 Hz), 6.06 (1 H, dtt, J=16.0, 6.6, 3.3 Hz); ^{13}C NMR (CDCl$_3$): δ 14.1, 22.6, 28.2, 28.8, 31.6, 31.8, 115.6 (t, J=232.6 Hz), 132.2 (t, J=23.9 Hz), 140.3 (t, J=11.0 Hz); ^{19}F NMR (CDCl$_3$): δ -110.0 (ddtd, J=56, 8, 4, 3 Hz); IR (neat): ν 2920, 2850, 1675, 1460, 1385, 1125, 1020, 960, 915 cm^{-1}.

Note

For the preparation of the starting monofluorinated allylic alcohol, see page 69 of this volume.

During the fluorination reaction, formation of the byproduct was observed (S$_N$2':S$_N$2=95:5), while due to the instability of the latter product, isolation furnished the material only *via* the S$_N$2' pathway.   See also Tellier, F.; Sauvêtre, R. *Tetrahedron Lett.* **1995**, *36*, 4223.   The same procedure for 3,3-difluoroallylic alcohols realized the construction of the corresponding CF$_3$ compounds.   See Tellier, F.; Sauvêtre, R. *J. Fluorine Chem.* **1993**, *62*, 183.

Tellier, F.; Sauvêtre, R. *J. Fluorine Chem.* **1996**, *76*, 181.

3.1.2 Electrophilic Fluorination

2', 3', 5'-Tri-*O*-acetyl-5,5-difluoro-5,6-dihydro-6-methoxycytidine

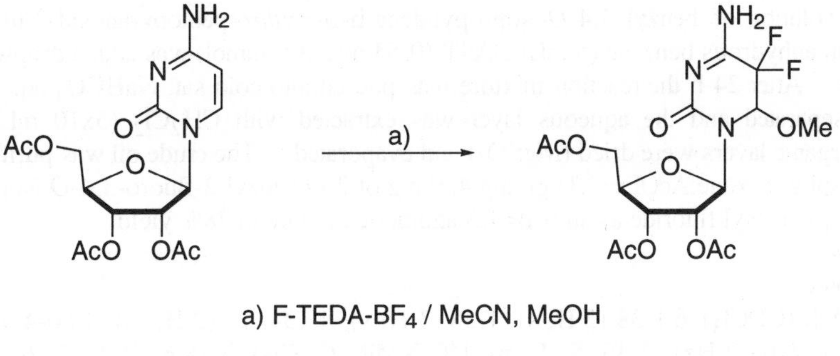

a) F-TEDA-BF$_4$ / MeCN, MeOH

Synthetic method

A solution of cytidine (337 mg, 0.91 mmol) in MeCN (10 mL) containing MeOH (2 mL) was treated with 1-(chloromethyl)-4-fluoro-1,4-diazabicyclo[2.2.2]octane bis(tetrafluoroborate) (F-TEDA-BF$_4$) (644 mg, 1.82 mmol) and refluxed for 4 h.   On cooling, the solution was poured into AcOEt (50 mL), washed with H$_2$O (25 mL) and sat. NaHCO$_3$ (2x25 mL), dried (MgSO$_4$), filtered, and evaporated *in vacuo*.   Flash chromatography on silica gel (hexane:AcOEt=2:3) afforded 313 mg of 2',3',5'-tri-*O*-acetyl-5,5-difluoro-5,6-dihydro-6-methoxycytidine in 82% yield.

Spectral data

^{1}H NMR (CDCl$_3$): δ 2.15 (9 H, br s), 3.45-3.60 (3 H, m), 4.20-4.50 (3 H, m), 4.85-5.00 (1 H, m), 5.15-5.50 (2 H, m), 5.90 (1 H, dd, J=36, 7 Hz), 9.15 (1 H, dd); ^{19}F NMR (CDCl$_3$): δ -109 (dd), -131 (dd).

Note

The present fluorinating reagent was also employed for the substitution of a tri-*n*-butyltin group for fluorine as described in the next page.   See Matthews, D. P., *et al. Tetrahedron Lett.* **1993**,*34*, 3057.   See also the following recent reviews on this class of fluorinating agents: i)

Banks, R. E. *J. Fluorine Chem.* **1998**, *87*, 1, ii) Pez, G. P., *et al. Chem. Rev.* **1996**, *96*, 1737.

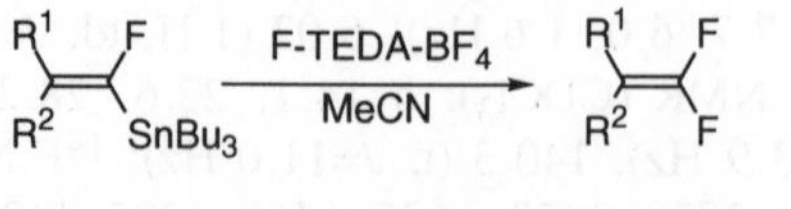

Lal, G. S., *et al. J. Org. Chem.* **1995**, *60*, 7340.

3.1.3 Miscellaneous

2-*O*-Benzyl-2-fluoro-3,4-*O*-isopropylidene-D-*ribo*-pentopyranosyl Fluoride

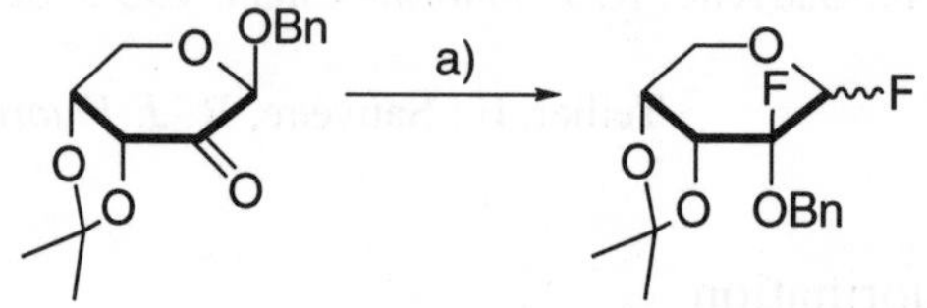

a) DAST / benzene

Synthetic method

To a solution of benzyl 3,4-*O*-isopropylidene-β-D-*erythro*-pentopyranosid-2-ulose (0.5 g, 1.8 mmol) in anhydrous benzene (5 mL) DAST (0.53 mL, 3.9 mmol) was added dropwise at room temperature.   After 24 h the reaction mixture was poured into cold sat. NaHCO$_3$ aq., the organic layer was separated and the aqueous layer was extracted with CH$_2$Cl$_2$ (3x10 mL); then, the combined organic layers were dried (MgSO$_4$) and evaporated.   The crude oil was purified by flash chromatography (hexane:AcOEt=2:1) giving 420 mg of 2-*O*-benzyl-2-fluoro-3,4-*O*-isopropylidene-D-*ribo*-pentopyranosyl fluoride as an α:β=1:3 anomeric mixture in 78% yield.

Spectral data

α-anomer

^{1}H NMR (CDCl$_3$): δ 1.38 (3 H, s), 1.58 (3 H, s), 4.13-4.22 (2 H, m), 4.60-4.93 (5 H, m), 5.31 (1 H, d, *J*=63.9 Hz), 7.36 (5 H, m); ^{13}C NMR (CDCl$_3$): δ 25.6, 25.9, 71.6, 72.4, 78.3, 79.3 (d, *J*=19.6 Hz), 106.8 (dd, *J*=223.0, 45.0 Hz), 112.2 (dd, *J*=239.9, 28.0 Hz), 115.4, 128.3-128.7 (Ph); ^{19}F NMR (CDCl$_3$): δ -123.5 (1 F, m), -143.7 (1 F, dd, *J*=63.9, 3.1 Hz).

β-anomer

^{1}H NMR (CDCl$_3$): δ 1.4 (3 H, s), 1.6 (3 H, s), 4.1-4.2 (2 H, m), 4.6-4.9 (5 H, m), 5.4 (1 H, dd, *J*=64.1, 2.5 Hz), 7.3-7.5 (5 H, m); ^{13}C NMR (CDCl$_3$): δ 25.6, 25.9, 71.8, 72.5, 78.3, 79.1 (d, *J*=19.2 Hz), 107.1 (dd, *J*=223.8, 45.6 Hz), 112.4 (dd, *J*=236.6, 34.3 Hz), 115.3, 128.3-128.7 (Ph); ^{19}F NMR (CDCl$_3$): δ -123.3 (1 F, dt, *J*=13.0, 2.5 Hz), -141.6 (1 F, dd, *J*=64.1, 13.0 Hz).

Note

DAST fluorination sometimes suffers from unexpected rearrangements by neighboring group participation.   See the following articles on sugar fluorination: i) Mori, Y.; Morishima, N. *Chem. Pharm. Bull.* **1991**, *39*, 1088, ii) Cabrera-Escribano, F., *et al. Tetrahedron Lett.* **1997**, *38*, 1231, iii) Kitano, K., *et al. Tetrahedron* **1997**, *53*, 13315.

Castillón, S., *et al. Tetrahedron* **1994**, *50*, 9125.

3.2 Difluoromethylenation

2,5-Anhydro-6-O-(tert-butyldimethylsilyl)-1-deoxy-1,1-difluoro-3,4-O-isopropylidene-D-ribo-hex-1-enitol

a) CBr_2F_2, $(Me_2N)_3P$, Zn / THF

Synthetic method

To a solution of 5-O-(tert-butyldimethylsilyl)-2,3-O-isopropylidene-D-ribonolactone (2.0 g, 6.6 mmol) in THF (65 mL), cooled to -20 °C, was added dibromodifluoromethane (2.73 mL, 30.0 mmol) using a cooled syringe. To the vigorously stirred solution was added tris(dimethylamino)-phosphine (6.50 mL, 30.0 mmol), and a dense white precipitate was formed immediately. The mixture was stirred at room temperature for 30 min, and then zinc powder (1.95 g, 30.0 mmol) and tris(dimethylamino)phosphine (0.3 mL) were added and the mixture heated to reflux for 18 h. The dark brown reaction mixture was allowed to cool to room temperature and Et_2O (20 mL) added. The Et_2O layer was decanted and the residue washed with Et_2O (10 mL). The combined Et_2O extracts were washed with a sat. $CuSO_4$ solution until it remained blue, H_2O (50 mL) and brine (50 mL), and dried over $MgSO_4$. The solvent was removed under reduced pressure to give a yellow oil. Column chromatography (petroleum ether (40-60 °C):Et_2O=5:1) afforded 1.50 g of 2,5-anhydro-6-O-(tert-butyldimethylsilyl)-1-deoxy-1,1-difluoro-3,4-O-isopropylidene-D-ribo-hex-1-enitol in 68% yield as a colorless oil.

Spectral data

$[\alpha]_D^{20}$ -104.6 (c 1.03, $CHCl_3$); ^1H NMR ($CDCl_3$): δ 0.03 (3 H, s), 0.05 (3 H, s), 0.87 (9 H, s), 1.39 (3 H, d, J=0.5 Hz), 1.50 (3 H, s), 3.72 (1 H, dd, J=11.2, 2.4 Hz), 3.77 (1 H, dd, J=11.2, 2.7 Hz), 4.45 (1 H, m), 4.81 (1 H, ddd, J=6.1, 1.7, 1.0 Hz), 5.28 (1 H, dd, J=6.1, 3.2 Hz); ^{13}C NMR ($CDCl_3$): δ -5.9, -5.8, 18.1, 25.6, 26.6, 64.3, 78.3, 81.4, 87.2, 112.9, 121.4 (dd, J=19, 13 Hz), 149.9 (dd, J=285, 270 Hz); ^{19}F NMR ($CDCl_3$): δ -105.6 (1 F, d, J=98 Hz), -122.6 (1 F, d, J=98 Hz); IR (neat): ν 2934, 1792, 1471, 1383,1214, 1164, 1087, 836 cm^{-1}.

Note

Ishikawa and his co-workers reported the difluoromethylenation by way of *in situ* prepared difluoromethylene ylide (CBr_2F_2, PPh_3, followed by Zn dust). Their method was found to be comparable or superior to Burton's original work *via* the isolated difluoromethylene ylide. See the following articles: i) Ishikawa, N., *et al. Chem. Lett.* **1979**, 983, ii) Naae, D. G.; Burton, D. J. *Synth. Commun.* **1983**, *3*, 197. Slight modification of the original procedure by Motherwell *et al.* as described above proved to be considerably more reliable.

Dolbier recently reported the alternative route for the construction of the difluorovinyl structure starting from α,α-difluoro-β-hydroxyesters: Dolbier, Jr., W. R., *et al. J. Fluorine Chem.* **1998**, *88*, 41.

Motherwell, W. B., *et al. Tetrahedron* **1993**, *49*, 8087.

1,1-Difluoro-3-[(trimethylsilyl)oxy]-3-phenylbut-1-ene

a) CBr$_2$F$_2$, (Me$_2$N)$_3$P / THF

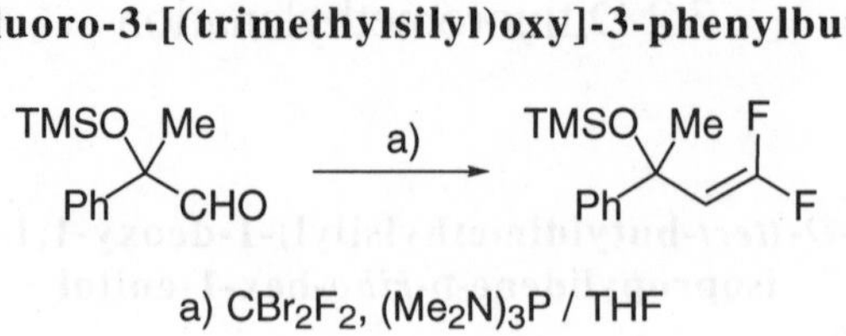

Synthetic method

Dibromodifluoromethane (approximately 38 g, 181 mmol) was condensed into 250 mL of THF at -78 °C with the help of a dry ice condenser.   Tris(dimethylamino)phosphine (54 g, 330 mmol) was then added, and the mixture was allowed to warm to room temperature before 2-phenyl-2-[(trimethylsilyl)oxy]propanal (15.31 g, 68.87 mmol) was added.   The aldehyde, as shown by gas chromatography, was consumed within 30 min.   The reaction slurry was transferred to a separatory funnel using 300 mL of H$_2$O and 500 mL of pentane.   The layers were separated, and the aqueous layer was washed with two further 200-mL aliquots of pentane.   The combined organic extracts, washed with brine and dried (MgSO$_4$), yielded the crude product, which was purified by LPLC (hexane:AcOEt=19:1) to afford 1,1-difluoro-3-[(trimethylsilyl)oxy]-3-phenylbut-1-ene in 65% yield.

Spectral data

^{1}H NMR (CDCl$_3$): δ 0.13 (9 H, s), 1.73 (3 H, d, *J*=3 Hz), 4.63 (1 H, dd, *J*=26, 6 Hz), 7.17-7.67 (5 H, m); ^{19}F NMR (CDCl$_3$): δ -83.27 (1 F, ddq, *J*=41.5, 26, 3 Hz), -85.92 (1 F, dd, *J*=41.5, 6 Hz): IR (film): ν 1740 cm^{-1}.

Note

The terminally difluorinated silyl ethers thus obtained were further transformed into the corresponding acetates, which, on treatment with a catalytic amount of *p*-toluenesulfonic acid in acetic anhydride, led to the acetoxy-group transposition to the carbon atom possessing two fluorines.   See the following article: Tellier, F.; Sauvêtre, R. *J. Fluorine Chem.* **1996**, *76*, 79.

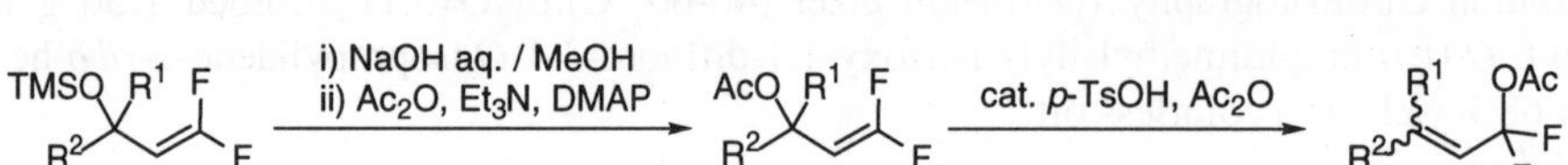

Ortiz de Montellano, P. R., *et al. J. Org. Chem.* **1983**, *48*, 4661.

Octyl 3,3-difluoroacrylate

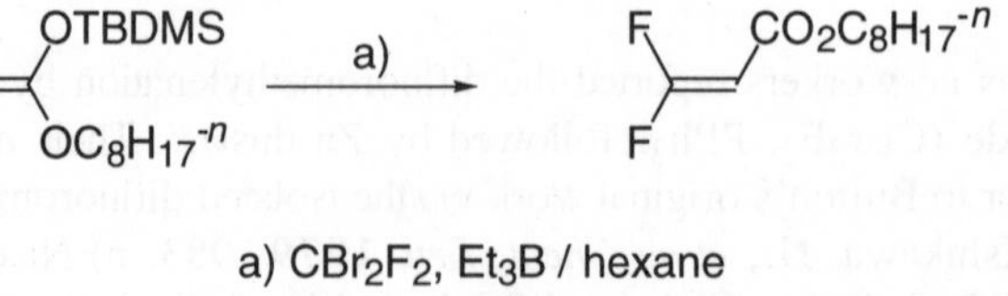

a) CBr$_2$F$_2$, Et$_3$B / hexane

Synthetic method

Under an argon atmosphere, Et$_3$B (a 0.96 *M* hexane solution, 0.21 mL, 0.20 mmol) was added to a solution of CBr$_2$F$_2$ (1.0 mmol) and ketene *tert*-butyldimethylsilyl octyl acetal (0.57 g, 2.0 mmol) in hexane (5 mL) at room temperature (25±3 °C).   After stirring for 30 min, sat. NaHCO$_3$ aq. (5 mL) was added to the reaction mixture.   The mixture was stirred vigorously for 1 h, then poured into H_2O (20 mL), and extracted with hexane (3x20 mL). The combined organic layer was dried over anhydrous Na_2SO_4 and concentrated *in vacuo*. The residual oil was purified by silica gel column chromatography (hexane:Et_2O=40:1) to give octyl 3,3-difluoroacrylate in 84% yield along with 5% of octyl 3-bromo-3,3-difluoropropionate as a byproduct.

Spectral data

bp. 69-73 °C (bath temperature)/9 mmHg; ^1H NMR ($CDCl_3$): δ 0.89 (3 H, t, J=6.5 Hz), 1.28 (10 H, bs), 1.58-1.72 (2 H, m), 4.15 (2 H, q, J=6.7 Hz), 4.98 (1 H, dd, J=21.8, 2.6 Hz); ^{13}C NMR ($CDCl_3$): δ 13.97, 22.58, 25.80, 28.47, 29.12 (two peaks), 31.72, 64.98, 77.10 (dd, J=28.6, 9.1 Hz), 161.87 (dd, J=311.7, 298.5 Hz), 162.97 (dd, J=17.1, 7.5 Hz); ^{19}F NMR ($CDCl_3$): δ -64.74 (1 F, dd, J=21.7, 15.8 Hz), -70.85 (1 F, dd, J=15.8, 2.0 Hz); IR (neat): ν 2954, 2926, 2856, 1749, 1734, 1711, 1357, 1279, 1137 cm^{-1}.

Note

Upon employment of dibromofluoromethane and treatment of the reaction mixture with triethylamine before purification, the corresponding (E)-3-fluoroacrylate was obtained in good yield. In the absence of triethylamine, the bromofluoromethylated ester was isolated.

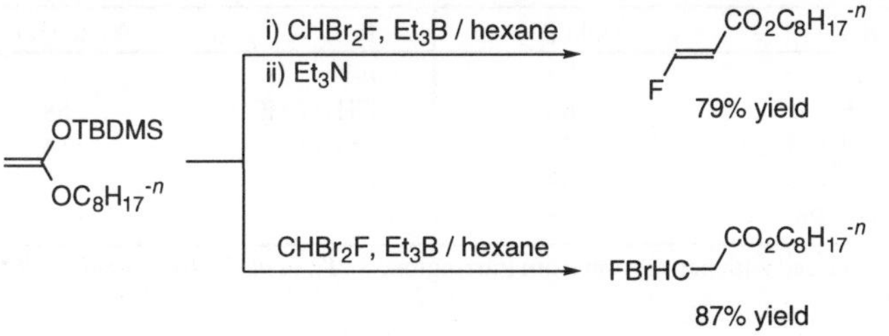

Utimoto, K., *et al. Bull. Chem. Soc. Jpn.* **1992**, 65, 1513.

3.3 Introduction of Difluoromethyl and Related Groups

Diethyl 2-(bromodifluoromethyl)-2-methylmalonate

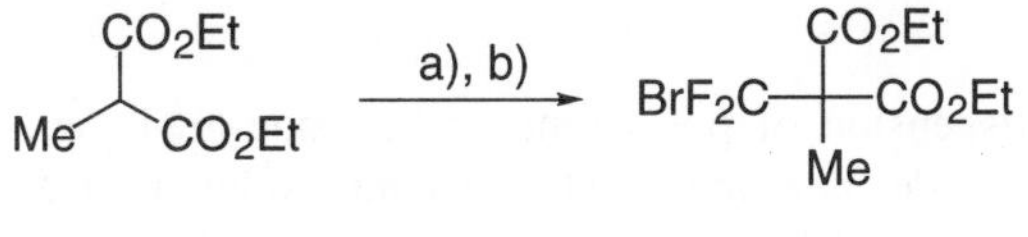

a) NaH / THF, b) CF_2Br_2

Synthetic method

Sodium hydride (50% oil dispersion, 0.10 mol) was transferred to an anhydrous reaction flask under nitrogen and washed free of the oil with hexane. Dry THF (100 mL) was added, then diethyl 2-methylmalonate (0.095 mol) was pipetted into the NaH/THF suspension at room temperature and stirred for 0.5 h. Heat was generated and H_2 evolved as the sodium salt of the malonic ester formed. Dibromodifluoromethane (0.10 mol, bp. 24 °C) was condensed with a dry ice-acetone cold finger and collected in a graduated receiver, with the amount either weighed or determined by volume (d=2.3 g/mL). This was joined to the reaction pot by a U tube type connection, which allowed the contents to be poured quickly by inverting the tube.   This addition at room temperature produced noticeable heating in some cases and within minutes an obvious change in appearance could be observed as NaBr precipitated from the solution.   The reaction vessel was then tightly sealed to prevent loss of the low boiling methane component and the reaction mixture stirred overnight at room temperature.

The reaction was worked up by evaporating THF and taking up the residue in an Et_2O/H_2O extraction.   Separation of the two phases followed by drying and concentration of the organic portion yielded the crude reaction mixture, which after vacuum distillation gave diethyl 2-(bromodifluoromethyl)-2-methylmalonate in 76% yield in 90+% purity.

Spectral data

bp. 66 °C/0.2 mmHg; ^{1}H NMR (CDCl$_3$): δ 1.3 (3 H, t), 1.8 (3 H, s), 4.3 (2 H, q); ^{19}F NMR (CDCl$_3$): δ 50 (s).

R	Yield (%)	R	Yield (%)
Et	73	*n*-Pr	74
n-Bu	66	CH$_2$=CHCH$_2$-	68
CH$_2$CH$_2$CN	80	CO$_2$Et	14
CH$_2$CO$_2$Et	53	Ph	85
CH$_2$Ph	75		

(Reproduced with permission from Purrington, S. T., *et al. J. Org. Chem.* **1984**, *49*, 3702)

Purrington, S. T., *et al. J. Org. Chem.* **1984**, *49*, 3702.

Diethyl 2-(difluoromethyl)-2-(methylsulfenyl)malonate

a) KOBu^{-t} / THF,   b) CHClF$_2$ / THF

Synthetic method

To a well-stirred suspension of potassium *tert*-butoxide (181 mg, 1.6 mmol) in THF (0.7 mL) was added at -78 °C under nitrogen a THF (0.7 mL) solution of diethyl 2-(methylsulfenyl)-malonate (103 mg, 0.5 mmol), and the temperature was gradually raised to 10 °C and maintained for 30 min.   Next, a THF solution containing chlorodifluoromethane in large excess was added all at once and the mixture was kept well stirred for 30 min at room temperature to complete the reaction.   The reaction mixture was then poured into 10% NH$_4$Cl aq., saturated with saline, and extracted with AcOEt.   The organic layer was washed with H$_2$O three times, dried MgSO$_4$, filtered, and evaporated *in vacuo*, leaving an oily residue.   Chromatography of this residue over Merck silica gel Lobar column (type B, benzene) afforded 54 mg of diethyl 2-(difluoromethyl)-2-(methyl-sulfenyl)malonate in 42% yield as an oily substance.

Spectral data

^{1}H NMR (CDCl$_3$): δ 1.30 (6 H, t, *J*=7.0 Hz), 2.23 (3 H, s), 4.32 (4 H, q, *J*=7.0 Hz), 6.32 (1 H, t, *J*=54.6 Hz); ^{19}F NMR (CDCl$_3$): δ 37.83 (d, *J*=54.6 Hz) from C$_6$F$_6$; IR (film): ν 2960,

2925, 1730, 1310-1170, 1140 cm^{-1}.

Tsushima, T., *et al. Tetrahedron* **1988**, *44*, 5375.

1-Benzyloxy-4,4-difluorobut-2-yne

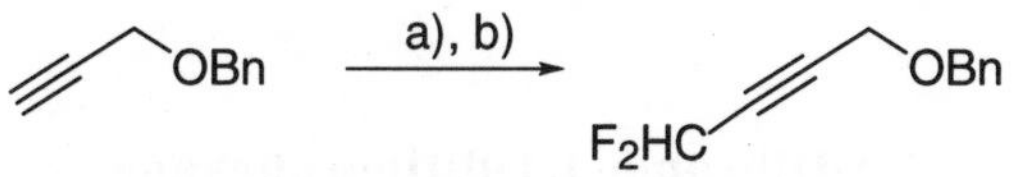

a) *n*-BuLi / THF,   b) CHClF$_2$

Synthetic method
To a THF (5 ml) solution of 3-(benzyloxy)prop-1-yne (1 mmol) was added 0.75 mL (1.2 mmol) of *n*-BuLi (a 1.6 *M* hexane solution) at -78 °C and the whole was stirred for 30 min.   To this solution was added excess CHF$_2$Cl (gas) at -78 °C and stirring was continued at that temperature for 1 h.   Quenched with a sat. NH$_4$Cl solution, the organic material was extracted with Et$_2$O, washed with H$_2$O and brine, dried over anhydrous MgSO$_4$ and the evaporation of the solvent afforded the crude oil.   Column chromatography gave the pure 1-benzyloxy-4,4-difluorobut-2-yne in 72% yield.
Spectral data
^{1}H NMR (CDCl$_3$): δ 4.27 (2 H, td, *J*=4.98, 0.98 Hz), 4.61 (2 H, s), 6.24 (1 H, tt, *J*=54.69, 1.22 Hz), 7.2-7.5 (5 H, m); ^{13}C NMR (CDCl$_3$): δ 56.55 (t, *J*=2.4 Hz), 72.04, 77.48 (t, *J*=24.1 Hz), 85.10 (t, *J*=7.1 Hz), 103.45 (t, *J*=232.4 Hz), 128.09, 128.50, 136.67; ^{19}F NMR (CDCl$_3$): δ 55.53 (dt, *J*=54.9, 4.57 Hz) from C$_6$F$_6$; IR (neat): ν 3055, 3040, 3032, 3000, 2950, 2858, 2252 cm^{-1}.

Konno, T.; Kitazume, T. *Chem. Commun.* **1996**, 2227.

Diethyl (bromodifluoromethyl)phosphonate

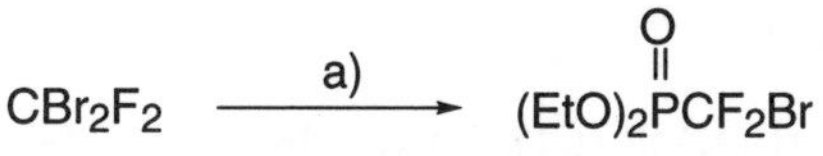

a) (EtO)$_3$P / Et$_2$O

Synthetic method
Dibromodifluoromethane (134 g, 0.64 mol) was added to a magnetically stirred cold solution of triethylphosphite (100 g, 0.60 mol) in 300 mL of dry Et$_2$O at 0 °C.   Under a nitrogen atmosphere the colorless solution was warmed to room temperature and then refluxed for 24 h.   After concentration of the reaction mixture, vacuum distillation yielded 152 g (0.57 mol, 95% yield) of diethyl (bromodifluoromethyl)phosphonate.
Spectral data
bp. 99-102 °C/16 mmHg; ^{13}C NMR (CDCl$_3$): δ 116.8 (td, *J*=330, 238 Hz); ^{19}F NMR (CDCl$_3$): δ -61.9 (d, *J*=93 Hz); ^{31}P NMR (CDCl$_3$): δ -1.16 (dt, *J*=238, 93 Hz).
Note
A similar process enabled the formation of the corresponding dibromofluoro- or difluoro-

methyl derivatives by using CFBr$_3$ or CClF$_2$H, respectively.   For the latter case, see Berkowitz, D. B.; Sloss, D. G. *J. Org. Chem.* **1995**, *60*, 7047 (phosphonate) or Edwards, M. L., *et al.* *Tetrahedron Lett.* **1990**, *31*, 5571 (phosphine oxide).   See also the following recent articles on the monofluorinated phosphonates: i) Nieschalk, J.; O'Hagan, D. *J. Chem. Soc., Chem. Commun.* **1995**, 719, ii) Waschbüsch, R., *et al.* *J. Chem. Soc., Perkin Trans. 1* **1997**, 1135.

Burton, D. J.; Flynn, R. M. *J. Fluorine Chem.* **1977**, *10*, 329.

1,3-Dibromo-1,1-difluorohexane, 1-Bromo-1,1-difluorohexane

a) cat. CuCl, H$_2$NCH$_2$CH$_2$OH / *tert*-BuOH,   b) NaBH$_4$ / DMSO

Synthetic method

   1,3-Dibromo-1,1-difluorohexane

   A Carius tube of approximately 200 mL capacity equipped with a small magnetic stir bar was charged with 0.14 g (1.42 mmol) copper(I) chloride, 4.36 g (71.2 mmol) ethanolamine, 12 mL *tert*-butanol, 10.0 g (142 mmol) 1-pentene, and 59.84 g (285 mmol) dibromodifluoromethane. The tube was flushed with nitrogen and flame-sealed; upon swirling, a deep blue coloration was observed.   The tube was immersed halfway into a silicon oil bath preheated at 85 °C and allowed to stir for 48 h, during which time the coloration turned from deep blue to olive green to brown. The tube was cooled in an ice bath, opened, and the contents transferred to a 250-mL Erlenmeyer flask (at this point, unreacted dibromodifluoromethane may be recovered by distillation) and the tube rinsed with three 50-mL portions of hexanes.   All organic materials (which consisted of a cloudy yellow-green supernatant and a brown resin) were filtered through 50 mL of silica gel, which was rinsed with two additional 50-mL portions of hexanes.   The resulting colorless filtrate was concentrated by rotary evaporation and subjected to reduced pressure fractional distillation through a 15-cm Vigreux column.   A total of 22.79 g (57.3%) of 1,3-dibromo-1,1-difluorohexane was obtained as a colorless liquid.

Spectral data

   bp. 80-85 °C/25 mmHg; ^{1}H NMR (CDCl$_3$): δ 0.96 (3 H, t, *J*=7.42 Hz), 1.50 (2 H, m), 1.86 (2 H, m), 3.01 (2 H, m), 4.24 (1 H, m); ^{13}C NMR (CDCl$_3$): δ 13.2, 20.4, 40.5, 46.6, 52.7 (t, *J*=21.8 Hz), 120.6 (t, *J*=306.9 Hz); ^{19}F NMR (CDCl$_3$): δ -43.2 (m).

Caution

   The reaction tube should be heated behind a safety shield.

Synthetic method

   1-Bromo-1,1-difluorohexane

   A 250-mL three-necked round-bottomed flask equipped with an ice-H$_2$O condenser, an argon inlet, and a magnetic stir bar was charged with 20.0 g (71.4 mmol) 1,3-dibromo-1,1-difluoro-hexane dissolved in 100 mL anhydrous DMSO.   A total of 10.8 g (285 mmol) of sodium boro-hydride was then added in small portions with vigorous stirring over the course of 1 h, during which time the flask became warm and a semisolid gel was observed to form.   After the addition was complete, the bath temperature was raised to 70 °C over the course of 1 h and heating continued for an additional 6 h (^{19}F NMR analysis of a small aliquot of the reaction mixture at this time showed complete consumption of starting material). The flask was cooled to room temperature, the contents transferred to a 1-liter Erlenmeyer flask, and the reaction quenched with chips of ice. The resulting mixture was carefully acidified with conc HCl aq., 100 mL Et_2O was added, and the aqueous DMSO layer extracted with three 50-mL portions of Et_2O. The combined Et_2O layers were washed with three 25 mL portions of H_2O, dried over $MgSO_4$, and subjected to ambient pressure fractional distillation through a 15-cm Vigreux column. After concentration in this way, 8.91 g (62.1%) 1-bromo-1,1-difluorohexane (contaminated with a small amount of 1,1-difluoro-hexane) was obtained as a colorless liquid.

Spectral data

bp. 125-128 °C; ^{1}H NMR ($CDCl_3$): δ 0.92 (3 H, t, J=7.5 Hz), 1.35 (4 H, m), 1.62 (2 H, m), 2.33 (2 H, m); ^{13}C NMR ($CDCl_3$): δ 13.8, 22.3, 23.6, 30.6, 44.3 (t, J=21.1 Hz), 123.3 (t, J=303.5 Hz); ^{19}F NMR ($CDCl_3$): δ -43.9 (t, J=14.7 Hz).

1,1-Difluorohexane

^{1}H NMR ($CDCl_3$): δ 0.91 (3 H, t, J=6.9 Hz), 1.34 (4 H, m), 1.45 (2 H, m), 1.80 (2 H, m), 5.79 (1 H, tt, J=57, 4.5 Hz); ^{13}C NMR ($CDCl_3$): δ 13.8, 21.8 (t, J=5.5 Hz), 22.4, 31.2, 34.1 (t, J=20.6 Hz), 117.5 (t, J=237.4 Hz); ^{19}F NMR ($CDCl_3$): δ -116.3 (dt, J=56.2, 17.1 Hz).

Note

A similar type of addition reaction of CBr_2F_2 to olefins was reported under the action of trimethylaluminum in the presence of a catalytic amount of tetrakis(triphenylphosphino)palladium. See Yamamoto, H., *et al. Chem. Lett.* **1985**, 1689.

Dolbier, Jr., W. R., *et al. Tetrahedron* **1997**, *53*, 9857.

Ethyl difluoroiodoacetate,
Ethyl 2,2-difluoro-3-iodooctanoate

$$CBrF_2CO_2Et \xrightarrow{\text{a), b)}} CF_2ICO_2Et \xrightarrow{\text{c)}} n\text{-}C_4H_9\text{-CH(I)-CH}_2\text{-}CF_2CO_2Et$$

a) Zn, cat. $HgCl_2$ / triglyme,   b) I_2,   c) 1-hexene, cat. Cu

Synthetic method

Ethyl difluoroiodoacetate

A flask fitted with a stirring bar and an N_2 inlet was charged with 19.5 g (0.3 mol) of zinc, 3.3 g (12 mmol) of mercury(II) chloride, and 250 mL of triglyme. Then 50.8 (0.25 mol) of ethyl bromodifluoroacetate was added slowly *via* a syringe with stirring at 25 °C over 30 min. After the addition was completed, the reaction mixture was stirred for 3 h. A 101.6-g (0.4 mol) portion of I_2 was added and the solution was stirred overnight. The reaction was flash distilled (<70 °C/ 0.1 mmHg) to give a dark mixture of ethyl difluoroiodoacetate and the solvent, which was poured into a beaker with a Na_2SO_3 solution. The light yellow organic lower layer was separated, washed with H_2O, dried over molecular sieves, and distilled to give 41.3 g (64%) of difluoroiodoacetate with 98% purity by CG analysis.

Spectral data

^{1}H NMR ($CDCl_3$): δ 1.38 (3 H, t, J=7.0 Hz), 4.40 (2 H, q, J=7.0 Hz); ^{19}F NMR ($CDCl_3$): δ -57.9 (s); IR (CCl_4): ν 2986, 1774, 1289, 1162, 1116, 930 cm[-1].

Synthetic method

Ethyl 2,2-difluoro-3-iodooctanoate

A heterogeneous mixture of 0.1 g (1.5 mmol) of copper, 0.82 g (10 mmol) of 1-hexene, and 1.25 g (5 mmol) of ethyl difluoroiodoacetate was stirred at 55 °C under nitrogen for 6 h. The reaction mixture was distilled at reduced pressure to give 1.1 g (65%) of ethyl 2,2-difluoro-3-iodo-octanoate.

Spectral data

bp. 102-104 °C/3.4 mmHg; ^{1}H NMR (CDCl$_3$): δ 0.93 (3 H, t, J=6.9 Hz), 1.37 (3 H, t, J=7.1 Hz), 1.27-1.55 (4 H, m), 1.70-1.84 (2 H, m), 2.69-2.96 (2 H, m), 4.22 (1 H, m), 4.34 (2 H, q, J=7.1 Hz); ^{13}C NMR (CDCl$_3$): δ 13.88, 21.71, 22.27, 23.22 (t, J=4.1 Hz), 31.63, 40.21, 45.43 (t, J=23.2 Hz), 63.18, 115.24 (t, J=251.6 Hz), 163.43 (t, J=32.0 Hz); ^{19}F NMR (CDCl$_3$): δ -102.2 (1 F, ddd, J=264, 17.1, 14.7 Hz), -107.6 (1 F, dt, J=264, 17.1 Hz); IR (CCl$_4$): ν 2963, 1774, 1764, 1301, 1288, 1191, 1079 cm^{-1}.

Note

Burton *et al.* also reported a similar addition reaction of difluoroiodoacetate to alkenes proceeding in a single electron transfer (SET) mechanism. See Yang, Z.-Y.; Burton, D. J. *J. Org. Chem.* **1992**, *57*, 5144.

$$CF_2ICO_2Et + \textit{n-}C_4H_9 \longrightarrow \xrightarrow[\text{72\% yield}]{\text{Zn, cat. NiCl}_2\cdot6H_2O \text{ / THF}} \textit{n-}C_4H_9 \longrightarrow CO_2Et$$

Yang, Z.-Y.; Burton, D. J. *J. Org. Chem.* **1991**, *56*, 5125.

(*S*)-3-[(*S*)-3-(Ethoxycarbonyl)-3,3-difluoro-2-methylpropionyl]-4-(prop-2-yl)oxazolidin-2-one

a) LDA / THF,   b) CF$_2$ICO$_2$Et, Et$_3$B / THF

Synthetic method

(*S*)-3-Propionyl-4-(prop-2-yl)oxazolidin-2-one (370 mg, 2.0 mmol) in THF (6 mL) was added at -78 °C to a solution of LDA (2.2 mmol) in THF (4 mL). After 1 h at the same temperature, the enolate solution was added to a solution of ethyl difluoroiodoacetate (0.34 mL, 2.6 mmol) and triethylborane (1 *M* in hexane, 2.0 mL, 2.0 mmol) in THF (4 mL) at 4 to 5 °C with a cannula over 12 min. The mixture was stirred at the same temperature for 3 min, then the reaction was quenched with sat. NH$_4$Cl aq. and the whole system was extracted with Et$_2$O. The combined ethereal extracts were washed with brine, dried over anhydrous MgSO$_4$ and filtered. Diastereo-meric excess was determined to be 88% by capillary GC analysis. After evaporation of the solvent, chromatography of the residue (hexane:CH$_2$Cl$_2$=1:4 and hexane:AcOEt=5:1) gave the starting material (0.043 g, 12%) and (*S*)-3-[(*S*)-3-(ethoxycarbonyl)-3,3-difluoro-2-methyl-propionyl]-4-(prop-2-yl)oxazolidin-2-one (0.455 g, 74%) as a mixture of stereoisomers. The isomers were separated by HPLC (hexane:AcOEt=7:1).

Spectral data

$[\alpha]_D^{24}$ +45.1 (*c* 1.02, CHCl$_3$); ^{1}H NMR (CDCl$_3$): δ 0.90 (3 H, d, *J*=6.9 Hz), 0.91 (3 H, d, *J*=6.9 Hz), 1.35 (3 H, t, *J*=7.1 Hz), 1.41 (3 H, d, *J*=7.0 Hz), 2.37 (1 H, qqd, *J*=7.1, 6.9, 3.7 Hz), 4.20-4.46 (5 H, m), 4.72 (2 H, ddq, *J*=14.4, 11.9, 7.0 Hz); ^{19}F NMR (CDCl$_3$): δ -106.71 (1 F, dd, *J*=270.0, 11.9 Hz), -110.85 (1 F, dt, *J*=270.0, 14.4 Hz); IR (neat): ν 1779, 1704, 1391, 1206, 1123 cm^{-1}.

(*S*)-3-[(*R*)-3-(ethoxycarbonyl)-3,3-difluoro-2-methyl-propionyl]-4-(prop-2-yl)oxazolidin-2-one

$[\alpha]_D^{25}$ +71.7 (*c* 0.70, CHCl$_3$); ^{1}H NMR (CDCl$_3$): δ 0.87 (3 H, d, *J*=7.0 Hz), 0.91 (3 H, d, *J*=7.0 Hz), 1.35 (3 H, t, *J*=7.2 Hz), 1.50 (3 H, d, *J*=7.2 Hz), 2.32 (1 H, qqd, *J*=7.0, 7.0, 4.1 Hz), 4.20-4.51 (5 H, m), 4.62 (2 H, ddq, *J*=15.6, 10.5, 7.2 Hz); ^{19}F NMR (CDCl$_3$): δ -106.15 (1 F, dd, *J*=272.9, 10.5 Hz), -112.63 (1 F, dt, *J*=272.9, 15.6 Hz); IR (neat): ν 1783, 1704, 1390, 1207, 1122 cm^{-1}.

Note

For the preparation of ethyl difluoroiodoacetate, see page 89 of this volume.

Iseki, K., *et al.* Chem. Pharm. Bull. **1996**, *44*, 1314.

3.4 Carbon-Carbon Bond-forming Reactions

3.4.1 Reformatsky Type Reactions

Ethyl 2,2-difluoro-3-hydroxy-12-methyltridecanoate

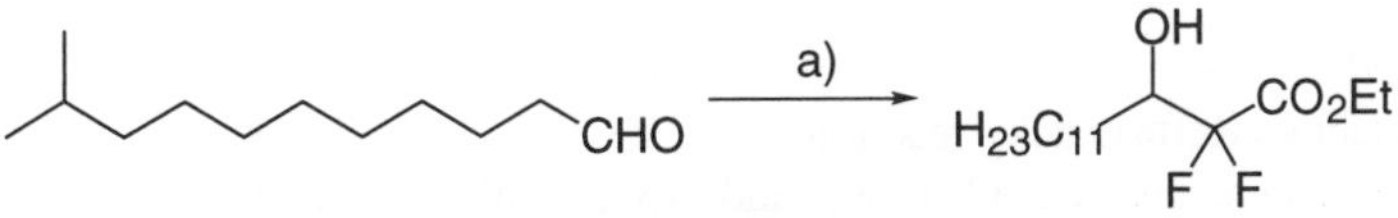

a) CBrF$_2$CO$_2$Et, Zn / THF

Synthetic method

Ethyl bromodifluoroacetate (1.15 mL, 8.97 mmol) was added to a solution of 10-methyl-undecanal (1.1 g, 5.98 mmol) and zinc (586 mg, 8.97 mmol) in THF (30 mL) at room temperature under a nitrogen atmosphere. After stirring for 2.5 h at that temperature, the reaction was quenched with 1*N* HCl aq. The organic layer was separated and the aqueous layer was extracted with CH$_2$Cl$_2$. The combined organic extracts were dried over MgSO$_4$ and evaporated *in vacuo*. The residual oil was purified by silica gel column chromatography (hexane:AcOEt=4:1) to give ethyl 2,2-difluoro-3-hydroxy-12-methyltridecanoate (1.68 g, 5.45 mmol) in 91% yield.

Spectral data

^{1}H NMR (CDCl$_3$): δ 0.86 (6 H, d, *J*=6.54 Hz), 1.07-1.74 (20 H, m), 1.37 (3 H, t, *J*=7.14 Hz), 1.91-2.09 (1 H, br s), 3.90-4.16 (1 H, m), 4.36 (2 H, q, *J*=7.14 Hz); ^{13}C NMR (CDCl$_3$): δ 13.96, 22.67, 25.23, 27.41, 27.99, 29.17-29.26, 29.34, 29.50, 29.61, 29.91, 39.06, 63.04, 71.81 (dd, *J*=27.04, 24.82 Hz), 114.72 (dd, *J*=255, 253 Hz), 163.76 (t, *J*=30.85 Hz); ^{19}F NMR

(CDCl$_3$): δ -123.5 (dd, J=266, 15.3 Hz), -116.2 (dd, J=266, 7.63 Hz); IR (neat): ν 3450, 1760 cm^{-1}.

Note

For related work, see the following articles: i) Kobayashi, Y., *et al. Chem. Pharm. Bull.* **1989**, *37*, 813, ii) Takayama, H., *et al. Chem. Pharm. Bull.* **1992**, *40*, 1120, iii) Fried, J., *et al. J. Org. Chem.* **1993**, *58*, 5724, iv) Kitazume, T., *et al. Tetrahedron* **1996**,*52*, 157.

A similar reaction with aldehydes was carried out with ethyl chlorodifluoroacetate and Zn dust in DMF, affording the desired products in comparable or better yields. See Lang, R. W.; Schaub, B. *Tetrahedron Lett.* **1988**, *29*, 2943.

Fukuda, H.; Kitazume, T. *J. Fluorine Chem.* **1995**,*74*, 171.

Ethyl (*S*)-2,2-difluoro-3-hydroxy-3-phenylpropanoate

a) Zn / THF, b) (1*S*,2*R*)-*N*-Methylephedrine, c) PhCHO

Synthetic method

Preparation of Reformatsky reagent

Chlorotrimethylsilane (0.037 mL, 0.29 mmol) was added to a suspension of zinc dust (261 mg, 4 mmol) in anhydrous THF (1 mL). The mixture was refluxed for 15 min, heating was stopped, and a solution of ethyl bromodifluoroacetate (0.53 mL, 4 mmol) in anhydrous THF (7 mL) was added. The mixture was heated at 50-55 °C for 20 min, then the dark green solution was cooled to room temperature to afford the Reformatsky reagent in 70-85% yield.

Synthetic method

Enantioselective Reformatsky reaction

A solution of benzaldehyde (1 mmol) and (1*S*,2*R*)-*N*-methylephedrine (1 mmol, 1 equiv) in anhydrous THF (4 mL) was cooled to 0 °C and stirred for 20 min. Then the Reformatsky reagent prepared as described above was added *via* a syringe, and the mixture was stirred at that temperature until the reaction was finished (TLC) and then quenched with a 10% solution of HCl (8 mL). The organic layer was separated and the aqueous phase extracted with AcOEt (3x10 mL). The combined organic layers were washed with brine, then dried (Na$_2$SO$_4$). The solvents were eliminated on Rotavapor and the residue was purified by flash column chromatography to afford ethyl (*S*)-2,2-difluoro-3-hydroxy-3-phenylpropanoate in 65% yield. 82% ee was confirmed by integration of diastereomeric peaks by ^{1}H and ^{19}F NMR after derivatization into the corresponding (*R*)-MTPA ester.

Spectral data

[α]$_D$^{23} -9.5 (*c* 1, CHCl$_3$); ^{1}H NMR (CDCl$_3$): δ 1.29 (3 H, t, J=7.1 Hz), 2.78 (1 H, d, J=5.3 Hz), 4.30 (2 H, q, J=7.1 Hz), 5.17 (1 H, ddd, J=15.4, 7.9, 5.3 Hz), 7.35-7.45 (5 H, m); ^{13}C NMR (CDCl$_3$): δ 13.7, 63.1, 73.6 (t, J=26 Hz), 113.7 (t, J=256 Hz), 127.6, 128.3, 129.1, 134.4, 163.6 (t, J=32 Hz); ^{19}F NMR (CDCl$_3$): δ -121.0 (1 F, dd, J=260.9, 15.4 Hz), -114.2 (1 F, dd, J=260.9, 7.9 Hz); IR (film): ν 3420, 1745 cm^{-1}.

Note

A similar type of reaction was also reported by Braun and his co-workers. See Braun, M., *et al. Liebigs Ann.* **1995**, 1447.

Pedrosa, R., *et al. Synthesis* **1996**, 1070.

4,4-Difluoro-3-hydroxyundecan-5-one

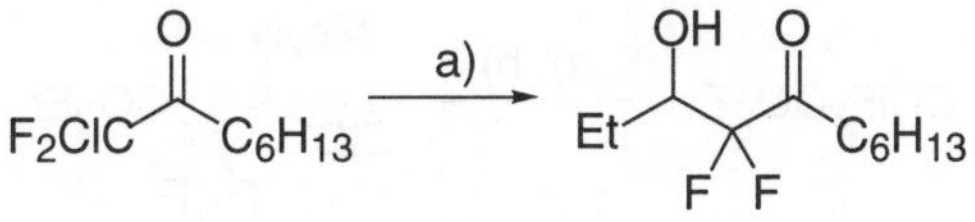

a) EtCHO, Zn, cat. CuCl / THF

Synthetic method

In a 50-mL three-necked flask purged with argon were placed acid-washed zinc dust (0.196 g, 3.0 mmol), copper(I) chloride (0.030 g, 0.3 mmol), and THF (5 mL). After the suspension was stirred for 0.5 h at room temperature, propanal (0.064 g, 1.1 mmol) and 1-chloro-1,1-difluoro-octan-2-one (0.199 g, 1.0 mmol) were added by a syringe. Then the mixture was refluxed with stirring for 4 h. After being cooled to room temperature, the reaction mixture was filtered through a pad of Celite 545 and the filtrate was concentrated *in vacuo*. Column chromatography of the crude product on silica gel (hexane:AcOEt=3:1) provided 4,4-difluoro-3-hydroxyundecan-5-one (0.220 g) in quantitative yield.

Spectral data

^{1}H NMR (CDCl$_3$): δ 0.89 (3 H, t, *J*=6.2 Hz), 1.04 (3 H, t, *J*=6.2 Hz), 1.2-2.0 (10 H, m), 2.69 (2 H, t, *J*=6.0 Hz), 2.3-2.9 (1 H, m), 3.5-4.3 (1 H, m); ^{19}F NMR (CDCl$_3$): δ -114.32 (1 F, dd, *J*=274.7, 6.1 Hz), -125.73 (1 F, dd, *J*=274.7, 15.9 Hz): IR (neat): ν 3422, 2956, 2932, 2858, 1740, 1648, 1637, 1459, 1401, 1379, 1307, 1221, 1110, 1074, 1046, 986, 893, 810, 757, 713, 664 cm^{-1}.

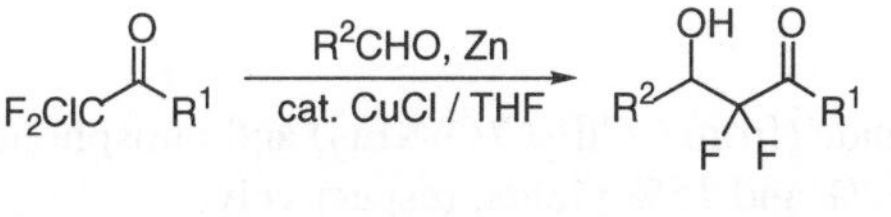

R^1	R^2	Yield (%)	R^1	R^2	Yield (%)
n-C$_6$H$_{13}$	Me$_2$CHCH$_2$	81	*n*-C$_6$H$_{13}$	Me$_3$C	60
n-C$_6$H$_{13}$	(*E*)-MeCH=CH	85	*n*-C$_6$H$_{13}$	Ph	90
c-C$_6$H$_{11}$	Pr	81	*c*-C$_6$H$_{11}$	Me$_2$CHCH$_2$	86
c-C$_6$H$_{11}$	(*E*)-MeCH=CH	77	*c*-C$_6$H$_{11}$	Ph	93
Bn	Pr	70	Bn	(*E*)-MeCH=CH	93
Bn	Ph	90	Ph[a]	Pr	80
Ph[b]	(*E*)-MeCH=CH	100	Ph[b]	Ph	88

[a] Reaction was conducted at -20 °C in the presence of 1 equiv of BF$_3$·OEt$_2$ in THF:Et$_2$O=1:4.
[b] Reaction was carried out in THF:Et$_2$O=1:4.
(Reproduced with permission from Kuroboshi, M.; Ishihara, T. *Bull. Chem. Soc. Jpn.* **1990**, *63*, 428)

Note

For the preparation of the starting chlorodifluorinated ketone, see page 102 of this volume.

The intermediary zinc enolates were successfully trapped in refluxing MeCN when TMSCl was employed as an electrophile instead of aldehydes in the above case. See Ishihara, T., *et al.*

Tetrahedron Lett. **1983**,*24*, 507.

Kuroboshi, M.; Ishihara, T. *Bull. Chem. Soc. Jpn.* **1990**, *63*, 428.

Ethyl 3-ethoxy-2,2-difluoro-3-(dimethylamino)propionate

$$CClF_2CO_2Et \xrightarrow{a), b)} \underset{\underset{F\ \ F}{EtO}}{\overset{Me_2N}{\diagup}}CO_2Et$$

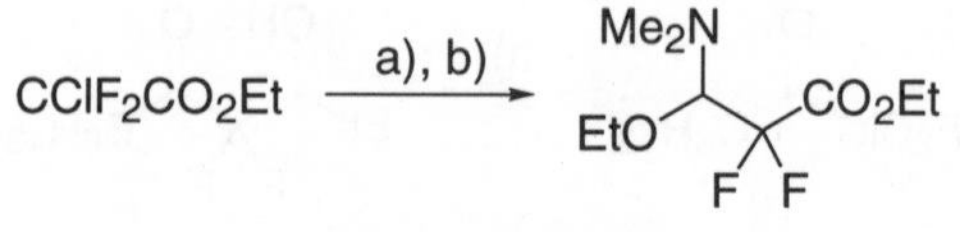

a) Et$_2$SO$_4$ / DMF,   b) Zn

Synthetic method

A solution of diethyl sulfate (2.31 g, 15 mmol) in DMF (4 mL) was stirred at 90 °C for 2 h. Zinc powder (1.31 g, 20 mmol) and ethyl chlorodifluoroacetate (1.59 g, 10 mmol) were added to the above solution at 70 °C, and the mixture was stirred at that temperature for 4 h.   The mixture was filtered to remove excess zinc, diluted with pentane (30 mL), and poured into NH$_4$Cl aq. (30 mL).   The organic layer was separated, and the aqueous layer extracted with pentane.   The combined organic extracts were washed with NaHCO$_3$ aq. (100 mL), dried over MgSO$_4$, and evaporated.   The residual oily materials were distilled to give ethyl 3-ethoxy-2,2-difluoro-3-(dimethylamino)propionate in 84% yield.

Spectral data

bp. 90-92 °C/12 mmHg; ^{1}H NMR (CDCl$_3$): δ 1.22 (3 H, t, *J*=7.0 Hz), 1.34 (3 H, t, *J*=7.1 Hz), 2.49 (6 H, s), 3.63 (1 H, dq, *J*=9.5, 7.0 Hz), 3.73 (1 H, dq, *J*=9.5, 7.0 Hz), 4.33 (2 H, q, *J*=7.1 Hz), 4.36 (1 H, dd, *J*=12.4, 10.7 Hz); ^{13}C NMR (CDCl$_3$): δ 14.0, 15.2, 40.2 (t, *J*=2.6 Hz), 62.8, 66.9, 92.0 (t, *J*=25 Hz), 115.4 (dd, *J*=261, 259 Hz), 164.3 (t, *J*=31.3 Hz); ^{19}F NMR (CDCl$_3$): δ -116.6 (1 F, dd, *J*=258, 12.2 Hz), -118.2 (1F, dd, *J*=258, 9.9 Hz); IR (neat): ν 1780, 1760 cm^{-1}.

Note

The corresponding amide (from CClF$_2$C(O)NEt$_2$) and phosphonate (from CBrF$_2$P(O)(OEt$_2$)) were similarly obtained in 52% and 15% yields, respectively.

Tsukamoto, T.; Kitazume, T. *J. Chem. Soc. Perkin Trans. 1* **1993**, 1177.

Diethyl (*E*)-(1,1-difluoro-3-phenylprop-2-en-1-yl)phosphonate

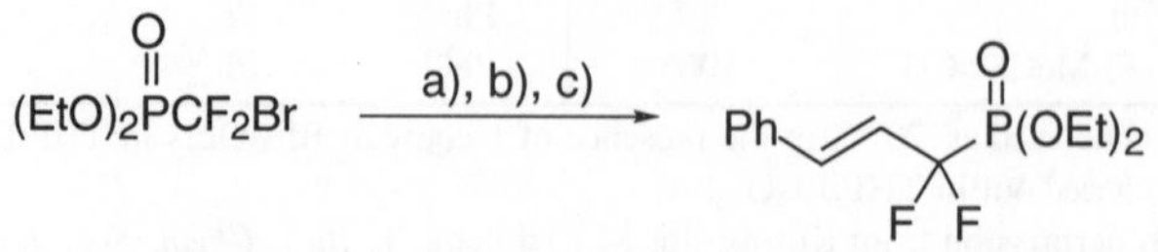

$$\underset{(EtO)_2PCF_2Br}{\overset{O}{\overset{\|}{}}} \xrightarrow{a), b), c)} \underset{F\ \ F}{Ph\diagdown\diagup\diagup}\underset{}{\overset{O}{\overset{\|}{P(OEt)_2}}}$$

a) Zn / DMF,   b) CuBr,   c) (*E*)-PhCH=CHBr

Synthetic method

To a stirred suspension of zinc dust (1.3 g, 20 mmol) in dry DMF (10 mL) was slowly added a solution of diethyl (bromodifluoromethyl)phosphonate (5.34 g, 20 mmol) in DMF (10 mL).

During the addition, an exothermic reaction occurred. The addition was controlled so that the internal temperature was maintained at 50-60 °C. After addition was completed, the solution was stirred at room temperature for an additional 3 h, and then copper(I) bromide (3.02 g, 20 mmol) was added in one portion. The mixture was stirred at the same temperature for 30 min to give an organocopper reagent in DMF. (E)-β-bromostyrene (10 mmol, (E):(Z)=86:14) was added dropwise at room temperature (an exothermic reaction occurred). After the mixture was stirred at room temperature for 60 h, 2% HCl aq. was added to quench the reaction. The biphasic mixture was passed through Celite and extracted with Et$_2$O. The extract was washed with sat. NaHCO$_3$ and brine, and dried over MgSO$_4$. Evaporation of the solvent, followed by chromatographic purification on silica gel (hexane:AcOEt=5:1 to 3:1) gave diethyl (E)-(1,1-difluoro-3-phenylprop-2-en-1-yl)phosphonate in 67% yield as an 86:14 (E):(Z) mixture.

Spectral data

^{1}H NMR (CDCl$_3$): δ 1.38 (6 H, t, J=7.1 Hz), 4.19-4.38 (4 H, m), 6.30 (1 H, dtd, J=16.2, 12.9, 2.9 Hz), 7.07 (1 H, ddt, J=16.2, 3.0, 2.8 Hz), 7.29-7.53 (5 H, m); ^{13}C NMR (CDCl$_3$): δ 16.3 (d, J=5.3 Hz), 64.7 (d, J=6.7 Hz), 117.4 (td, J=257.9, 221.0 Hz), 118.6 (td, J=21.1, 12.9 Hz), 127.3, 128.7, 129.4, 134.2, 136.9 (td, J=10.5, 5.9 Hz); ^{19}F NMR (CDCl$_3$): δ -45.6 (dd, J=114.3, 12.9 Hz) from PhCF$_3$; ^{31}P NMR (CDCl$_3$): δ 6.19 (t, J=114.3 Hz); IR (neat): ν 1270, 1039 cm^{-1}.

Note

For the preparation of the starting bromodifluoromethylated phosphonate, see page 87 of this volume.

This reaction was found to proceed in a stereospecific manner, and (Z)-β-bromostyrene ((E):(Z)=26:74) furnished the corresponding adduct in 50% yield as a 32:68 (E):(Z) mixture. When (E)-β-iodostyrene was employed, a shorter reaction time (15 h) and higher yield (85%) were realized.

For a representative related work, see i) Martin, S. F., et al. Tetrahedron Lett. **1992**, 33, 1839, ii) Nieschalk, J.; O'Hagan, D. J. Chem. Soc., Chem. Commun. **1995**, 719, iii) Berkowitz, D. B., et al. J. Org. Chem. **1996**, 61, 4666, iv) Percy, J. M., et al. Tetrahedron **1997**, 53, 10623.

Shibuya, S., et al. J. Org. Chem. **1996**, 61, 7207.

3.4.2 Aldol Reactions and Claisen Condensations

S-tert-Butyl (R)-4,4-difluoro-3-hydroxybutanethiolate

a) CHF$_2$CH(OEt)OH, (R)-BINOL-TiCl$_2$, MS 5 Å / toluene

Synthetic method

To a solution of the chiral titanium complex, (R)-BINOL-TiCl$_2$, (0.20 mmol) in toluene (3 mL) were added 1-(tert-butylsulfenyl)-1-(trimethylsiloxy)ethylene (2.0 mmol) and a toluene (0.5 mL) solution of difluoroacetaldehyde ethyl hemiacetal (1 mmol) separately but simultaneously over several minutes at 0 °C. After stirring for 2 h at that temperature, Et$_2$O (2 mL) and a sat. NaHCO$_3$

solution (2 mL) was added to the reaction mixture. The solution was filtered through a pad of Celite and the filtrate was extracted three times with Et_2O (totally 15 mL). The combined organic layer was washed with brine. The extract was dried over $MgSO_4$ and evaporated under reduced pressure to give the crude product as a silyl ether form. The crude product was treated with 10% HCl-MeOH. Separation of the resultant mixture by silica gel chromatography (hexane:AcOEt= 20:1) gave S-tert-butyl 4,4-difluoro-3-hydroxybutanethioate in 47% yield.

Spectral data

$[\alpha]_D^{28}$ +17.6 (c 1.15, CHCl$_3$, 95% ee); ^{1}H NMR (CDCl$_3$): δ 1.48 (9 H, s), 2.75 (1 H, dd, J=16.1, 8.1 Hz), 2.83 (1 H, dd, J=16.1, 4.5 Hz), 3.02 (1 H, d, J=4.9 Hz), 4.15-4.32 (1 H, m), 5.75 (1 H, td, J=55.7, 3.6 Hz); ^{13}C NMR (CDCl$_3$): δ 29.7, 43.3, 49.1, 68.4 (t, J=25 Hz), 115.3 (t, J=243 Hz), 199.2; IR (neat): ν 3460, 2970, 1670, 1460, 1370, 1260, 1070, 670, 660 cm^{-1}.

Note

For the preparation of difluoroacetaldehyde ethyl hemiacetal, see page 112 of this volume.

Mikami, K., *et al.* Tetrahedron **1996**, *52*, 85.

S-*tert*-Butyl difluoroethanethioate,
S-*tert*-Butyl 2,2-difluoro-3-hydroxy-3-phenylpropanethioate

a) (COCl)$_2$ / MeCN,   b) *tert*-BuSH,   c) CoCl$_2$,   d) LDA,   e) PhCHO

Synthetic method

S-*tert*-Butyl difluoroethanethioate

To a solution of oxalyl chloride (9.08 mL, 104 mmol) in MeCN (50 mL) at 25 °C was added dropwise difluoroacetic acid (6.55 mL, 104 mmol). After 3 h, 2-methyl-2-propanethiol (11.74 mL, 104 mmol) was added dropwise followed by cobalt(II) chloride (10 mg). After being stirred at room temperature for 18 h, an additional 2-methyl-2-propanethiol (4.0 mL, 35.4 mmol) was added. After 2 h, the solution was poured onto Et_2O (500 mL) and washed with sat. NaHCO$_3$ (2x300 mL) and H$_2$O (2x300 mL). The organic layer was dried over Na$_2$SO$_4$, and the solvent was removed under reduced pressure to give a red oil. Vacuum distillation gave 8.26 g of S-*tert*-butyl 2,2-difluoroethanethioate in 47% yield as a colorless oil.

Spectral data

bp. 59-63 °C/30 mmHg; ^{1}H NMR (CDCl$_3$): δ 1.48 (9 H, s), 5.67 (1 H, t, J=54 Hz); ^{13}C NMR (CDCl$_3$): δ 29.45, 49.32, 108.94 (t, J=255 Hz), 191.23 (t, J=28 Hz); ^{19}F NMR (CDCl$_3$): δ -123.65 (d, J=55 Hz): IR (film): ν 2969, 1684, 1154, 1093, 1064 cm^{-1}.

Synthetic method

S-*tert*-Butyl 2,2-difluoro-3-hydroxy-3-phenylpropanethioate

To a solution of S-*tert*-butyl difluoroethanethioate (0.152 mL, 1.00 mmol) in toluene (10 mL) at -78 °C was added dropwise over 1 min LDA (0.550 mL of a 2.0 *M* solution in heptane/THF/ benzene, 1.10 mmol). After 2 min, benzaldehyde (0.112 mL, 1.10 mmol) was added dropwise. After 1 h, the reaction was warmed to 25 °C over 1 h and quenched by the addition of a phosphate buffer (5 mL, 0.5 *M*, pH=7). The resulting organic layer was concentrated under reduced pressure. The crude product was purified by radial chromatography using Merck TLC grade 7749 silica gel (hexane:AcOEt=90:10) to give 192 mg of *S*-*tert*-butyl 2,2-difluoro-3-hydroxy-3-phenyl-propanethioate in 70% yield as a colorless oil.

Spectral data

^{1}H NMR (CDCl$_3$): δ 1.46 (9 H, s), 2.52 (1 H, br s), 5.16 (1 H, dd), 7.4 (5 H, m); ^{13}C NMR (CDCl$_3$): δ 29.31, 49.05, 73.49 (t, *J*=24 Hz), 114.78 (t, *J*=260 Hz), 127.80, 128.32, 129.10, 134.56, 193.47 (t, *J*=30 Hz); ^{19}F NMR (CDCl$_3$): δ -112.88 (1 F, dd, *J*=259, 9 Hz), -119.22 (1 F, dd, *J*=259, 15 Hz): IR (film): ν 3442, 2967, 1673, 1660, 1456, 1367, 1194, 1078, 917 cm^{-1}.

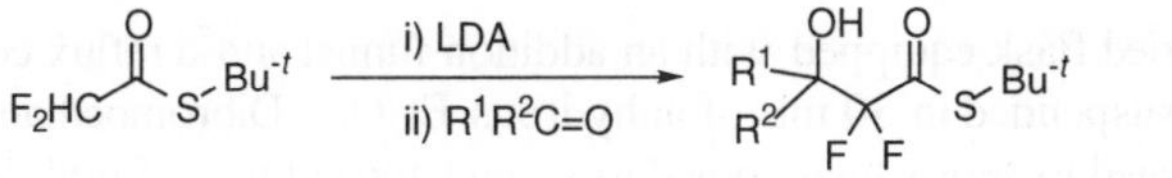

R^1	R^2	Yield (%)	R^1	R^2	Yield (%)
p-MeOC$_6$H$_4$-	H	62	*n*-C$_5$H$_{11}$-	H	61
Me$_2$CH-	H	44	Me$_3$C-	H	42
Ph-	CH$_3$	69	Ph-	Cl	77[a]
-C$_3$H$_6$CH=CH-		65			

a) Two molecules of *S*-*tert*-butyl 2,2-difluorothiolacetate were added (in the product structure, R^2=CF$_2$C(O)SBu-^{t}).

(Reproduced with permission from Weigel, J. A. *J. Org. Chem.* **1997**, *62*, 6108)

Note

This method is also applicable for preparing the corresponding ketene silyl thioacetals. For recent reports on the usage of the terminally difluorinated ketene silyl acetals, see i) Taguchi, T., *et al. Chem. Lett.* **1990**, 1307, ii) Iseki, K., *et al. Tetrahedron* **1997**, *53*, 10271.

In quite sharp contrast to the result shown above, subjection of the corresponding amides or esters to lithium cyclohexylisopropylamide (LCIA) resulted in self-condensation as the major reaction path.   Sterically hindered amide, however, realized the aldol reaction with benzaldehyde but only in a moderate (40%) yield.   See Kobayashi, Y., *et al. Tetrahedron Lett.* **1986**, *27*, 6103.

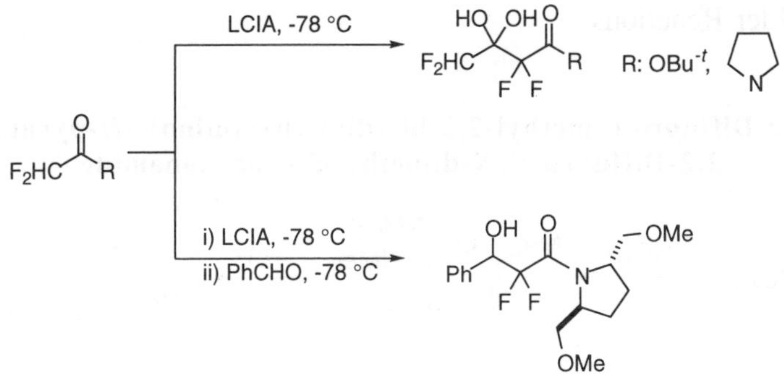

Weigel, J. A. *J. Org. Chem.* **1997**, *62*, 6108.

Ethyl 4,4-difluoro-3-oxobutyrate

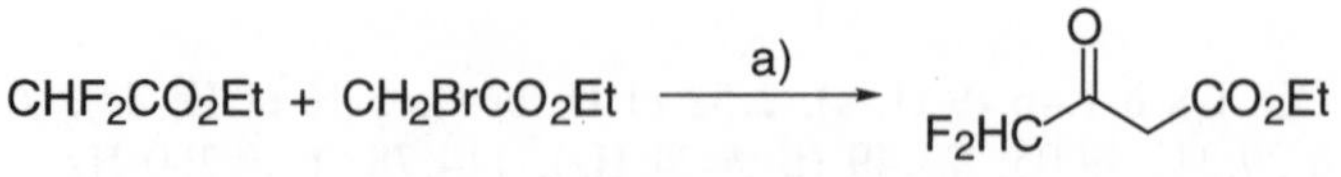

$$CHF_2CO_2Et + CH_2BrCO_2Et \xrightarrow{a)}$$

a) Zn / Et$_2$O

Synthetic method

In an oven-dried flask equipped with an addition funnel and a reflux condenser, zinc (8.80 g, 135.0 mmol) was suspended in 30 mL of anhydrous Et$_2$O. Dibromoethane (0.5 mL) was added; the mixture was heated to reflux then cooled to room temperature. Ethyl difluoroacetate (5.4 mL, 54.0 mmol) was added and ethyl bromoacetate (15.0 mL, 135 mmol) in 70 mL of Et$_2$O was added dropwise. The addition was controlled so that a gentle reflux was maintained. After addition was complete (1 h), the solution was heated at reflux for an additional hour, at which time almost all the Zn was consumed. The light green solution was cooled in an ice bath and quenched with 1 N HCl with stirring. The Et$_2$O layer was separated, washed with 1 N HCl, H$_2$O, dried, and concentrated to afford 7.8 g (88%) of a yellow oil as a mixture of keto and enol tautomers (35:65).

Spectral data

bp. 153-154 °C

keto form

^{1}H NMR (CDCl$_3$): δ 1.29 (3 H, t, J=7 Hz), 3.70 (2 H, t, J=1 Hz), 4.23 (2 H, q, J=7 Hz), 5.91 (1 H, t, J=54 Hz); ^{19}F NMR (CDCl$_3$): δ 34.0 (d, J=55 Hz) from C$_6$F$_6$.

enol form

^{1}H NMR (CDCl$_3$): δ 1.32 (3 H, t, J=7 Hz), 4.26 (2 H, q, J=7 Hz), 5.49 (1 H, s), 6.03 (1 H, t, J=54 Hz), 11.80 (1 H, s); ^{19}F NMR (CDCl$_3$): δ 35.3 (d, J=55 Hz) from C$_6$F$_6$.

Dolence, J. M.; Poulter, C. D. *Tetrahedron* **1996**, *52*, 119.

3.4.3 Diels-Alder Reactions

3,3-Difluoro-6-methyl-2,2-bis(dimethylamino)-2*H*-pyran, 2,2-Difluoro-*N*,*N*-dimethyl-5-oxohexanamide

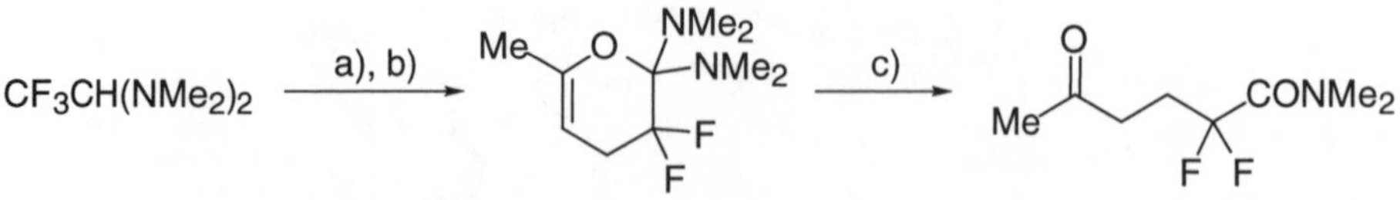

$$CF_3CH(NMe_2)_2 \xrightarrow{a), b)}$$

a) *n*-BuLi / Et$_2$O, b) CH$_2$=CHC(O)CH$_3$ / hexane, c) 3% HCl / CH$_2$Cl$_2$

Synthetic method

3,3-Difluoro-6-methyl-2,2-bis(dimethylamino)-2*H*-pyran

To a solution of 2,2,2-trifluoro-1,1-bis(dimethylamino)ethane (1.7 g, 10 mmol) in hexane (10 mL) at -78 °C under nitrogen was added *n*-BuLi (2.5 *M* in hexanes, 4.5 mL). The reaction mixture was then allowed to warm to room temperature and stirred for 10 h. Then the reaction mixture was distilled at reduced pressure into a receiving flask cooled with a dry ice-acetone bath

(100 mL, round bottom) equipped with a rubber septum, a magnetic stir bar, and an H_2O-cooled condenser. Before and after distillation, the apparatus was maintained under a dry nitrogen atmosphere. But-3-en-2-one (14 mmol) was added *via* a syringe to this distillate, which contained 2,2-difluoro-1,1-bis(dimethylamino)ethene, and the mixture was stirred at room temperature for 5 min. 3,3-Difluoro-6-methyl-2,2-bis(dimethylamino)-2*H*-pyran was isolated by distillation under reduced pressure in 72% yield.

Spectral data

bp. 62-64 °C/1.5 mmHg; ^1H NMR (CDCl$_3$): δ 1.59 (3 H, m), 2.35 (2 H, m), 2.49 (12 H, t, J=1.2 Hz), 4.14 (1 H, m); ^{13}C NMR (CDCl$_3$): δ 18.8, 32.7 (t, J=25.7 Hz), 38.7, 93.4, 123.1 (t, J=256.3 Hz), 99.3, 149.0; ^{19}F NMR (CDCl$_3$): δ -102.6 (t, J=14.2 Hz); IR (neat): ν 1691, 1192 cm^{-1}.

Synthetic method

2,2-Difluoro-*N*,*N*-dimethyl-5-oxohexanamide

3,3-Difluoro-6-methyl-2,2-bis(dimethylamino)-2*H*-pyran (1 mmol) was dissolved in 20 mL of CH_2Cl_2, and 0.5 mL of a 3% HCl solution was added to the solution. The reaction mixture was stirred for 2 h at room temperature, then 30 mL of Et_2O was added. The organic layer was washed with H_2O and dried. Then Et_2O was removed, and the residue was distilled at reduced pressure to give *N*,*N*-dimethyl-2,2-difluoro-5-oxohexanamide in 89% yield.

Spectral data

bp. 95-97 °C/1.5 mmHg; ^1H NMR (CDCl$_3$): δ 1.57 (3 H, s), 2.30-2.58 (7 H, m), 2.59 (3 H, t, J=3.2 Hz); ^{13}C NMR (CDCl$_3$): δ 29.1, 29.4, 29.7, 36.0 (m), 119.9 (t, J=253.8 Hz), 162.4 (t, J=14.1 Hz), 204.1; ^{19}F NMR (CDCl$_3$): δ -99.5 (t, J=19.5 Hz); IR (neat): ν 1722, 1667, 1190 cm^{-1}.

Note

For the preparation of the starting material, see page 176 of this volume.

Use of α,β-unsaturated esters or nitriles has led to the formation of four-membered ring compounds in good yields by [2+2] cycloaddition, which were similarly hydrolyzed to acyclic amides with carboalkoxy or cyano functionalities at the γ position.

Xu, Y.-L.; Dolbier, Jr., W. R. *J. Org. Chem.* **1997**, *62*, 6503.

cis-5,6-Difluoro-5-(phenylsulfonyl)bicyclo[2.2.1]hept-2-ene

a) Cyclopentadiene / MeCN

Synthetic method

160 µL (1.96 mmol) of freshly cracked cyclopentadiene was added to a solution of (*E*)-1,2-difluoro-1-(phenylsulfonyl)ethene (100 mg, 0.49 mmol) in MeCN (1 mL). The solution was stirred at room temperature for 6 d. Evaporation of the solvent and chromatography of the residue on silica gel (petroleum ether:CH_2Cl_2=1:1) gave 85 mg of *cis*-5,6-difluoro-*exo*-5-(phenylsulfonyl)-bicyclo[2.2.1]hept-2-ene and 30 mg of a mixture of *exo*- and *endo*-isomers (87% total yield, *exo*:*endo*=19:1).

Spectral data

   exo(5-SO₂Ph)-isomer

Rf 0.40 (petroleum ether:CH₂Cl₂=1:1); mp. 90.4-91.5 °C; ¹H NMR (CDCl₃): δ 1.78 (1 H, dtm, *J*=10.28, 8.18, 2.17, 1.92 Hz), 2.10 (1 H, dm, *J*=10.60, 1.62, 1.58 Hz), 3.28 (1 H, m), 3.46 (1 H, m), 5.45 (1 H, dt, *J*=52.89, 3.50 Hz), 6.21 (1 H, dd, *J*=5.68, 3.22 Hz), 6.42 (1 H, dd, *J*=5.68, 2.95 Hz); ¹³C NMR (CDCl₃): δ 41.40, 45.79 (d, *J*=18.20 Hz), 47.68 (d, *J*=19.10 Hz), 90.33 (dd, *J*=209.40, 13.10 Hz), 108.56 (dd, *J*=240.20, 15.60 Hz), 129.27, 129.93, 134.12, 134.71, 134.90, 137.37; ¹⁹F NMR (CDCl₃): δ -194.77 (1 F, dt, *J*=52.60, 8.50 Hz), -161.59 (1 F, t, *J*=8.50 Hz); IR (CH₂Cl₂): ν 1330, 1160 cm⁻¹.

   endo(5-SO₂Ph)-isomer (representative data are shown)

Rf 0.38 (petroleum ether:CH₂Cl₂=1:1); ¹H NMR (CDCl₃): δ 1.99 (1 H, dm, *J*=9.80 Hz), 2.27 (1 H, dm, *J*=9.76 Hz), 3.02 (1 H, m), 3.10 (1 H, m), 5.00 (1 H, dt, *J*=53.35, 2.96 Hz), 6.2 (1 H (masked by a proton of the *exo*-adduct)), 6.32 (1 H, dd, *J*=5.81, 2.98 Hz), 7.59-7.95 (5 H, m); ¹⁹F NMR (CDCl₃): δ -191.55 (1 F, ddd, *J*=52.30, 20.40, 6.43 Hz), -156.11 (1 F, dm, *J*=20.40 Hz).

Note

For the preparation of the starting 1,2-difluorinated sulfone, see page 119 of this volume.

When the corresponding Z-olefin was employed, the reaction became faster but the selectivity was drastically decreased.

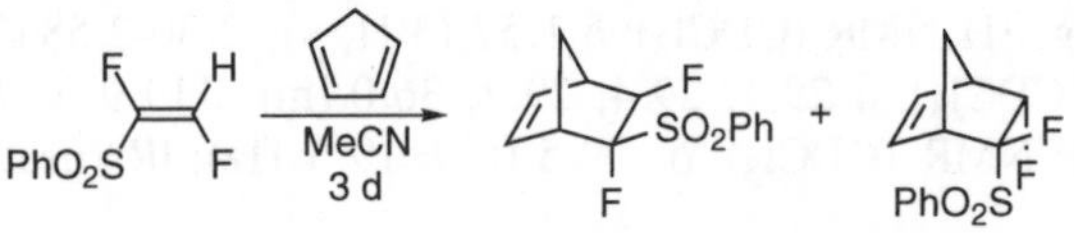

95% yield (1:1.4)

de Tollenaere, C.; Ghosez, L. *Tetrahedron* **1997**, *53*, 17127.

3.4.4 Claisen Rearrangement

3-(Allyloxy)-1,1-difluoro-2-[(2-methoxyethoxy)methoxy]pent-1-ene,
(*E*)-4,4-Difluoro-5-[(2-methoxyethoxy)methoxy]octa-1,5-dien-3-ol

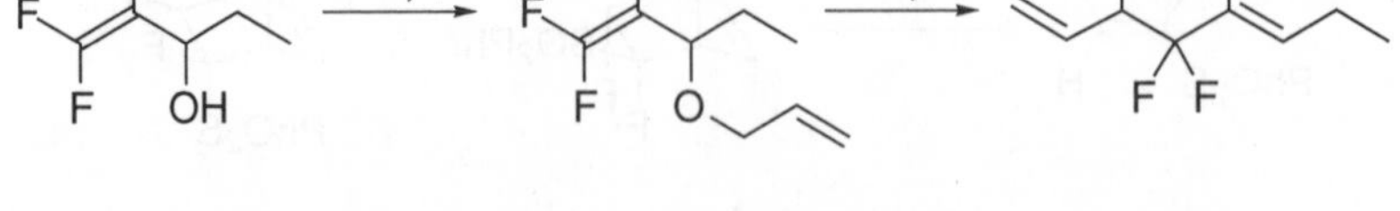

a) Allyl-Br, NaOH, cat. *n*-BuN₄HSO₄ / H₂O,   b) LDA / THF

Synthetic method

   3-(Allyloxy)-1,1-difluoro-2-[(2-methoxyethoxy)methoxy]pent-1-ene

A mixture of 1,1-difluoro-2-[(2-methoxyethoxy)methoxy]pent-1-en-3-ol (1.50 g, 6.63 mmol), allyl bromide (0.70 mL, 9.63 mmol), 50% NaOH aq. (3.70 mL, 46.4 mmol), and tetra-butylammonium hydrogen sulfate (0.11 g, 0.33 mmol) was stirred at 0 °C for 30 min. The mixture was allowed to warm to room temperature and stirred overnight. Sat. NH₄Cl aq. (10 mL) was added, and the mixture was extracted with Et₂O (3x30 mL). The combined organic extracts were washed with H_2O (10 mL), dried $MgSO_4$, and concentrated *in vacuo* to give 3-(allyloxy)-1,1-difluoro-2-[(2-methoxyethoxy)methoxy]pent-1-ene as a pale yellow oil (1.65 g, 93%), which was subjected to rearrangement without further purification.

Spectral data

^{1}H NMR (CDCl$_3$): δ 0.87 (3 H, t, J=7.3 Hz), 1.59-1.80 (2 H, m), 3.36 (3 H, s), 3.58 (2 H, t, J=4.2 Hz), 3.70-3.91 (3 H, m), 4.05 (1 H, ddt, J=13.0, 5.0, 1.5 Hz), 4.08 (1 H, tdd, J=12.7, 4.9, 1.5 Hz), 4.89 (1 H, d, J=6.1 Hz), 5.00 (1 H, d, J=6.1 Hz), 5.14 (1 H, dq, J=10.2, 1.5 Hz), 5.23 (1 H, dq, J=14.0, 1.5 Hz), 5.79-5.90 (1 H, m); ^{13}C NMR (CDCl$_3$): δ 9.8, 24.9, 58.9, 68.2, 69.2, 71.6, 76.0, 97.0, 112.3 (dd, J=36.7, 9.7 Hz), 117.1, 134.4, 156.2 (dd, J=284.3, 284.3 Hz); ^{19}F NMR (CDCl$_3$): δ -110.5 (1 F, d, J=64.1 Hz), -98.3 (1 F, d, J=64.1 Hz); IR (film): ν 1750, 1647, 1458 cm^{-1}.

R^1	R^2	G	Yield (%)	R^1	R^2	G	Yield (%)
Et	H	Ph	65[a]	Et	H	C(Me)=CH$_2$	69[a]
Et	H	C≡CH	51	Et	H	C≡CSiPr$^i_3$	83[b]
H	H	CH=CH$_2$	93	H	H	Ph	96
H	H	C(Me)=CH$_2$	48	H	H	C≡CH	82
H	H	C≡CSiPr$^i_3$	78	Et	Et	CH=CH$_2$	61[c]
Et	Et	Ph	57[c]	CH=CH$_2$	H	CH=CH$_2$	74

a) BnEt$_3$NCl was used as the PTC.   b) Silylation was carried out at the terminal acetylene.   c) Formed *via* the sodium alkoxide in the presence of *n*-BuN$_4$I as the PTC.

(Reproduced with permission from Percy, J. M., *et al. J. Org. Chem.* **1996**, *61*, 166)

Synthetic method

(*E*)-4,4-Difluoro-5-[(2-methoxyethoxy)methoxy]octa-1,5-dien-3-ol

A solution of 3-(allyloxy)-1,1-difluoro-2-[(2-methoxyethoxy)methoxy]pent-1-ene (4.5 mmol) in THF (*ca.* 3 mL) was added dropwise over 5 min to an LDA solution (2 equiv) in THF (*ca.* 10 mL) at -78 ℃. The solution became red-brown in color instantaneously. After stirring at -78 ℃ for 2 h, the solution was allowed to warm slowly to -30 ℃ and was maintained at that temperature for a further 4 h. The solution was quenched with a methanolic solution of NH$_4$Cl (10 mL) and washed with H$_2$O (20 mL), and the aqueous layer was extracted with Et$_2$O (3x30 mL). The combined organic extracts were dried over MgSO$_4$ and concentrated *in vacuo*. Purification by flash chromatography (petroleum ether:AcOEt=4:1) gave (*E*)-4,4-difluoro-5-[(2-methoxyethoxy)-methoxy]octa-1,5-dien-3-ol as a yellow oil (1.03 g, 86%).

Spectral data

Rf 0.21 (petroleum ether:AcOEt=4:1); ^{1}H NMR (CDCl$_3$): δ 1.00 (3 H, t, J=7.6 Hz), 2.10-2.22 (2 H, m), 2.80 (1 H, br s), 3.39 (3 H, s), 3.58 (2 H, t, J=5.0 Hz), 3.80-3.90 (2 H, m), 4.48-4.52 (1 H, m), 5.00(2 H, s), 5.33 (1 H, dt, J=10.5, 1.5 Hz), 5.49 (1 H, dt, J=17.1, 1.5 Hz), 5.50 (1 H, td, J=7.6, 1.5 Hz), 5.77-5.91 (1 H, m); ^{13}C NMR (CDCl$_3$): δ 13.7, 18.8, 59.0, 68.9, 71.6, 72.5, 98.3, 118.2 (t, J=248.6 Hz), 118.9, 122.0 (t, J=5.1 Hz), 132.5, 144.7 (t, J=26.3 Hz); ^{19}F NMR (CDCl$_3$): δ -116.9 (1 F, dd, J=250.2, 14.0 Hz), -109.6 (1 F, d, J=250.2 Hz); IR (film): ν 3424, 1750, 1674, 1460 cm^{-1}.

Note

For the preparation of the starting difluorinated allylic alcohol, see page 110 of this volume.

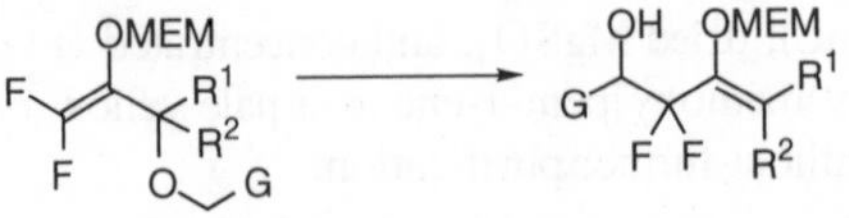

R^1	R^2	G	Yield (%)	R^1	R^2	G	Yield (%)
Et	H	Ph	76	Et	H	$C(Me)=CH_2$	84
Et	H	$C\equiv CSiPr^{-i}_3$	68	H	H	$CH=CH_2$	61
H	H	Ph	52	H	H	$C(Me)=CH_2$	58
H	H	$C\equiv CSiPr^{-i}_3$	52	Et	Et	$CH=CH_2$	51
Et	Et	Ph	70	$CH=CH_2$	H	$CH=CH_2$	74

(Reproduced with permission from Percy, J. M., *et al. J. Org. Chem.* **1996**, *61*, 166)

Percy, J. M., *et al. J. Org. Chem.* **1996**, *61*, 166.

3.4.5 Alkylations and Acylations

1-Chloro-1,1-difluorooctan-2-one

a) n-$C_6H_{13}MgBr$ / Et_2O

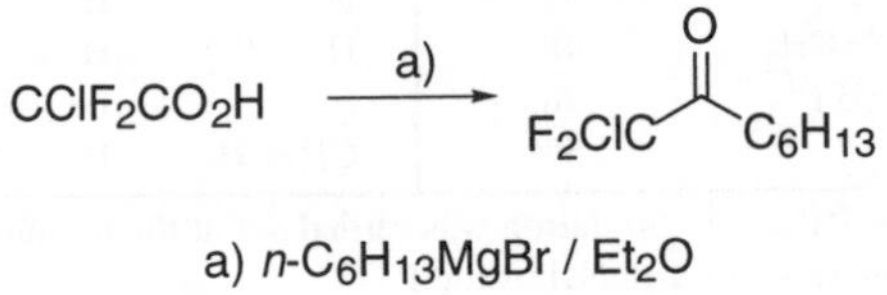

Synthetic method

Hexylmagnesium bromide was prepared from 1-bromohexane (20.6 g, 125 mmol) and magnesium (3.65 g, 150 mmol) in Et_2O (50 mL). A solution of chlorodifluoroacetic acid (6.52 g, 50 mmol) in Et_2O (15 mL) was gradually added under argon to the Grignard reagent at such a rate that the temperature did not rise above -10 °C. After stirring for 12 h at -20 °C, the reaction mixture was hydrolyzed with 6 N HCl aq. (50 mL) below 0 °C and stirred for 1 h at room temperature. The resulting mixture was extracted with Et_2O (3x50 mL), and the ethereal extracts were washed with sat. $NaHCO_3$ aq. (2x50 mL) and brine (2x50 mL), dried over anhydrous Na_2SO_4, and concentrated *in vacuo*. The oily residue was distilled under reduced pressure to give 1-chloro-1,1-difluorooctan-2-one (8.63 g) in 87% yield.

Spectral data

bp. 86-87 °C/25 mmHg; ^{1}H NMR ($CDCl_3$): δ 0.88 (3 H, t, $J=5.4$ Hz), 1.1-1.5 (6 H, m), 1.67 (2 H, t, $J=7.5$ Hz), 2.72 (2 H, t, $J=7.5$ Hz); ^{19}F NMR ($CDCl_3$): δ -68.80 (s): IR (neat): ν 2956, 2864, 1760, 1462, 1408, 1380, 1210, 1148, 1116, 1042, 912, 720, 660 cm^{-1}.

Note

When cyclohexyl, benzyl, and phenyl Grignard reagents were used in a similar manner, the products were obtained in 85%, 80%, and 84% yields, respectively. Keeping the temperature at -20 °C for 12 h was found to be essential for attaining a high yield of product ketones. The corresponding alkynyl ketones were also prepared by the reaction of lithiated alkynes with methyl 2-chloro-2,2-difluoroacetate at -78 °C for 2 h.

Kuroboshi, M.; Ishihara, T. *Bull. Chem. Soc. Jpn.* **1990**, *63*, 428.

1,1-Difluoronon-1-en-3-yl acetate, (E)-5,5-Difluorotridec-6-ene

$F_2C=CH_2$ $\xrightarrow{\text{a), b), c)}}$ [structure: F F / OAc on vinyl, C_6H_{13}-n] $\xrightarrow{\text{d), e)}}$ [structure: F F, n-Bu / C_6H_{13}-n]

a) *sec*-BuLi / THF-Et$_2$O, b) n-C$_6$H$_{13}$CHO / Et$_2$O, c) Ac$_2$O, DMAP / CH$_2$Cl$_2$,
d) CuCN, LiCl / THF, e) n-BuMgBr / THF

Synthetic method

1,1-Difluoronon-1-en-3-yl acetate

To a solution of 1,1-difluoroethylene (60 mmol) in 80 mL of THF and 20 mL of Et$_2$O was added 50 mmol of *sec*-BuLi in cyclohexane at -100 °C. The reaction mixture was stirred for 20 min at -90 °C, then a solution of heptanal (40 mmol) in Et$_2$O (10 mL) was added at -100 °C. After 30 min stirring at -90 °C, the temperature was raised to -20 °C over 30 min. 60 mmol of acetic anhydride was added at -50 °C and the whole was stirred for 30 min at -50 °C to 0 °C. The mixture was then hydrolyzed by the addition of cold 25% H$_2$SO$_4$. After neutralization by a sat. NaHCO$_3$ solution and washing with brine, the organic phase was dried over MgSO$_4$, the solvent was evaporated, and distillation from NaHCO$_3$ afforded 1,1-difluoronon-1-en-3-yl acetate in 60% yield.

Spectral data

bp. 46-48 °C/0.2 mmHg; ^{1}H NMR (CDCl$_3$): δ 0.9 (3 H, t), 1.3 (6 H, m), 1.6 (2 H, m), 1.7 (2 H, m), 2.0 (3 H, s), 4.3 (1 H, ddd, J=24.2, 9.3, 2.2 Hz), 5.4 (1 H, dq, J=9.3, 7.2 Hz); ^{13}C NMR (CDCl$_3$): δ 14.1, 21.1, 22.6, 25.0, 29.0, 31.8, 34.7, 67.7 (d, J=9.2 Hz), 79.6 (dd, J=23.0, 17.5 Hz), 157.3 (t, J=291.3 Hz), 170.2; ^{19}F NMR (CDCl$_3$): δ -84.2 (1 F, dd, J=33.6, 24.2 Hz), -86.2 (1 F, d, J=33.6 Hz); IR (neat): ν 2920, 2850, 1740, 1460, 1365, 1300, 1230, 1020, 805, 730 cm^{-1}.

Synthetic method

(E)-5,5-Difluorotridec-6-ene

To a THF (40 mL) solution of 1,1-difluoronon-1-en-3-yl acetate (5 mmol) were added at -20 °C, 1.35 g of copper(I) cyanide (15 mmol, 3 equiv) and 0.63 g of lithium chloride (15 mmol, 3 equiv), and after some minutes stirring at -15 °C n-butylmagnesium bromide (15 mmol) was added. After 8 h stirring, the reaction mixture was hydrolyzed by a solution of NH$_4$Cl:NH$_4$OH=3:1, filtered through Celite, and extracted with Et$_2$O. The organic phase was washed with brine, and dried over MgSO$_4$. The solvent was evaporated, and the residue was chromatographed by silica gel column chromatography (pentane) to afford (E)-5,5-difluorotridec-6-ene in 95% yield as a 98:2 (E):(Z) mixture.

Spectral data

^{1}H NMR (CDCl$_3$): δ 0.9-0.95 (6 H, two sets of t), 1.2-1.4 (12 H, m), 1.9 (2 H, m), 2.1 (2 H, m), 5.5 (1 H, dtt, J=15.9, 11.0, 1.6 Hz), 6.0 (1 H, dtt, J=15.9, 7.1, 2.7 Hz); ^{13}C NMR (CDCl$_3$): δ 13.8, 14.0, 22.5, 22.6, 24.6 (t, J=4.6 Hz), 28.5, 28.8, 31.6, 31.8, 37.3 (t, J=26.6 Hz), 121.7 (t, J=238.1 Hz), 125.2 (t, J=26.6 Hz), 135.7 (t, J=9.2 Hz); ^{19}F NMR (CDCl$_3$): δ -91.3 (1 F, m, J=33.6 Hz; (Z) form), -95.1 (1 F, dt, J=12, 11 Hz; (E) form); IR (neat): ν 2910, 2840, 1675, 1460, 1150, 1000, 960, 900 cm^{-1}.

Note

Acidic hydrolysis (1 N H$_2$SO$_4$ aq.) instead of the reaction with acetic anhydride in the first procedure afforded the corresponding allylic alcohols in good yields. Such alcohols were reported

$$F_2C=CH_2 \xrightarrow[\text{iii) Ac}_2\text{O, DMAP}]{\begin{array}{l}\text{i) } s\text{-BuLi}\\ \text{ii) R}^1\text{R}^2\text{C=O}\end{array}} \quad \text{[structure]} \xrightarrow[\text{v) R}^3\text{MgX}]{\text{iv) CuCN, LiCl}} \quad \text{[structure]}$$

| | | | Reaction | | | | | |
R^1	R^2	R^3	Temp (°C)	Time (h)	Yield$^{a)}$ (%)	(E):(Z) ratio$^{b)}$		
n-C$_6$H$_{13}{}^{e)}$	H$^{d)}$	Me	20	24	75$^{c)}$	97	:	3
n-C$_6$H$_{13}{}^{e)}$	H	i-Pr	-20	2	94	97	:	3
n-C$_6$H$_{13}{}^{d)}$	H	t-Bu	-20	3	95	95	:	5
n-C$_6$H$_{13}{}^{d)}$	H	CH$_2$=CH	20	24	0			
n-C$_6$H$_{13}{}^{e)}$	H	Allyl	20	24	0			
(E)-MeCH=CH$^{d)}$	H	t-Bu	-20	1	25	65	:	35
Thienyl$^{d)}$	H	t-Bu	-20	1	70	100	:	0
Thienyl$^{e)}$	H	c-C$_6$H$_{11}$	-20	2	80	100	:	0
-(CH$_2$)$_5$-$^{e)}$		i-Pr	-30	1	85			

a) The yield of the second step.   b) Determined by ^{19}F NMR.   c) 15% of byproduct (further S$_N$2' reaction of MeMgBr to the above final product) was detected.   d) X=Cl.   e) X=Br.
(Reproduced with permission from Tellier, F.; Sauvêtre, R. *J. Fluorine Chem.* **1995**, *70*, 265)

to be unstable in the pure state, but able to be stored without any problem in an Et$_2$O solution in the presence of a small amount of NaHCO$_3$.   See, for example, Tellier, F.; Sauvêtre, R. *J. Fluorine Chem.* **1993**, *62*, 183.

See also the following report on the alternative route to access this type of γ,γ-difluorinated allylic alcohols from 1,1-difluoro-1-chloroalkan-2-ones: Percy, J. M., *et al. J. Chem. Soc., Chem. Commun.* **1995**, 1857.

Tellier, F.; Sauvêtre, R. *J. Fluorine Chem.* **1995**, *70*, 265.

2,2-Difluoro-1-phenylbut-3-en-1-ol

a) PhCHO, cat. TASF / DMPU

Synthetic method

TASF (20 mg, 0.06 mmol) was added to a mixture of 1,1-difluoro-3-(dimethylphenylsilyl)-propene (0.208 mL, 1.0 mmol) and benzaldehyde (0.102 mL, 1.0 mmol) dissolved in DMPU (2 mL) at room temperature.   The resulting mixture was stirred at room temperature overnight, treated with 1 N HCl and extracted with Et$_2$O.   The ethereal extract was washed with sat. NaCl aq., dried over MgSO$_4$, and concentrated *in vacuo*.   Purification by preparative TLC gave 2,2-difluoro-1-phenyl-3-buten-1-ol (0.172 g, 93% yield).

Spectral data

Rf 0.55 (hexane:AcOEt=3:1); ^{1}H NMR (CCl$_4$): δ 3.59 (1 H, d, J=5 Hz), 4.79 (1 H, td, J=10, 5 Hz), 5.34-6.18 (3 H, m), 7.20-7.40 (5 H, m); ^{19}F NMR (CCl$_4$): δ -105.87 (1 F, dt, J=248, 10 Hz), -110.65 (1 F, dt, J=248, 10 Hz); IR (film): ν 3440, 1650, 1500, 1421, 1200, 1160, 995, 955, 852, 703, 637 cm^{-1}.

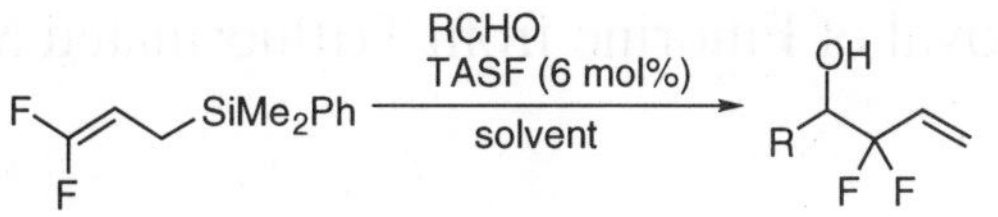

RCHO
TASF (6 mol%)
solvent

F-\/\-SiMe₂Ph → R-CH(OH)-C(F)(F)-CH=CH₂

R	Solvent	Yield (%)	R	Solvent	Yield (%)
Ph	DMPU	93	Ph	THF	100
4-Cl-C₆H₄-	DMPU	100	PhCH=CH	DMPU	52
n-C₁₀H₂₁-	DMPU	44	n-C₁₀H₂₁-	THF	42

(Reproduced with permission from Hiyama, T., *et al. Tetrahedron* **1988**, *44*, 4135)

Note

For the preparation of the starting material, see page 109 of this volume.

The regioisomeric allylic silanes containing two fluorines α or γ to the silyl group afforded the same product as the result of the new carbon-carbon bond formation of a carbonyl carbon atom with the fluorine-attached carbon atom when reacted with benzaldehyde.

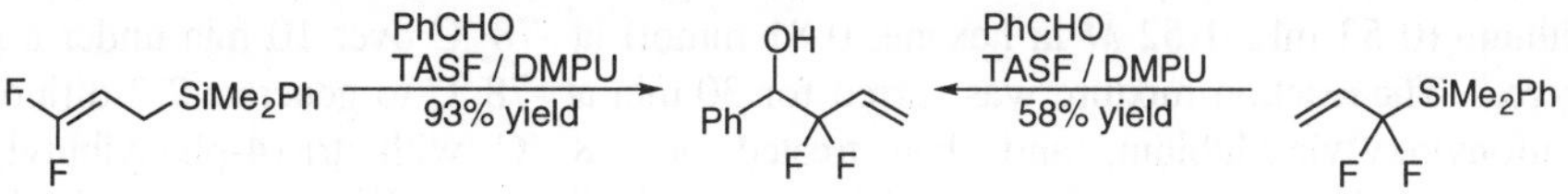

PhCHO
TASF / DMPU
93% yield
→

OH

←
PhCHO
TASF / DMPU
58% yield

Hiyama, T., *et al. Tetrahedron* **1988**, *44*, 4135.

(*E*)-1-Chloro-4-ethoxy-1,1-difluorobut-3-en-2-one

(CClF₂CO)₂O →(a)→ F₂ClC-C(O)-CH=CH-OEt

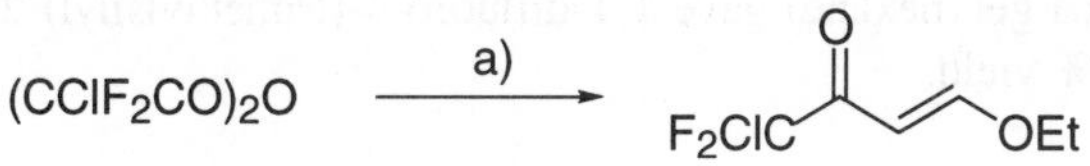

a) EtOCH=CH₂, pyridine / CH₂Cl₂

Synthetic method

A solution of ethyl vinyl ether (1.0 g, 13.9 mmol) and pyridine (0.55 g, 6.9 mmol) in CH₂Cl₂ (12.5 mL) was cooled to 0 °C, then chlorodifluoroacetic anhydride (4.86 g, 20 mmol) was added. The solution was stirred overnight at room temperature, then washed with 5% NaOH aq., with 1 *N* HCl and with H₂O. After drying (Na₂SO₄) the solvent was removed under reduced pressure to give 1.49 g of (*E*)-1-chloro-4-ethoxy-1,1-difluorobut-3-en-2-one as an oil in 58% yield.

Spectral data

^{1}H NMR (CDCl₃): δ 1.41 (3 H, t, *J*=7 Hz), 4.12 (2 H, q, *J*=7 Hz), 5.87 (1 H, d, *J*=11 Hz), 7.91 (1 H, d, *J*=11 Hz); ^{19}F NMR (CDCl₃): δ -68.55 (s); IR (film): ν 1710 cm^{-1}.

Note

For the original study using trifluoroacetic anhydride furnishing the adduct in quantitative yield. See Hojo, M., *et al. Chem. Lett.* **1976**, 499. The same compound was also prepared: Taguchi, T., *et al. Carbohydr. Res.* **1993**, *249*, 243 and the boiling point of this material was reported to be 66 °C at 8 mmHg.

Bravo, P., *et al. J. Chem. Soc. Perkin Trans. 1* **1995**, 1667.

3.5 Removal of Fluorine from Trifluorinated Materials

1,1-Difluoro-4-(trimethylsilyl)-2-(4-phenylbutyl)but-1-en-3-yne

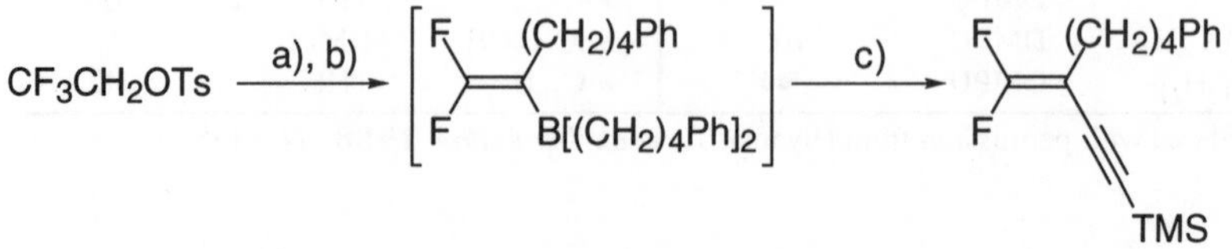

a) 2 equiv *n*-BuLi / THF,   b) B[(CH₂)₄Ph]₃,   c) HMPA, TMS-≡-I, CuCl·SMe₂

Synthetic method

To 2,2,2-trifluoroethyl *p*-toluenesulfonate (98 mg, 0.39 mmol) in THF (2.5 mL) was added *n*-butyllithium (0.53 mL, 1.52 *M* in hexane, 0.81 mmol) at -78 °C over 10 min under a nitrogen atmosphere.   The reaction mixture was stirred for 30 min at -78 °C to generate 2,2-difluoro-1-(*p*-toluenesulfonyloxy)vinyllithium,   and   then   treated   at   -78 °C   with   tris(4-phenylbutyl)borane, generated from 4-phenylbutene (168 mg, 1.27 mmol) and a borane-THF complex (0.42 mL, 1.0 *M* in THF, 0.42 mmol).   After 1 h of stirring at -78 °C, the mixture was warmed to room temperature and stirred for an additional 3 h.   To the resulting solution of vinylborane were added successively HMPA (0.9 mL),   (iodoethynyl)trimethylsilane (78 mg, 0.35 mmol) and copper(I) chloride/ dimethyl sulfide (1:1) (62 mg, 0.38 mmol).   After the mixture had been stirred for 1 h at room temperature, H₂O was added to quench the reaction.   The usual work-up followed by column chromatography on silica gel (hexane) gave 1,1-difluoro-4-(trimethylsilyl)-2-(4-phenylbutyl)but-1-en-3-yne (61 mg) in 60% yield.

Spectral data

^{1}H NMR (CDCl₃): δ 0.22 (9 H, s), 1.36-1.89 (4 H, m), 1.89-2.40 (2 H, m), 2.66 (2 H, m), 7.21 (5 H, br s); ^{13}C NMR (CDCl₃): δ 0.0, 26.7, 27.2 (t, *J*=2 Hz), 30.5, 35.7, 78.7 (dd, *J*=34, 15 Hz), 96.7 (dd, *J*=8, 4 Hz), 99.3 (t, *J*=5 Hz), 125.8, 128.4, 128.5, 142.4, 159.6 (dd, *J*=296, 292 Hz); ^{19}F NMR (CDCl₃): δ 76.9 (d, *J*=17 Hz), 82.4 (d, *J*=17 Hz) from C₆F₆: IR (neat): ν 3270, 2910, 1715, 1600, 1495, 1450, 1295, 1145, 1095, 695 cm^{-1}.

Ichikawa, J., *et al. J. Fluorine Chem.* **1993**, *63*, 281.

1,1-Difluoro-2-phenylbut-1-ene

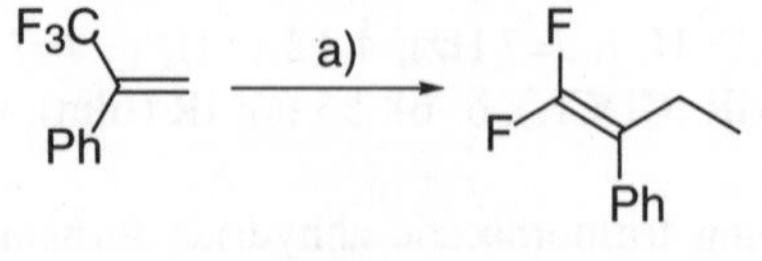

a) MeLi / THF

Synthetic method

Methyllithium (a 1.6 *M* solution in Et₂O, 2.5 mL, 4 mmol) was added at -78 °C over a period of 2 min to a solution of 3,3,3-trifluoro-2-phenylpropene (0.7 g, 4 mmol) in THF (20 mL).   The pale yellow solution was stirred for 30 min and then allowed to warm to room temperature during 1 h. The brown solution was poured into sat. NH$_4$Cl aq., the layers were separated, and the aqueous phase was extracted with Et$_2$O (3x50 mL). The combined organic extracts were dried and evaporated to afford a brown oil, which was purified by passage through a short column of silica gel (pentane) to give 1,1-difluoro-2-phenylbut-1-ene (0.61 g, 90%) as an oil.

Spectral data

^{1}H NMR (CDCl$_3$): δ 1.0 (3 H, t, *J*=7.5 Hz), 2.4-2.5 (2 H, m), 7.1-7.4 (5 H, m); ^{13}C NMR (CDCl$_3$): δ 12.9, 21.4, 94.0 (dd, *J*=18, 15 Hz), 127.4, 128.4, 128.7, 133.8 (d, *J*=1.9 Hz), 153.5 (t, *J*=288 Hz); ^{19}F NMR (CDCl$_3$): δ -91.9 (s); IR (neat): ν 1737 cm^{-1}.

Note

For the preparation of the starting β-(trifluoromethyl)styrene, see page 168 of this volume.

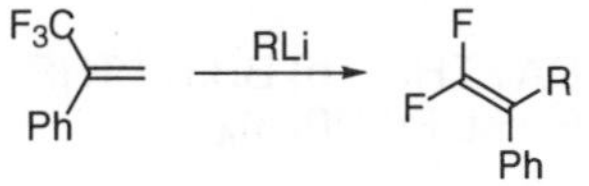

R	Yield (%)	R	Yield (%)
Bu	93	*t*-Bu	92
Ph	90	(EtO)$_2$P(O)CH$_2$	85
PhS(O)CH$_2$	83	PhS(O)$_2$CH$_2$	80
1,3-dithian-2-yl	90	(*i*-Pr)$_2$N	90
(allyl)$_2$N	85	2-phenylethylamino	60

(Reproduced with permission from Bonnet-Delpon, D., *et al. J. Chem. Soc., Perkin Trans. 1* **1996**, 1409)

Bonnet-Delpon, D., *et al. J. Chem. Soc., Perkin Trans. 1* **1996**, 1409.

Ethyl 3,3-difluoro-2-methoxypropenoate

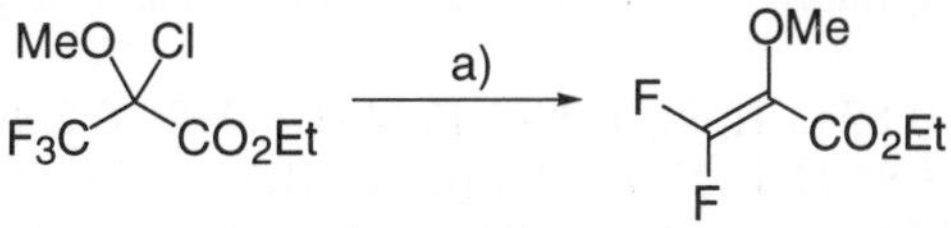

a) Zn, cat. CuI / DMF

Synthetic method

A mixture of ethyl 2-chloro-3,3,3-trifluoro-2-methoxypropionate (11.0 g, 50 mmol), freshly activated zinc powder (12.8 g, 0.20 mol), and a small amount of copper(I) iodide (*ca.* 50 mg) in DMF (100 mL) was vigorously stirred at 25 °C under nitrogen. After the heat evolution ceased, the reaction mixture was diluted with Et$_2$O (100 mL) and filtered. The filtrate was washed with H$_2$O (2x50 mL) and dried over Na$_2$SO$_4$. Distillation under reduced pressure afforded 7.1 g of ethyl 3,3-difluoro-2-methoxypropenoate in 85% yield as a colorless liquid.

Spectral data

bp. 52 °C/24 mmHg; ^{1}H NMR (CCl$_4$): δ 1.31 (3 H, t, *J*=7.2 Hz), 3.65 (3 H, s), 4.25 (2 H, q, *J*=7.2 Hz); ^{19}F NMR (CCl$_4$): δ -5.3 (1 F, d, *J*=3.0 Hz), -9.8 (1 F, d, *J*=3.0 Hz) from CF$_3$CO$_2$H.

Note

For the preparation of the starting material, see page 178 of this volume.

It was reported that the corresponding 2-benzyloxy compound reacted with a variety of nucleophiles in an addition-elimination manner.   See Shi, G.-Q.; Cao, Z.-Y. *J. Chem. Soc., Chem. Commun.* **1995**, 1969.

Shi, G.-Q., *et al. J. Org. Chem.* **1995**, *60*, 6608.

1,1-Difluoro-2-phenylhex-1-ene

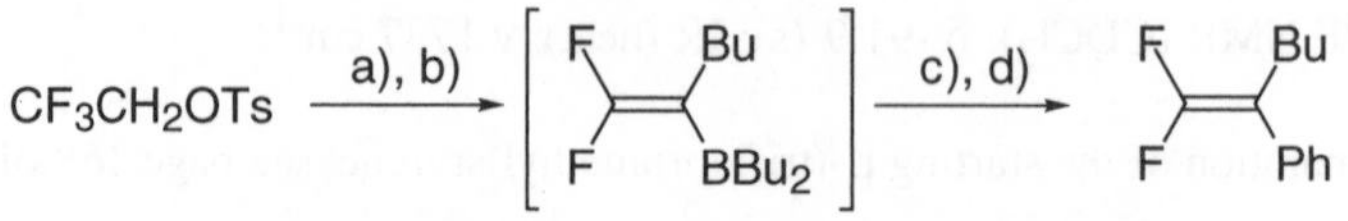

a) 2 equiv LDA / THF,   b) BBu₃ / THF,   c) Me₃NO,
d) PhI, TBAF, cat. Pd(PPh₃)₄

Synthetic method

2,2,2-Trifluoroethyl *p*-toluenesulfonate (160 mg, 0.63 mmol) in THF (1 mL) was added dropwise to a THF solution of LDA (1.32 mmol) at -78 °C under an argon atmosphere.   The reaction mixture was stirred for 30 min at -78 °C to generate 2,2-difluoro-1-(*p*-toluenesulfonyl-oxy)vinyllithium, and then treated with tributylborane (0.69 mL, 1.0 *M* in THF, 0.69 mmol) at -78 °C.   After being stirred for 1 h at -78 °C, the mixture was brought to room temperature and stirred for an additional 3 h.   The vinylborane solution thus obtained was treated with solid trimethylamine *N*-oxide (117 mg, 1.56 mmol) at 0 °C and stirred for 1 h at that temperature.   To the resulting solution were added the palladium catalyst generated from dichlorobis(triphenyl-phosphine)palladium(II) (44 mg, 0.063 mmol) and butyllithium (0.077 mL, 1.63 *M* in hexane, 0.126 mmol) in THF (1 mL), iodobenzene (53 µL, 0.47 mmol), and TBAF (1.89 mL, 1.0 *M* in THF, 1.89 mmol) successively at room temperature.   After stirring for 12 h at 50 °C, the reaction mixture was quenched with H_2O.   The usual work-up followed by thin-layer chromatography on silica gel (hexane) gave 1,1-difluoro-2-phenylhex-1-ene (80 mg) in 86% yield.

Spectral data

^{1}H NMR (CDCl₃): δ 0.87 (3 H, s), 1.08-1.46 (4 H, m), 2.20-2.49 (2 H, m), 7.31 (5 H, br s); ^{13}C NMR (CDCl₃): δ 13.8, 22.1, 27.4, 29.9 (t, *J*=2 Hz), 92.5 (t, *J*=18 Hz), 127.2, 128.3 (t, *J*=3 Hz), 128.4, 133.9, 153.6 (t, *J*=288 Hz); ^{19}F NMR (CDCl₃): δ 69.7 (t, *J*=2 Hz) from C₆F₆; IR (neat): ν 1725, 1230, 690 cm⁻¹.

Ichikawa, J. *et al. Chem. Lett.* **1991**, 961.

1,1-Difluoro-2-[(triphenylsilyl)oxy]propene

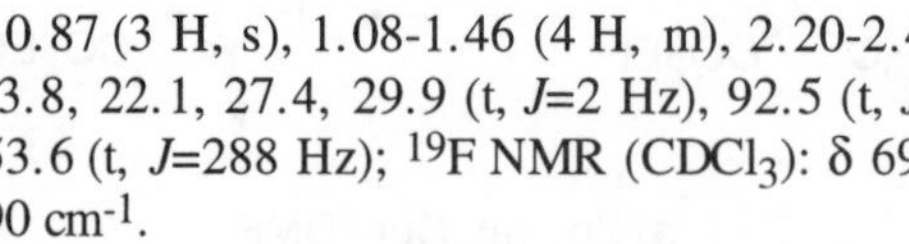

a) MeMgI / THF

Synthetic method

To a solution of (trifluoroacetyl)triphenylsilane (1 mmol) in dry THF (10 mL), cooled with dry ice at -30 ℃, was added *via* a syringe MeMgI (1.2 mmol) and stirring was continued at -30 ℃ for 15 min, and then at room temperature for 1 h. Hexane (30 mL) was added and the solid portion was filtered off. The filtrate was concentrated under reduced pressure and the residue chromatographed on silica gel (light petroleum ether:AcOEt=99:1) to afford 1,1-difluoro-2-[(triphenylsilyl)oxy]propene in 99% yield.

Spectral data

mp. 51-52.5 °C; ^{1}H NMR (CCl$_4$): δ 1.60 (3 H, t, J=4 Hz), 7.10-7.80 (15 H, m); ^{19}F NMR (CCl$_4$): δ -28.5 (1 F, dq, J=90, 4 Hz), -45.5 (1 F, dq, J=90, 4 Hz) from CF$_3$CO$_2$H; IR (KBr): ν 1780, 1590, 1480, 1250, 1120, 740, 700 cm^{-1}.

Note

The above enol silyl ether was found to undergo the Lewis acid-catalyzed reaction with the steroidal aldehyde shown below, affording the product in 71% yield with complete diastereo-selectivity at the newly created stereogenic center.

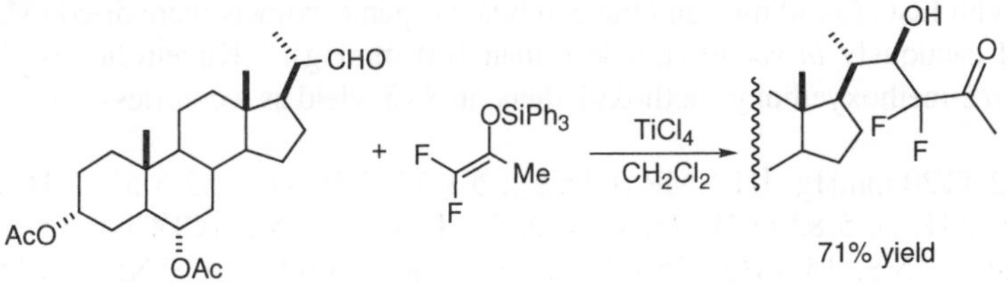

Xu, Y.-Y., *et al. J. Chem. Soc. Perkin Trans. 1* **1993**, 795.

1,1-Difluoro-3-(dimethylphenylsilyl)propene

a) Me$_3$SiSiMe$_2$Ph, cat. TBAF / THF

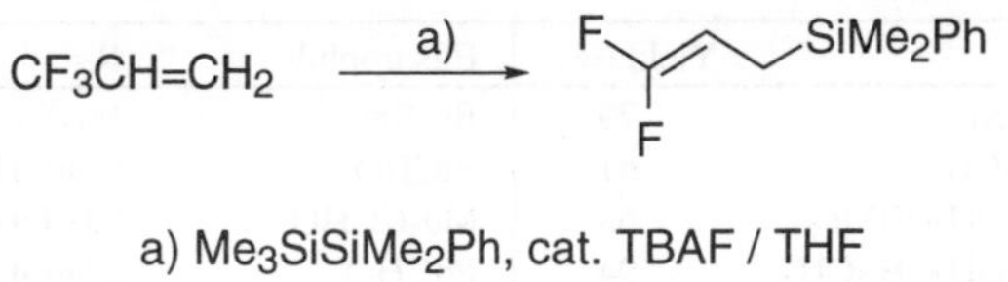

Synthetic method

A THF (2 mL) solution of pentamethylphenylsilane (0.238 mL, 1 mmol) was treated with TBAF (a 0.5 M THF solution, 0.2 mL, 0.1 mmol), and the mixture was stirred for 5 min. Then 3,3,3-trifluoropropene (26.9 mL at 1 atm, 1.2 mmol) was bubbled at room temperature. The resulting reaction mixture was stirred for 6 h, then washed with H$_2$O, and concentrated. Distillation gave 1,1-difluoro-3-(dimethylphenylsilyl)propene (0.181 g, 85% yield) as a colorless oil.

Spectral data

bp. 120-130 ℃/23 mmHg; ^{1}H NMR (CCl$_4$): δ 0.40 (6 H, s), 1.55 (2 H, d, J=9 Hz), 4.10 (1 H, dtd, J=25, 8, 2 Hz), 7.20-7.70 (5 H, m); ^{19}F NMR (CCl$_4$): δ -89.85 (1 F, dd, J=50.8, 2.0 Hz), -93.00 (1 F, dd, J=50.8, 25.1 Hz).

Note

As an alternative method, 1,2-diphenyltetramethylsilane (0.284 mL, 1 mmol) in the presence of HMPA (2 mL) can be used instead of pentamethylphenylsilane, affording the same product in 85% yield.

Hiyama, T., *et al. Tetrahedron* **1988**, *44*, 4135.

1,1-Difluoro-2-[(2-methoxyethoxy)methoxy]ethene

$$CF_3CH_2OMEM \xrightarrow{a),\ b)} \begin{array}{c} \text{OMEM} \\ F \diagup \diagdown H \\ F \end{array}$$

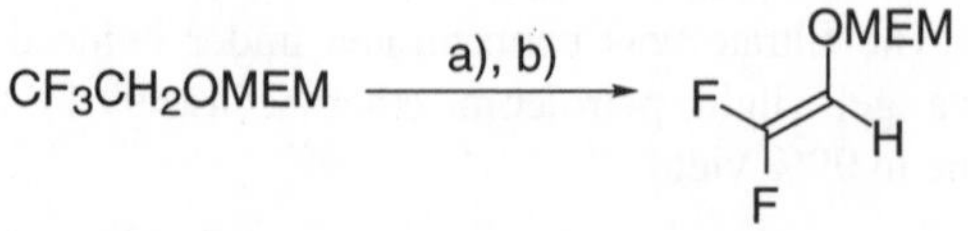

a) 2 LDA / THF,   b) NH$_4$Cl / MeOH

Synthetic method

1,1,1-Trifluoro-2-[(2-methoxyethoxy)methoxy]ethane (0.94 g, 5.0 mmol) was added slowly to a stirred solution of LDA (10.5 mmol) in dry THF (5 mL) at -78 °C. The dark orange suspension was stirred at -78 °C for 30 min then treated with methanolic NH$_4$Cl (5.0 mL). The mixture was allowed to warm to room temperature and poured into H$_2$O (50 mL). The mixture was extracted with Et$_2$O (3x50 mL) and the combined organic extracts were dried (MgSO$_4$), filtered and evaporated cautiously *in vacuo* (not less than 100 mmHg). Kügelrohr distillation afforded 1,1-difluoro-2-[(2-methoxyethoxy)methoxy]ethene in 88% yield as a colorless oil.

Spectral data

bp. 70-72 °C/20 mmHg; ^{1}H NMR (CDCl$_3$): δ 3.37 (3 H, s), 3.52-3.57 (2 H, m), 3.71-3.75 (2 H, m), 4.82 (2 H, s), 5.82 (1 H, dd, *J*=16.0, 3.0 Hz); ^{13}C NMR (CDCl$_3$): δ 58.9, 67.8, 71.5, 96.1, 105.3 (dd, *J*=15.8, 15.4 Hz), 155.4 (dd, *J*=287.8, 275.6 Hz); ^{19}F NMR (CDCl$_3$): δ -100.6 (1 F, dd, *J*=76.3, 16.0 Hz), -119.7 (1 F, dd, *J*=76.3, 3.0 Hz).

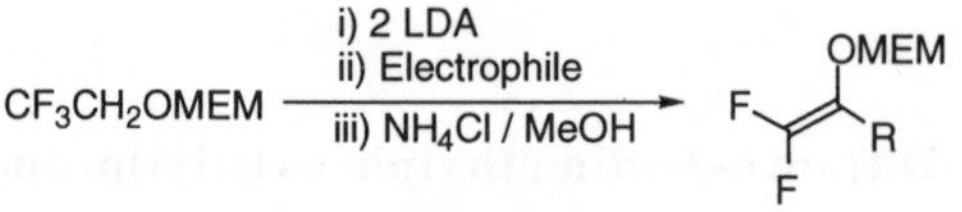

$$CF_3CH_2OMEM \xrightarrow[\text{iii) NH}_4\text{Cl / MeOH}]{\substack{\text{i) 2 LDA} \\ \text{ii) Electrophile}}} \begin{array}{c} \text{OMEM} \\ F \diagup \diagdown R \\ F \end{array}$$

Electrophile	R	Yield (%)	Electrophile	R	Yield (%)
Me$_3$SiCl	Me$_3$Si	79	Bu$_3$SnCl	Bu$_3$Sn	70
HCHO	CH$_2$OH	61	EtCHO	CH(OH)Et	83
Me$_2$CHCHO	CH(OH)CHMe$_2$	68	Me$_3$CCHO	CH(OH)CMe$_3$	74
CH$_2$=CHCHO	CH(OH)CH=CH$_2$	94	PhCHO	CH(OH)Ph	67
MeC(O)Me	C(OH)Me$_2$	68	EtC(O)Et	C(OH)Et$_2$	79

(Reproduced with permission from Percy, J. M., *et al. Tetrahedron* **1995**, *51*, 9201)

Note

For the preparation of the starting MEM ether, see page 200 of this volume.

The same type of reaction was also reported by Percy *et al.*, starting from the carbamoyl-

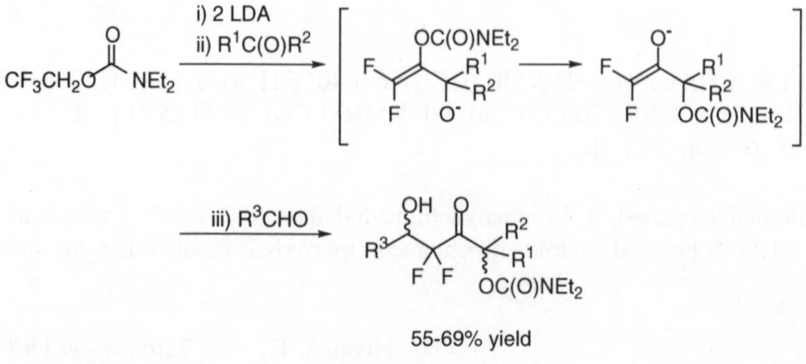

55-69% yield

protected 2,2,2-trifluoroethanol. However, this moiety was found to migrate to the neighboring alkoxide when reacted with an appropriate carbonyl compound after subjection to 2 equiv of LDA, and the resultant enolate was able to be reacted with different carbonyl compounds as shown in the previous page. See Percy, J. M., *et al. Tetrahedron* **1995**, *51*, 10289.

Percy, J. M., *et al. Tetrahedron* **1995**, *51*, 9201.

3.6 Reactions of Hemiacetals

Chlorodifluoroacetaldehyde ethyl hemiacetal,
Ethyl (E)-4-chloro-4,4-difluorobut-2-enoate

a) LAH / Et$_2$O, b) (EtO)$_2$P(O)CH$_2$CO$_2$Et, Et$_3$N, LiBr / THF

Synthetic method

To a suspension of LAH (0.57 g, 15.0 mmol) in Et$_2$O (50 mL) was added a solution of ethyl chlorodifluoroacetate (7.93 g, 50.0 mmol) in Et$_2$O (50 mL) at -78 °C, and the mixture was stirred at that temperature for 30 min. The reaction mixture was then diluted with Et$_2$O (100 mL) and poured into 0.5 N HCl (200 mL). The organic layer was separated and the aqueous layer extracted with Et$_2$O (2x150 mL). The combined organic extracts were dried over Na$_2$SO$_4$ and evaporated to give the crude chlorodifluoroacetaldehyde ethyl hemiacetal.

To a suspension of LiBr (5.56 g, 64 mmol) in dry THF (50 mL) was added triethyl phosphonoacetate (9.9 ml, 50 mmol) and triethylamine (7.7 mL, 55 mmol) at 0 °C. After being stirred at room temperature for 10 min, the above crude 2-chloro-1-ethoxy-2,2-difluoroethanol was added dropwise at 0 °C and the mixture was stirred at room temperature for 3 h. The reaction mixture was then diluted with Et$_2$O and poured into 0.5 N HCl (150 mL). The organic layer was separated and the aqueous layer extracted with Et$_2$O (2x150 mL). The combined organic extracts were washed with brine (400 mL), dried over MgSO$_4$ and evaporated. The residual oil was distilled to give ethyl (E)-4-chloro-4,4-difluorobut-2-enoate in 64% overall yield.

Spectral data

bp. 147-148 °C; ^{1}H NMR (CDCl$_3$): δ 1.33 (3 H, t, J=7.2 Hz), 4.28 (2 H, q, J=7.2 Hz), 6.39 (1 H, dt, J=15.6, 1.9 Hz), 6.93 (1 H, dt, J=15.6, 9.2 Hz); ^{13}C NMR (CDCl$_3$): δ 14.5, 62.1, 124.1 (t, J=287 Hz), 126.7 (t, J=6.5 Hz), 137.3 (t, J=28.7 Hz), 164.6; ^{19}F NMR (CDCl$_3$): δ -54.9 (d, J=9.2 Hz); IR (neat): ν 3000, 1740 cm^{-1}.

Note

A similar Höner-Wittig reaction was possible with ethyl 3-ethoxy-2,2-difluoro-3-hydroxy-propionate. See also the following report on the remarkable reversal of stereoselectivity: Kakinuma, *et al. Tetrahedron Lett.* **1992**, *33*, 5545.

Ethyl (E)-4-chloro-4,4-difluorobut-2-enoate was employed for the Reformatsky-type reaction described earlier to afford the products in good yields with high α-regioselectivity.

$$\text{F}_2\text{ClC}\diagdown\diagup\text{CO}_2\text{Et} \xrightarrow{\text{RCHO, Zn-Cu(OAc)}_2} \text{F}\diagup\diagdown\text{CO}_2\text{Et}$$

Tsukamoto, T.; Kitazume, T. *Synlett* **1992**, 977.
The detailed experimental procedures were obtained from the following dissertation.
See Tsukamoto, T. PhD dissertation, Tokyo Institute of Technology (1993).

Difluoroacetaldehyde ethyl hemiacetal,
4,4-Difluoro-3-hydroxy-1-phenylbutan-1-one

$$\text{CHF}_2\text{CO}_2\text{Et} \xrightarrow{\text{a), b)}} \text{F}_2\text{HC}\diagup\diagdown\text{OEt} \xrightarrow{\text{c)}} \text{F}_2\text{HC}\diagup\diagdown\diagup\text{Ph}$$

a) LAH / Et$_2$O,   b) H$_2$SO$_4$,   c) CH$_2$=C(OTMS)Ph, ZnCl$_2$ / THF

Synthetic method

Difluoroacetaldehyde ethyl hemiacetal

To a solution of ethyl difluoroacetate (20 mL, 200 mmol) in dry Et$_2$O (30 mL) was added a solution of lithium aluminum hydride in THF (1.0 *M*, 50 mL, 50 mmol) at -78 °C.   After the solution was stirred at that temperature for 3 h, EtOH (95%, 5.0 mL) was added and the whole solution was allowed to warm to room temperature.   The mixture was poured into a solution of crushed ice and conc H$_2$SO$_4$ (15 mL), and then extracted with Et$_2$O.   The extracts were dried over anhydrous MgSO$_4$.   On removal of the solvent, the residual oil was purified by distillation to afford 1-ethoxy-2,2-difluoroethanol (15.1 g, 120 mmol) in 60% yield.

Spectral data

bp. 45-47 °C/27 mmHg; ^{1}H NMR (CDCl$_3$): δ 1.26 (3 H, t, *J*=7.1 Hz), 3.2-3.4 (1 H, br), 3.63 (1 H, dq, *J*=9.6, 7.1 Hz), 3.91 (1 H, dq, *J*=9.6, 7.1 Hz), 4.70 (1 H, ddd, *J*=8.0, 5.7, 2.4 Hz), 5.60 (1 H, ddd, *J*=55.5, 54.9, 2.7 Hz); ^{13}C NMR (CDCl$_3$): δ 15.05, 64.55, 93.77 (dd, *J*=29.2, 26.3 Hz), 113.40 (t, *J*=241 Hz); ^{19}F NMR (CDCl$_3$): δ 25.15 (1 F, ddd, *J*=291, 55.7, 7.6 Hz), 30.11 (1 F, ddd, *J*=291, 54.9, 6.1 Hz) from C$_6$F$_6$; IR (neat): ν 3400 cm^{-1}.

Synthetic method

4,4-Difluoro-3-hydroxy-1-phenylbutan-1-one

To a suspension of zinc chloride powder (90%, 0.91 g, 6.0 mmol) and difluoroacetaldehyde ethyl hemiacetal (0.25 g, 2.0 mmol) in dry THF (10 mL) was added α-[(trimethylsilyl)oxy]styrene (1.15 g, 6.0 mmol) at 0 °C.   After 1 h refluxing, the mixture was quenched with H$_2$O (10 mL) and extracted with Et$_2$O.   The extract was dried over anhydrous MgSO$_4$, and the solvent was removed.   Flash chromatography (hexane:AcOEt=5:1) gave 4,4-difluoro-3-hydroxy-1-phenylbutan-1-one in 75% yield.

Spectral data

^{1}H NMR (CDCl$_3$): δ 3.30 (2 H, dd, *J*=6.1, 0.5 Hz), 3.2-3.4 (1 H, br), 4.35 (1 H, dddt, *J*=13.6, 9.7, 6.1, 3.5 Hz), 5.83 (1 H, ddd, *J*=56.3, 55.4, 3.5 Hz), 7.2-8.0 (5 H, m); ^{13}C NMR (CDCl$_3$): δ 37.83 (t, *J*=3.0 Hz), 67.69 (t, *J*=23.6 Hz), 115.56 (t, *J*=243 Hz), 128.19, 128.80, 133.95, 136.21, 198.79; ^{19}F NMR (CDCl$_3$): δ 30.27 (1 F, ddd, *J*=288, 56.1, 13.7 Hz), 33.17 (1 F, ddd, *J*=288, 55.3, 9.2 Hz) from C$_6$F$_6$; IR (neat): ν 3440, 1680 cm^{-1}.

Note

Difluoroacetaldehyde ethyl hemiacetal thus formed by reduction of ethyl difluoroacetate was found to undergo various types of reactions with nucleophiles as described in the next page.   See also the following articles on similar types of reactions: i) Haas, A. M.; Hägele, G. *J. Fluorine Chem.* **1996**, *78*, 75, ii) Parisi, M. F., *et al. J. Org. Chem.* **1995**,*60*, 5174.

    For the case of the corresponding trifluorinated materials, see Kubota, T., *et al. Tetrahedron Lett.* **1992**, *33*, 1351.

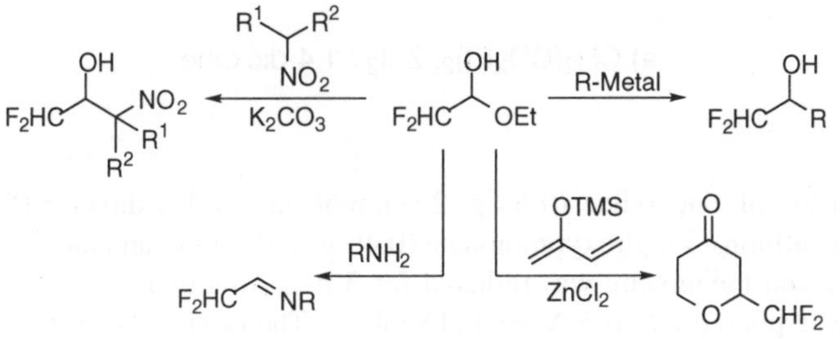

Kitazume, T., *et al. J. Org. Chem.* **1993**, *58*, 2302.

Ethyl 3-ethoxy-2,2-difluoro-3-hydroxypropionate

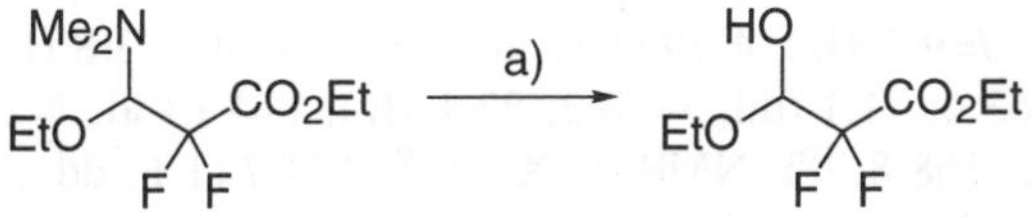

a) conc H_2SO_4 / EtOH

Synthetic method
    A drop of conc H_2SO_4 was added to a solution of ethyl 3-ethoxy-2,2-difluoro-3-(dimethyl-amino)propionate (0.45 g, 2 mmol) in EtOH (2.5 mL) and H_2O (0.5 mL), and the mixture was stirred at ambient temperature for 30 min.   After diluting with Et_2O (30 mL), the reaction mixture was poured into H_2O (30 mL).   The organic layer was separated and the aqueous layer was extracted with Et_2O.   The combined organic extracts were dried over Na_2SO_4, and evaporated. The residual oily material was bulb-to-bulb distilled to give ethyl 3-ethoxy-2,2-difluoro-3-hydroxy-propionate in 96% yield.
Spectral data
    bp. 140 ℃ (bath temperature)/1 mmHg; ^1H NMR ($CDCl_3$): δ 1.22 (3 H, t, *J*=7.1 Hz), 1.37 (3 H, t, *J*=7.1 Hz), 2.90 (1 H, s), 3.65 (1 H, dq, *J*=9.8, 7.1 Hz), 3.92 (1 H, dq, *J*=9.8, 7.1 Hz), 4.37 (2 H, q, *J*=7.1 Hz), 4.91 (1 H, dd, *J*=7.0, 5.6 Hz); ^{13}C NMR ($CDCl_3$): δ 13.9, 14.9, 63.2, 64.7, 94.0 (dd, *J*=31.5, 26.5 Hz), 111.3 (t, *J*=256 Hz), 163.0 (t, *J*=31.1 Hz); ^{19}F NMR ($CDCl_3$): δ -120.0 (1 F, dd, *J*=266, 6.1 Hz), -125.1 (1 F, dd, *J*=266, 6.9 Hz); IR (neat): ν 3480, 1780, 1760 cm^{-1}.
Note
    For the preparation of the starting *N*,*O*-acetal, see page 94 of this volume.
    The corresponding amide was prepared similarly in 87% yield, while the corresponding phosphonate was constructed directly by the reaction of lithiated difluorinated phosphonate and formyl morpholine at -78 ℃ in 81% yield.

Tsukamoto, T.; Kitazume, T. *J. Chem. Soc. Perkin Trans. 1* **1993**, 1177.

Diethyl 4-carboethoxy-2,2-difluoro-3-hydroxyglutarate

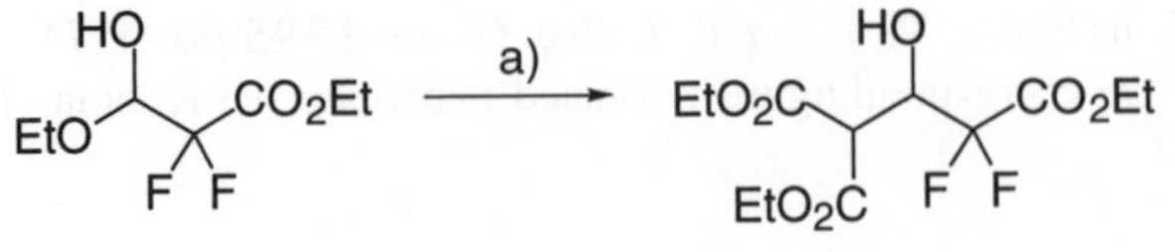

a) CH$_2$(CO$_2$Et)$_2$, ZnI$_2$ / 1,4-dioxane

Synthetic method

To a suspension of zinc iodide (0.64 g, 2.0 mmol) in dry 1,4-dioxane (5 mL) were added ethyl 3-ethoxy-2,2-difluoro-3-hydroxypropionate (0.40 g, 2.0 mmol) and diethyl malonate (0.48 g, 3.0 mmol) at 0 °C, and the mixture was refluxed for 3 h.   After dilution with Et$_2$O (15 mL), the reaction mixture was poured into 0.5 *N* HCl (15 mL).   The organic layer was separated and the aqueous layer was extracted with Et$_2$O.   The combined organic extracts were washed with brine (30 mL), dried over MgSO$_4$, and evaporated.   Diethyl 4-carboethoxy-2,2-difluoro-3-hydroxy-glutarate was obtained in 76% yield by column chromatography on silica gel (hexane:AcOEt=4:1).

Spectral data

^{1}H NMR (CDCl$_3$): δ 1.30 (3 H, t, *J*=7.1 Hz), 1.32 (3 H, t, *J*=7.1 Hz), 1.37 (3 H, t, *J*=7.1 Hz), 3.78 (1 H, d, *J*=3.7 Hz), 4.27 (2 H, q, *J*=7.1 Hz), 4.30 (2 H, q, *J*=7.1 Hz), 4.38 (2 H, q, *J*=7.1 Hz), 4.62 (1 H, d, *J*=9.2 Hz), 4.79 (1 H, ddt, *J*=20.4, 9.2, 3.8 Hz); ^{13}C NMR (CDCl$_3$): δ 13.9, 49.4, 62.5, 62.6, 63.4, 71.1 (dd, *J*=30.2, 23.1 Hz), 114.1 (dd, *J*=260, 255 Hz), 162.5 (dd, *J*=33.1, 29.2 Hz), 166.2, 168.8; ^{19}F NMR (CDCl$_3$): δ -124.7 (1 F, dd, *J*=261, 21.4 Hz), -110.9 (1 F, dd, *J*=261, 3.1 Hz); IR (neat): ν 3500, 1760 cm^{-1}.

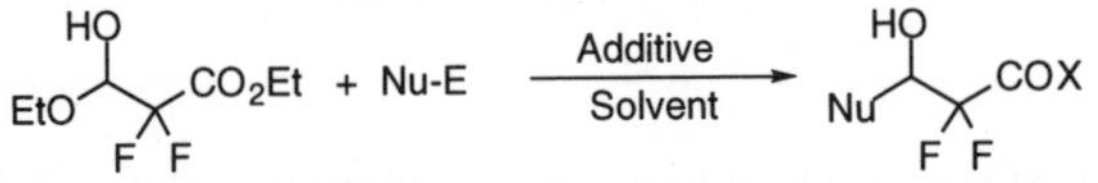

Nu-E	Additive	Solvent	X	Yield (%)
(MeC(O))$_2$CH$_2$	ZnI$_2$	1,4-dioxane	OEt	61
MeC(O)CH$_2$CO$_2$Et	ZnI$_2$	1,4-dioxane	OEt	63
(EtO$_2$C)$_2$CH$_2$	ZnI$_2$	1,4-dioxane	NEt$_2$	76
NC-TMS	ZnI$_2$	1,4-dioxane	OEt	82
NC-TMS	ZnI$_2$	1,4-dioxane	NEt$_2$	76
Allyl-TMS	ZnI$_2$	1,4-dioxane	OEt	68
EtO$_2$CCH$_2$NO$_2$	K$_2$CO$_3$	THF	NEt$_2$	68
O$_2$NMe	K$_2$CO$_3$	THF	OEt	73
O$_2$NMe	K$_2$CO$_3$	THF	NEt$_2$	88

Note

For the preparation of the starting hemiacetal, see page 113 of this volume.

Tsukamoto, T.; Kitazume, T. *J. Chem. Soc. Perkin Trans. 1* **1993**, 1177.

3.7 Oxidations and Reductions

anti-2,2-Difluoro-1-phenylhexane-1,3-diol

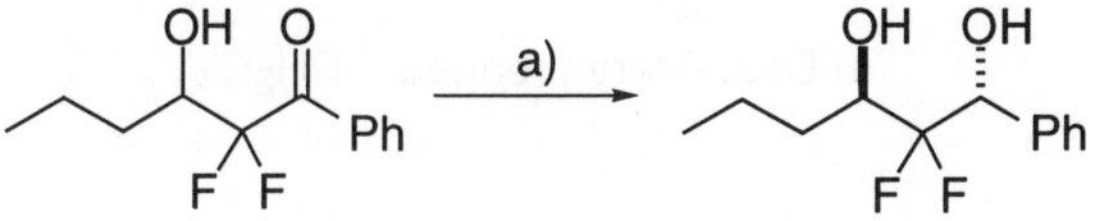

a) Al(OPr^{-i})$_3$ / benzene

Synthetic method

Aluminum isopropoxide (0.214 g, 1.05 mmol) was dissolved in benzene (5 mL) and to this solution was added 2,2-difluoro-3-hydroxy-1-phenylhexan-1-one (1.0 mmol) at room temperature. After being stirred for 18 h at room temperature, the reaction mixture was poured into a mixture of 6 N HCl (5 mL), sat. NH$_4$Cl aq. (5 mL) and ice.   The resultant was extracted with Et$_2$O (3x20 mL).   The combined extracts were dried over anhydrous Na$_2$SO$_4$, filtered and concentrated.   The ratio of *syn*- to *anti*-2,2-difluoro-1-phenylhexane-1,3-diol was determined by ^{19}F NMR before purification.   The diol was purified by silica gel column chromatography and the desired product, 2,2-difluoro-1-phenylhexane-1,3-diol, was obtained in 98% yield in a ratio of *syn:anti*=11:89.

Spectral data

^{1}H NMR (CDCl$_3$): δ 0.7-1.1 (3 H, m), 1.2-1.9 (4 H, m), 2.79 (1 H, br d, J=5.4 Hz), 3.4-3.7 (1 H, br s), 3.7-4.3 (1 H, br s), 4.99 (1 H, ddd, J=12.2, 9.8, 4.2 Hz), 7.26 (5 H, s); ^{19}F NMR (CDCl$_3$): δ -122.02 (ddd, J=12.2, 9.8, 9.8 Hz): IR (neat): ν 3580, 3378, 3056, 2962, 2932, 1203, 1177, 1109, 1086, 1059, 1028 cm^{-1}.

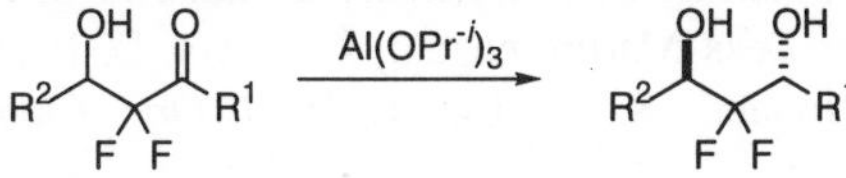

R^1	R^2	Yield (%)	syn:anti		
n-C$_6$H$_{13}$	Pr	99	9	:	91
n-C$_6$H$_{13}$	Me$_2$CH	92	0	:	100
n-C$_6$H$_{13}$	(*E*)-MeCH=CH	93	5	:	95
n-C$_6$H$_{13}$	Ph	95	0	:	100
Ph	Pr	98	11	:	89
Ph	(*E*)-MeCH=CH	95	8	:	92
Ph	Ph	91	0	:	100

(Reproduced with permission from Ishihara, T. *et al. Bull. Chem. Soc. Jpn.* **1990**, *63*, 1185)

Note

For the preparation of the starting hydroxyketone, see page 93 of this volume.

Kuroboshi, M.; Ishihara, T. *Bull. Chem. Soc. Jpn.* **1990**, *63*, 1185.

Ethyl 2,2-difluoro-12-methyl-3-oxotridecanoate

a) Dess-Martin reagent / CH_2Cl_2

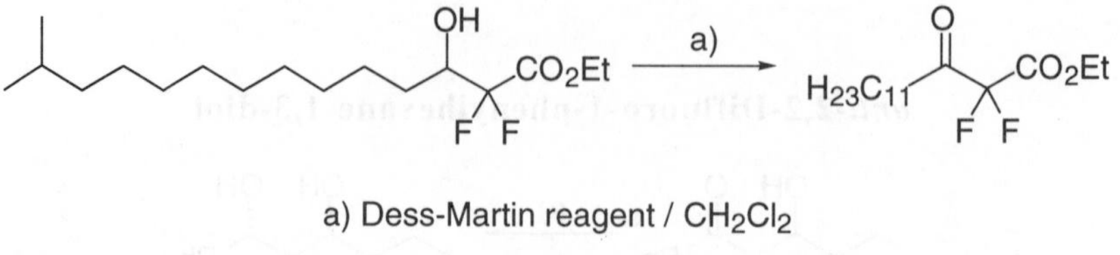

Synthetic method

Dess-Martin reagent (2.27 g, 5.36 mmol) was added to a solution of ethyl 2,2-difluoro-3-hydroxy-12-methyltridecanoate (1.1 g, 3.57 mmol) in CH_2Cl_2 (36 mL) at room temperature under argon.   After the solution was stirred for 30 min, the reaction was quenched with sat. NaHCO₃ aq. and sat. Na₂S₂O₃ aq.   The organic layer was separated and the aqueous layer was extracted with CH_2Cl_2.   The combined organic extracts were dried over MgSO₄, and evaporated *in vacuo*.   The residual oil was purified by silica gel column chromatography (hexane:AcOEt=9:1) to give ethyl 2,2-difluoro-12-methyl-3-oxotridecanoate (1.01 g, 3.30 mmol) in 92% yield.

Spectral data

^{1}H NMR (CDCl₃): δ 0.86 (6 H, d, *J*=6.56 Hz), 1.07-1.74 (18 H, m), 1.35 (3 H, t, *J*=7.14 Hz), 2.73 (2 H, tt, *J*=7.14, 1.04 Hz), 4.37 (2 H, q, *J*=7.14 Hz); ^{13}C NMR (CDCl₃): δ 13.88, 22.55, 22.70, 27.49, 28.07, 28.91, 29.37, 29.55, 29.94, 36.69, 39.13, 63.73, 108.33 (t, *J*=262.15 Hz), 161.54 (t, *J*=30.05 Hz), 197.53 (t, *J*= 27.70 Hz); ^{19}F NMR (CDCl₃): δ -115.00 (s); IR (neat): ν 1780, 1745 cm^{-1}.

Caution

The Dess-Martin precursor was reported to be explosive under excessive heating (>200 ℃) or impact.   See Plumb, J. B.; Harper, D. J. *Chem. Eng. News* **1990**, July 16, 3.

Note

For the preparation of the starting hydroxy ester, see page 91 of this volume.

For the preparation of the Dess-Martin reagent, see Dess, D. B.; Martin, J. C. *J. Am. Chem. Soc.* **1991**, *113*, 7277, and Ireland, R. E.; Liu, L.-B. *J. Org. Chem.* **1993**, *58*, 2899 for the improved procedure.

The following compound possessing a similar partial structure was nicely oxidized by the method of Pfitzner and Moffat, while other methods met with difficulty.   See Gelb, M. H., *et al. J. Med. Chem.* **1987**, *30*, 1617.

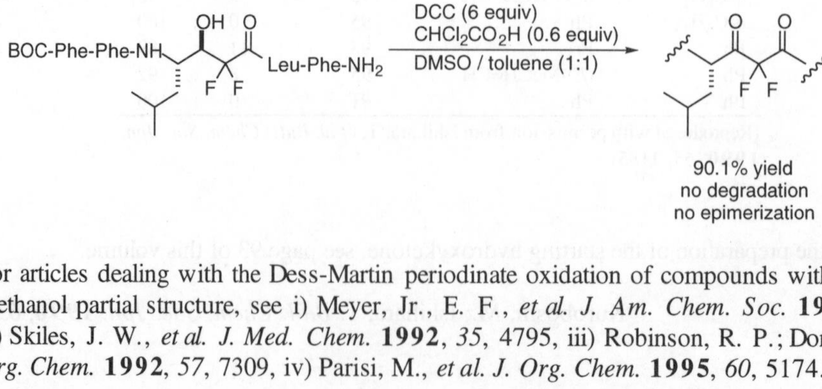

90.1% yield
no degradation
no epimerization

For articles dealing with the Dess-Martin periodinate oxidation of compounds with the 2,2-difluoroethanol partial structure, see i) Meyer, Jr., E. F., *et al. J. Am. Chem. Soc.* **1989**, *111*, 3368, ii) Skiles, J. W., *et al. J. Med. Chem.* **1992**, *35*, 4795, iii) Robinson, R. P.; Donahue, K. M. *J. Org. Chem.* **1992**, *57*, 7309, iv) Parisi, M., *et al. J. Org. Chem.* **1995**, *60*, 5174.

Fukuda, H.; Kitazume, T. *J. Fluorine Chem.* **1995**, *74*, 171.

syn-2,2-Difluoro-1-phenylhexane-1,3-diol

a) DIBAL-H, ZnCl₂·TMEDA / THF

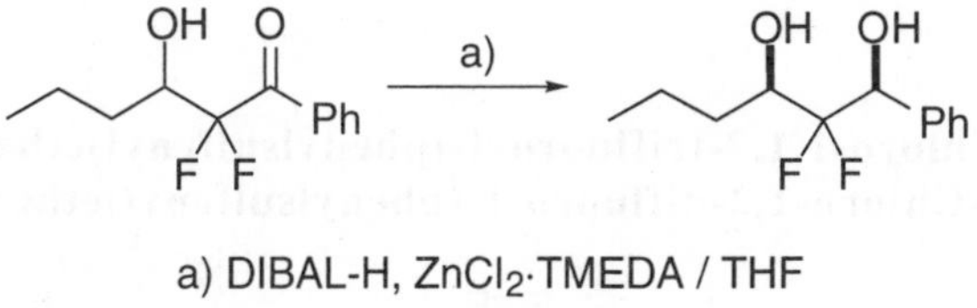

Synthetic method

To a mixture of 2,2-difluoro-3-hydroxy-1-phenylhexan-1-one (1.00 mmol), a $ZnCl_2 \cdot$ TMEDA complex (0.250 g, 1.00 mmol) and THF (5 mL) was added a solution of DIBAL-H (1 *M* in hexane, 3 mL, 3 mmol) at -78 °C under an argon atmosphere, and the resultant was stirred for 6 h at -78 °C.   The reaction mixture was poured into a mixture of 6 *N* HCl (5 mL) and sat. NH_4Cl aq. (5 mL), and the resultant was extracted with Et_2O (3x20 mL).   The combined extracts were dried over anhydrous Na_2SO_4 and concentrated.   After the *syn*- and *anti*-diol ratio was determined by ¹⁹F NMR, the residue was purified by silica gel column chromatography and the desired product, 2,2-difluoro-1-phenylhexane-1,3-diol, was obtained in 98% yield in a ratio of *syn:anti*= 91:9.

Spectral data

¹H NMR (CDCl₃): δ 0.90 (3 H, t, *J*=5.8 Hz), 1.2-2.0 (4 H, m), 2.2 (1 H, br d, *J*=7.2 Hz), 2.8-3.2 (1 H, br s), 3.2-4.1 (1 H, m), 5.00 (1 H, ddd, *J*=13.4, 13.4, 3.2 Hz), 7.28 (5 H, s); ¹⁹F NMR (CDCl₃): δ -120.00 (1 F, d, *J*=255.1 Hz), -129.23 (1 F, ddd, *J*=255.1, 13.4, 13.4 Hz): IR (neat): ν 3582, 3390, 2962, 2932, 1200, 1177, 1109, 1085, 1058, 1038, 1028 cm⁻¹.

Note

For the preparation of the starting hydroxyketone, see page 93 of this volume.

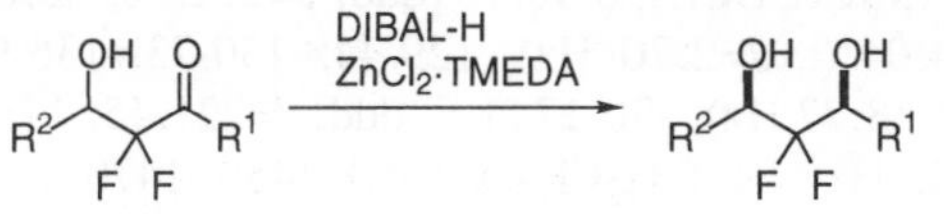

R¹	R²	Yield (%)	*syn:anti*		
n-C₆H₁₃	CH₃(CH₂)₂	93	92	:	8
n-C₆H₁₃	Me₂CH	99	88	:	12
n-C₆H₁₃	Me₃C	100	77	:	23
n-C₆H₁₃	(*E*)-MeCH=CH	91	94	:	6
n-C₆H₁₃	Ph	100	92	:	8
Ph	(*E*)-MeCH=CH	99	96	:	4
Ph	Ph	98	80	:	20

(Reproduced with permission from Ishihara, T., *et al. Bull. Chem. Soc. Jpn.* **1990**, *63*, 1185)

Kuroboshi, M.; Ishihara, T. *Bull. Chem. Soc. Jpn.* **1990**, *63*, 1185.

3.8 Miscellaneous Reactions

2-Chloro-1,1,2-trifluoro-1-(phenylsulfenyl)ethane, 2-Chloro-1,2-difluoro-1-(phenylsulfenyl)ethene

a) PhSNa / EtOH,   b) KOH, Aliquat 336®

Synthetic method

2-Chloro-1,1,2-trifluoro-1-(phenylsulfenyl)ethane

12.62 mL (158.67 mmol) of chlorotrifluoroethylene was rapidly added to a solution of 6.99 g (52.89 mmol) of sodium thiophenoxide in 97% EtOH (28 mL) in a 100-mL autoclave cooled to -78 ℃ *via* a dry ice-isopropanol bath.   The autoclave was sealed and warmed to room temperature for 15 h.   The autoclave was then cooled to -78 ℃ and opened.   The residue was concentrated under vacuum, poured into 90 mL of H_2O and extracted with CH_2Cl_2 (2x90 mL).   The combined organic layers were dried over $MgSO_4$, filtered and concentrated under vacuum.   The residue was distilled to give 9.10 g of 2-chloro-1,1,2-trifluoro-1-(phenylsulfenyl)ethane in 76% yield as a colorless liquid.

Spectral data

bp. 89-92 ℃/14 mmHg; ^{1}H NMR (CDCl$_3$): δ 6.09 (1 H, ddd, J=48.29, 7.74, 4.09 Hz), 7.37-7.67 (5 H, m); ^{13}C NMR (CDCl$_3$): δ 96.96 (ddd, J=252.80, 38.02, 33.77 Hz), 123.99 (dt, J=285.50, 27.3 Hz), 124.04 (t, J=2.70 Hz), 129.44, 130.73, 136.94; ^{19}F NMR (CDCl$_3$): δ -148.00 (1 F, dt, J=48.30, 18.77 Hz), -90.17 (1 F, ddd, J=221.45, 17.70, 7.80 Hz), -85.47 (1 F, ddd, J=221.40, 19.35, 4.03 Hz); IR (CH$_2$Cl$_2$): ν 1580, 1480, 1470 cm^{-1}.

Synthetic method

2-Chloro-1,2-difluoro-1-(phenylsulfenyl)ethene

4.56 g (20.12 mmol) of 2-chloro-1,1,2-trifluoro-1-(phenylsulfenyl)ethane was added to 1.35 g (24.13 mmol) of finely ground potassium hydroxide and 903 mg (1.99 mmol) of Aliquat 336®. After being vigorously stirred for 5 min, the mixture was heated for 8 h at 90-100 ℃.   After cooling, organic products were removed by filtration on Florisil after the addition of AcOEt (25 mL).   The crude mixture was purified by column chromatography on silica gel (petroleum ether) to give 3.12 g of 2-chloro-1,2-difluoro-1-(phenylsulfenyl)ethene in 75% yield as a colorless liquid in a 1.2:1 (E):(Z) ratio.

Spectral data

bp. 58 ℃/1 mmHg; Rf 0.25 (petroleum ether); ^{1}H NMR (CDCl$_3$): δ 7.20-7.50 (m); ^{13}C NMR (CDCl$_3$): δ 128.40, 128.44, 129.58, 130.16, 130.53, 130.83, 130.86, 139.93 (dd, J=295.60, 22.90 Hz), 141.09 (dd, J=289.00, 49.60 Hz), 142.52 (dd, J=291.80, 61.10 Hz), 143.36 (dd, J=314.00, 44.90 Hz); ^{19}F NMR (CDCl$_3$): δ -123.67 (1 F, d, J=140.20 Hz; (E)-isomer), -112.73 (1 F, d, J=17.30 Hz; (Z)-isomer), -104.17 (1 F, d, J=140.20 Hz; (E)-isomer), -86.94 (1 F, d, J=17.30 Hz; (Z)-isomer).

de Tollenaere, C.; Ghosez, L. *Tetrahedron* **1997**, *53*, 17127.

1,2-Difluoro-1-(phenylsulfenyl)ethene,
1,2-Difluoro-1-(phenylsulfonyl)ethene

a) t-BuLi / THF,   b) mCPBA / CH$_2$Cl$_2$

Synthetic method

1,2-Difluoro-1-(phenylsulfenyl)ethene

8.05 mL (12.08 mmol) of t-BuLi (1.5 M in pentane) was added dropwise to a solution of 1.92 g (9.29 mmol) of 2-chloro-1,2-difluoro-1-(phenylsulfenyl)ethene ((*E*):(*Z*)=1.3:1) in THF (22 mL) cooled to -105 °C *via* a liquid N$_2$-THF bath.   After being vigorously stirred at this temperature for 1 h, the mixture was cautiously treated with 7.50 mL (185.80 mmol) of MeOH and stirred again for 10 min.   8 mL of 3 M H$_2$SO$_4$ was added and the solution was allowed to slowly warm up to room temperature.   20 mL of H$_2$O was added and the organic material was extracted with CH$_2$Cl$_2$ (2x25 mL).   The combined organic layers were dried over MgSO$_4$, filtered and concentrated under vacuum.   The residue was purified by column chromatography on silica gel (petroleum ether) to give 1.29 g of 1,2-difluoro-1-(phenylsulfenyl)ethene in 81% yield as a colorless liquid in a 1:1.2 (*E*):(*Z*) ratio.

Spectral data

Rf 0.33 (petroleum ether); ^{1}H NMR (CDCl$_3$): δ 6.78 (1 H, dd, *J*=74.53, 13.80 Hz; (*E*)-isomer), 7.20-7.65 (5 H, m); ^{13}C NMR (CDCl$_3$): δ 127.84, 127.95, 129.10, 129.36, 130.21, 130.96, 131.63, 139.62 (dd, *J*=279.05, 20.00 Hz), 143.61 (dd, *J*=300.30, 14.10 Hz), 145.33 (dd, *J*=252.95, 71.30 Hz), 148.57 (dd, *J*=284.50, 39.70 Hz); ^{19}F NMR (CDCl$_3$): δ -157.44 (1 F, dd, *J*=145.70, 77.10 Hz; (*Z*)-isomer), -138.51 (1 F, d, *J*=145.50 Hz; (*Z*)-isomer), -133.70 (1 F, dd, *J*=74.70, 13.80 Hz; (*E*)-isomer), -114.83 (1 F, t, *J*=13.05 Hz; (*E*)-isomer); IR (CH$_2$Cl$_2$): ν 3050, 2960-2870, 1675, 1580 cm^{-1}.

Synthetic method

1,2-Difluoro-1-(phenylsulfonyl)ethene

1.78 g (7.20 mmol) of m-chloroperbenzoic acid (70%) was slowly added to a solution of 564 mg (3.27 mmol) of 1,2-difluoro-1-(phenylsulfenyl)ethene in CH$_2$Cl$_2$ (87 mL).   The mixture was refluxed for 29 h.   After cooling, it was washed with 5% NaHCO$_3$ aq. (2x80 mL).   The organic layers were dried over MgSO$_4$, filtered and concentrated under vacuum.   The residue was purified by column chromatography on silica gel (petroleum ether:CH$_2$Cl$_2$=1:1) to give 211 mg (31.5%) of (*E*)-1,2-difluoro-1-(phenylsulfenyl)ethene, 136 mg (20%) of a mixture of (*E*) and (*Z*), and 272 mg (41%) of (*Z*)-1,2-difluoro-1-(phenylsulfenyl)ethene as a colorless liquid.

Spectral data

(*E*)-isomer

Rf 0.39 (petroleum ether:CH$_2$Cl$_2$=1:1); ^{1}H NMR (CDCl$_3$): δ 7.51 (1 H, dd, *J*=69.00, 14.00 Hz), 7.57-7.99 (5 H, m); ^{13}C NMR (CDCl$_3$): δ 128.57, 129.62, 134.95, 136.86 (d, *J*=1.90 Hz), 142.15 (dd, *J*=284.55, 6.75 Hz), 147.42 (dd, *J*=296.80, 9.50 Hz); ^{19}F NMR (CDCl$_3$): δ -145.63 (1 F, dd, *J*=69.10, 4.60 Hz), -145.28 (1 F, dd, *J*=13.60, 4.00 Hz); IR (CH$_2$Cl$_2$): ν 3120-3060, 1695, 1590, 1460, 1350, 1160 cm^{-1}.

(*Z*)-isomer

Rf 0.27 (petroleum ether:CH$_2$Cl$_2$=1:1); ^{1}H NMR (CDCl$_3$): δ 7.32 (1 H, dd, *J*=71.05, 3.96 Hz), 7.56-8.04 (5 H, m); ^{13}C NMR (CDCl$_3$): δ 128.40, 129.57, 134.98, 137.85, 144.55 (dd,

J=272.70, 55.10 Hz), 149.93 (dd, J=281.35, 32.25 Hz); ^{19}F NMR (CDCl$_3$): δ -163.14 (1 F, dd, J=138.80, 4.20 Hz), -153.80 (1 F, dd, J=138.85, 70.40 Hz); IR (CH$_2$Cl$_2$): ν 3080-3060, 1670, 1590, 1450, 1350, 1160 cm^{-1}.

Note

For the preparation of the starting chlorodifluorinated olefin, see page 118 of this volume.

de Tollenaere, C.; Ghosez, L. *Tetrahedron* **1997**, *53*, 17127.

4 Preparation of Trifluorinated Materials

4.1 Introduction of a Trifluoromethyl Group

4.1.1 Introduction of a Trifluoromethyl Group (Nucleophilic)

α, α, α-Trifluorotoluene

$$CClF_2CO_2Me \xrightarrow{a)} PhCF_3$$

a) KF, CuI, PhI / DMF

Synthetic method

A mixture of dry KF (0.58 g, 10 mmol), copper(I) iodide (2.0 g, 10 mmol), PhI (2.0 g, 10 mmol) and DMF (30 mL) was placed in a 50-mL three-necked round-bottomed flask fitted with a stirrer bar, a dry ice condenser and a thermometer. The solution was heated to 120 °C under a nitrogen atmosphere for 0.5 h. Then methyl chlorodifluoroacetate (2.89 g, 10 mmol) was added over 3 h and the contents were stirred for a further 4 h. After reaction was completed, the reactants were poured into ice H_2O (200 mL). The solution was filtered and the residue was washed with Et_2O (3x20 mL) and the aqueous layer was extracted with Et_2O (3x20 mL). The combined extract was washed with H_2O (30 mL) and dried over Na_2SO_4. After evaporation of Et_2O, distillation gave 1.3 g of α,α,α-trifluorotoluene in 88% yield.

Spectral data

bp. 102-103 °C; ^{1}H NMR (CDCl_3): δ 6.57-6.88 (m); ^{19}F NMR (CDCl_3): δ -66 (s).

$$CClF_2CO_2Me \xrightarrow{KF, CuI, RX / DMF} RCF_3$$

RX	Temp (°C)	Yield (%)	RX	Temp (°C)	Yield (%)
PhI[a)]	100	80	PhBr	110-120	60
Allyl-Br	110-120	82	Allyl-I	110-120	85
(*E*)-PhCH=CHBr	110-120	81	BnBr	110-120	84
BnCl	110-120	46	1-Naph-I	110-120	94
p-Cl-C_6H_4-I[b)]	100	81	*p*-O_2N-C_6H_4-I	100	89
CH_2=CHBr[c)]	100	56	ClCH_2CO_2Et	110-120	5

[a)] In HMPA. [b)] No *p*-F_3C-C_6H_4-CF_3 was detected. [c)] Reaction was carried out for 8 h.
(Reproduced with permission from Chen, Q.-Y., *et al. J. Fluorine Chem.* **1993**, *61*, 279)

Chen, Q.-Y., *et al. J. Fluorine Chem.* **1993**, *61*, 279.

1-Phenyl-2,2,2-trifluoroethanol

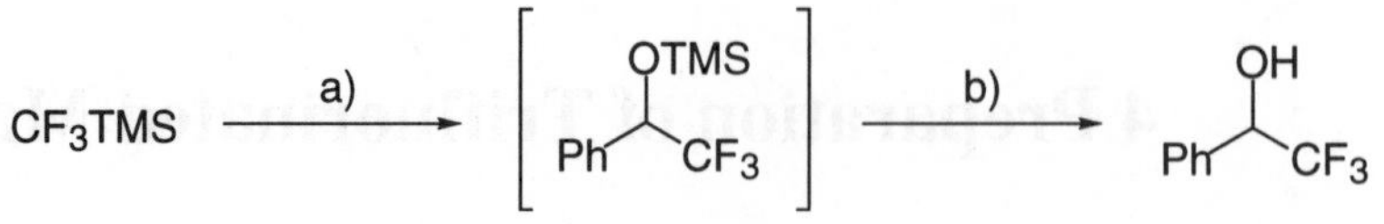

a) PhCHO, cat. TBAF / THF, b) HCl aq.

Synthetic method

A mixture of benzaldehyde (10 mmol) and (trifluoromethyl)trimethylsilane (12 mmol) in 10 mL of THF cooled to 0 °C was treated with a catalytic amount (*ca.* 20 mg) of tetra-*n*-butyl-ammonium fluoride. Instantaneously, a yellow color developed with the initial evolution of fluorotrimethylsilane, and the reaction mixture was brought to ambient temperature and stirred. The mixture was periodically analyzed by GC for the completion of the reaction. The resulting silylated compounds were then hydrolyzed with HCl aq. After the reaction, the mixture was extracted with Et_2O (75 mL), and concentrated. The residue was distilled to give 1-phenyl-2,2,2-trifluoroethanol in 85% yield as a colorless oil.

Spectral data

bp. 64-65 °C/5.0 mmHg; ^1H NMR ($CDCl_3$): δ 2.99 (1 H, br s), 4.89 (1 H, q, *J*=6.8 Hz), 7.34 (5 H, m); ^{13}C NMR ($CDCl_3$): δ 72.8 (q, *J*=31.9 Hz), 124.3 (q, *J*=282.0 Hz), 127.4, 128.6, 129.5, 134.0; ^{19}F NMR ($CDCl_3$): δ -79.2 (d, *J*=7.1 Hz); IR (film): ν 3405, 1267, 1172 cm^{-1}.

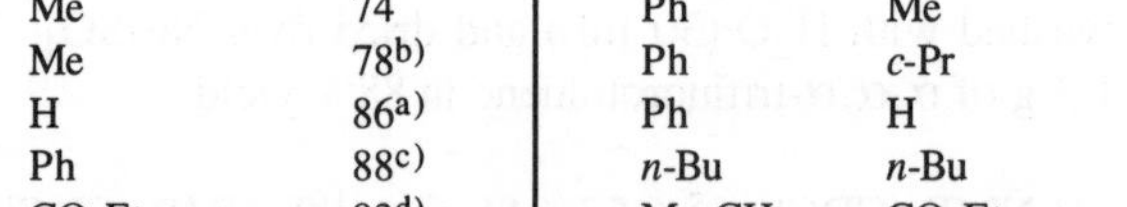

R^1	R^2	Yield (%)	R^1	R^2	Yield (%)
-(CH$_2$)$_5$-		77	-(CH$_2$)$_5$-		82[a]
-(CH$_2$)$_5$-		81[b]	c-C$_6$H$_{11}$	H	80
Ph	Me	74	Ph	Me	81[a]
Ph	Me	78[b]	Ph	c-Pr	81
Ph	H	86[a]	Ph	H	66[b]
Ph	Ph	88[c]	n-Bu	n-Bu	87
Me	CO$_2$Et	83[d]	Me$_2$CH	CO$_2$Et	76[d]
Ph	CO$_2$Et	70[d]			

[a] The corresponding (pentafluoroethyl)trimethylsilane was employed. [b] The corresponding (heptafluoropropyl)trimethylsilane was employed. [c] Yield of the corresponding trimethylsilyl ether. [d] See Ramaiah, P.; Surya Prakash, G. K. *Synlett* **1991**, 643.

Note

For the preparation of (trifluoromethyl)trimethylsilane, see page 206 of this volume.

The present versatile reagent (trifluoromethy)trimethylsilane was widely employed for the introduction of a CF_3 moiety, and for the reaction with cyclohex-2-en-1-one, 1,2-addition was found to predominate (>90%).

See the following review of the perfluoroalkylation using R_fSiR_3 compounds: Surya Prakash, G. K.; Yudin, A. K. *Chem. Rev.* **1997**, *97*, 757.

Surya Prakash, G. K., *et al. J. Org. Chem.* **1991**, *56*, 984.

1,2:5,6-Di-*O*-isopropylidene-3-*C*-trifluoromethyl-α-D-allofuranose, 3-Deoxy-1,2:5,6-di-*O*-isopropylidene-3-*C*-trifluoromethyl-α-D-allofuranose

a) CF₃TMS, cat. TBAF / THF, b) TBAF / MeOH, c) pyridine (3 equiv), ClC(O)CO₂Me (2 equiv) / CH₂Cl₂, d) *n*-Bu₃SnH (2 equiv), AIBN (0.5 equiv) / toluene

Synthetic method

1,2:5,6-Di-*O*-isopropylidene-3-*C*-trifluoromethyl-α-D-allofuranose

Before use, 1,2:5,6-di-*O*-isopropylidene-3-*C*-trifluoromethyl-α-D-*ribo*-hexofuranos-3-ulose was dehydrated by azeotropic distillation from a solution in toluene.   To a stirred solution of non-hydrated 1,2:5,6-di-*O*-isopropylidene-3-*C*-trifluoromethyl-α-D-*ribo*-hexofuranos-3-ulose (1 mmol) in anhydrous THF (5 mL) was added (trifluoromethyl)trimethylsilane (1.1 mmol) and a catalytic amount of tetra-*n*-butylammonium fluoride (20 mg for 10 mmol) at room temperature.   When the reaction was completed, the mixture was washed with a sat. NH₄Cl solution.   The aqueous layer was extracted with Et₂O (2x10 mL) and the organic layer was dried with MgSO₄, filtered and concentrated *in vacuo*.

To this crude solution dissolved in MeOH was added 1 equiv of TBAF at room temperature for desilylation.   The reaction was completed within 1 h, the solvent was evaporated and the residue was purified by flash chromatography (petroleum ether:AcOEt=80:20) to give 1,2:5,6-di-*O*-isopropylidene-3-*C*-trifluoromethyl-α-D-allofuranose in 88% yield as a white solid.

Spectral data

$[\alpha]_D^{20}$ +23.0 (*c* 1.8, CHCl₃); mp. 76 °C; ^{1}H NMR (CDCl₃): δ 1.36 (3 H, s), 1.39 (3 H, s), 1.45 (3 H, s), 1.61 (3 H, s), 3.31 (1 H, s), 3.94 (1 H, dd, *J*=8.8, 5.3 Hz), 4.00 (1 H, d, *J*=2.3 Hz), 4.11 (1 H, dd, *J*=8.8, 6.1 Hz), 4.34 (1 H, m), 4.60 (1 H, d, *J*=3.8 Hz), 5.83 (1 H, d, *J*=3.8 Hz); ^{13}C NMR (CDCl₃): δ 26.1 (2C), 26.8 (2C), 67.2, 72.8, 79.1, 80.1 (q, *J*=28.0 Hz), 81.4, 104.0, 109.8, 113.8, 124.2 (q, *J*=285.0 Hz); ^{19}F NMR (CDCl₃): δ -76.25 (s); IR (KBr): ν 3374, 2986, 1215, 1153, 1014, 877 cm⁻¹.

Synthetic method

3-Deoxy-1,2:5,6-di-*O*-isopropylidene-3-*C*-trifluoromethyl-α-D-allofuranose

To a solution of 1,2:5,6-di-*O*-isopropylidene-3-*C*-trifluoromethyl-α-D-allofuranose (1 g, 3 mmol) in CH₂Cl₂ (15 mL) and pyridine (0.73 mL, 9 mmol) was added dropwise, under argon, methyl oxalyl chloride (0.55 mL, 6 mmol).   The reaction mixture was stirred for 1 h at room temperature.   Then, the reaction was poured into a sat. NaHCO₃ solution.   The aqueous layer was extracted with CH₂Cl₂ and the combined organic layers were dried over Na₂SO₄.   After filtration, the solvent was removed under reduced pressure and the residue was dried under high vacuum to give the crude acylated material.   Without further purification, this compound was dissolved in anhydrous toluene (7 mL) and a solution of AIBN (246 mg, 1.5 mmol) and *n*-Bu₃SnH (1.61 mL, 6 mmol) in anhydrous toluene was added at room temperature.   The mixture was heated to 100 °C under an argon atmosphere.   When the reaction was complete (1 h), the solvent was removed under reduced pressure.   The residue was purified by flash chromatography (petroleum ether:AcOEt=95:5) to give 0.69 g of 3-deoxy-1,2:5,6-di-*O*-isopropylidene-3-*C*-trifluoromethyl-α-D-allofuranose in 73% yield containing a trace of the corresponding D-glucofuranose.

Spectral data

^{1}H NMR (CDCl$_3$): δ 1.35 (3 H, s), 1.43 (3 H, s), 1.56 (3 H, s), 1.72 (3 H, s), 2.68 (1 H, dqd, *J*=9.3, 7.6, 5.0 Hz), 3.91 (1 H, dd, *J*=8.4, 5.8 Hz), 4.07 (1 H, dd, *J*=8.4, 6.8 Hz), 4.21 (1 H, m), 4.37 (1 H, dd, *J*=9.3, 5.0 Hz), 4.85 (1 H, dd, *J*=5.0, 3.7 Hz), 5.83 (1 H, d, *J*=3.7 Hz); ^{13}C NMR (CDCl$_3$): δ 25.1 (2C), 26.2, 26.6, 50.3 (q, *J*=27.5 Hz), 65.7, 76.1, 77.0, 79.8, 104.5, 110.0, 113.6, 124.2 (q, *J*=278.0 Hz); ^{19}F NMR (CDCl$_3$): δ -61.02 (d, *J*=7.6 Hz), -65.05 (d, *J*=10.0 Hz for the minor isomer, D-glucofuranose); IR (film): ν 2990, 2941, 1386, 1024 cm^{-1}.

Note

For the preparation of (trifluoromethyl)trimethylsilane, see page 206 of this volume.

Diastereoselectivity of (trifluoromethyl)trimethylsilane was highly dependent on the structure of the carbonyl molecules, and stereorandom introduction of a CF$_3$ group usually occurred. On the other hand, for the compounds with constrained structure, a single isomer was sometimes obtained. For recent representative examples, see i) Wang, Z.-Q., *et al. Bioorg. Med. Chem. Lett.* **1995**, *5*, 1899, ii) Anker, D., *et al. J. Carbohydr. Chem.* **1996**, *15*, 739, iii) Terashima, S., *et al. Bioorg. Med. Chem. Lett.* **1996**, *6*, 1927.

Enantioselective introduction of a CF$_3$ group by (trifluoromethyl)trimethylsilane was also reported by use of cinchonine derivatives and up to 51% ee was realized with 2-methyl-1-phenyl-propan-1-one. See Iseki, K., *et al. Tetrahedron Lett.* **1994**, *35*, 3137.

Portella, C., *et al. J. Carbohydr. Chem.* **1996**, *15*, 361.

1-Phenyl-2,2,2-trifluoroethanol

$$CF_3Br \xrightarrow{\text{a)}} \underset{Ph}{\overset{OH}{\diagup}} CF_3$$

a) PhCHO, Zn / pyridine

Synthetic method

A thick glass flask containing benzaldehyde (10 mL, 98.5 mmol) and commercial zinc powder (6.5 g, 100 mmol) in pyridine (25 mL) was placed in a Parr apparatus. The air was evacuated and then the flask was shaken for 1 or 2 h under a pressure of trifluoromethyl bromide maintained between 2.5-4 atoms. The solution was filtered and hydrolyzed with ice-cold 20% HCl. This solution was extracted with Et$_2$O and the unreacted benzaldehyde was removed using a 38% NaHSO$_3$ solution. Distillation under vacuum gave 1-phenyl-2,2,2-trifluoroethanol in a yield of 52%.

Spectral data

bp. 75 °C/12 mmHg; ^{1}H NMR (CDCl$_3$): δ 3.5 (1 H, br s), 5 (1 H, q), 7.5 (5 H, m); ^{19}F NMR (CDCl$_3$): δ -77.8 (d, *J*=7.1 Hz).

Note

See the following articles on similar types of reactions: i) Ishikawa, N., *et al. Chem. Lett.* **1984**, 517, ii) Kitazume, T.; Ishikawa, N. *J. Am. Chem. Soc.* **1985**, *109*, 5186, iii) Utimoto, K., *et al. Tetrahedron Lett.* **1987**, *28*, 5857, iv) Fuchikami, T., *et al. Chem. Lett.* **1987**, *521*.

$$R^1 \overset{O}{\underset{}{\overset{\|}{C}}} R^2 \xrightarrow[\text{20 °C}]{\text{Zn, CF}_3\text{Br / pyridine}} R^1 \overset{OH}{\underset{F_3C}{\overset{|}{C}}} R^2$$

R^1	R^2	Yield (%)	R^1	R^2	Yield (%)
4-F-C$_6$H$_5$-	H	60	9-Anthryl	H	37[a]
2-Pyridyl	H	65[a]	n-Pr	H[b]	30
PhCH=CH-	H	55	Ph	Me[c]	20
-(CH$_2$)$_5$-		20[c]	Me	CO$_2$Me[b]	35

[a] Yield after sublimation.   [b] At 0-10 °C.   [c] Yield after preparative GLC.
(Reproduced with permission from Wakselman, C., *et al. J. Chem. Soc., Chem. Commun.*
1987, 642)

Wakselman, C., *et al. J. Chem. Soc., Chem. Commun.* **1987**, 642.

5-(Trifluoromethyl)-4-oxo-4*H*-1,3-dioxine-2-spirocyclohexane, Benzyl 2-(trifluoromethyl)-3-oxopropionate (enol form)

a) CF$_3$I, Cu / HMPA,   b) BnOH / toluene

Synthetic method
    5-(Trifluoromethyl)-4-oxo-4*H*-1,3-dioxine-2-spirocyclohexane
    A suspension of trifluoromethyl iodide (8.82 g, 45 mmol), copper powder (4.57 g, 72 mmol), and HMPA (18 mL) was stirred at 120-125 °C in a sealed tube for 3 h under an argon atmosphere. To the mixture was added 5-iodo-4-oxo-4*H*-1,3-dioxine-2-spirocyclohexane (4.43 g, 15 mmol) and the resultant solution was stirred at 55 °C for 1 h.   The reaction mixture was then cooled and poured into ice H$_2$O, and the insoluble material was removed by filtration through Celite.   After extraction of the filtrate with Et$_2$O, the organic layer was washed with H$_2$O and dried over MgSO$_4$. The residue obtained after evaporation of the solvent was chromatographed on silica gel (hexane: CH$_2$Cl$_2$=1:1) to give 3.14 g of 5-(trifluoromethyl)-4-oxo-4*H*-1,3-dioxine-2-spirocyclohexane as needles in 89% yield.
Spectral data
    mp. 77.5-78.0 °C (hexane:Et$_2$O); ^{1}H NMR (CDCl$_3$): δ 1.3-2.3 (10 H, m), 7.7 (1 H, br s); IR (CHCl$_3$): ν 2960, 1760, 1625, 1400, 1150 cm^{-1}.

Synthetic method
    Benzyl 2-(trifluoromethyl)-3-oxopropionate (enol form)
    A solution of 5-(trifluoromethyl)-4-oxo-4*H*-1,3-dioxine-2-spirocyclohexane (47.2 mg, 0.2 mmol) and benzyl alcohol (22.7 mg, 0.21 mmol) in toluene (0.5 mL) was refluxed for 30 min. The solvent and cyclohexanone were evaporated off under reduced pressure to give 42 mg of benzyl 2-(trifluoromethyl)-3-oxopropionate in 85% yield as an oil.

Spectral data
^{1}H NMR (CDCl$_3$): δ 5.37 (2 H, s), 7.40 (5 H, s), 7.76 (1 H, br d, J=13 Hz), 12.35 (1 H, d, J=13 Hz); IR (CHCl$_3$): ν 1680, 1420, 1190, 1140 cm^{-1}.

Note
The trimethylaluminum-mediated addition reaction of CF$_3$I to olefins has been reported.   See Yamamoto, H., *et al. Chem. Lett.* **1985**, 1689.

The above intermediary dioxin with a CF$_3$ group was also employed for the Diels-Alder reaction.   See Iwaoka, T., *et al. Chem. Pharm. Bull.* **1992**, *40*, 2319.

Kaneko, C., *et al. J. Chem. Soc., Perkin Trans. 1* **1992**, 1393.

N-(2,2,2-Trifluoro-1,1-diphenylethyl)acetamide

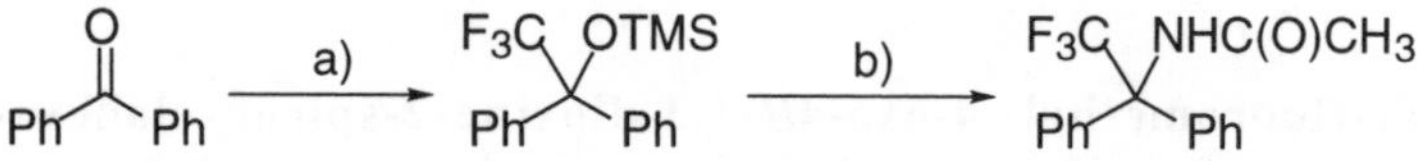

a) CF$_3$TMS, cat. TBAF / THF,   b) MeCN / H$_2$SO$_4$, AcOH

Synthetic method
To a solution of diphenylketone (1.25 mmol) in THF (2.5 mL) were added (trifluoromethyl)-trimethylsilane (0.25 mL, 1.5 mmol) and a catalytic amount of tetra-*n*-butylammonium fluoride (62.5 μL, 0.0625 mmol).   Conversion to 2,2,2-trifluoro-1,1-diphenyl-1-[(trimethylsilyl)oxy]-ethane was complete within several hours.   The solvent was then evaporated under vacuum, replaced with MeCN (10-20 mL), followed by the addition of excess H$_2$SO$_4$ and AcOH (0.25 mL each), and the reaction mixture set to reflux.   After 5 h the reaction mixture was carefully neutralized with NaHCO$_3$ aq., and extracted several times with CH$_2$Cl$_2$.   Evaporation of the solvent gave the crude amide, which was further purified through silica gel column chromatography (CH$_2$Cl$_2$:MeOH= 200:1).   *N*-(2,2,2-Trifluoro-1,1-diphenylethyl)acetamide was obtained in 68% yield.

Spectral data
mp. 187-188 °C; ^{1}H NMR (CDCl$_3$): δ2.05 (3 H, s), 6.37 (1 H, s), 7.36 (10 H, br s); ^{13}C NMR (CDCl$_3$): δ 24.2, 68.7 (q, J=27 Hz), 125.5 (q, J=288 Hz), 128.0, 128.1, 128.2, 128.5, 136.9, 168.4; ^{19}F NMR (CDCl$_3$): δ -67.9 (s).

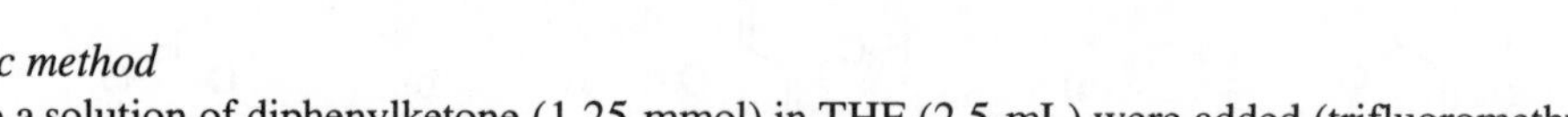

R^1	R^2	Reaction time (h)[a]	H$_2$SO$_4$:AcOH (mL)	Yield (%)
p-MeO-C$_6$H$_4$-	Ph	5	0.25:0.25	66
p-Me-C$_6$H$_4$-	Ph	5	0.25:0.25	81
p-F-C$_6$H$_4$-	Ph	5	0.25:0.25	57
Ph	Me	3	0.5:0.0	54
Ph	CO$_2$Et	48	1.5:1.5	59
n-Bu	*n*-Bu	24	1.0:1.0	32
-(CH$_2$)$_3$-CH=CH-		3	0.5:0.0	54
Bn	Et	15	0.25:0.25	49

[a] Reaction time at the second step.
(Reproduced with permission from Surya Prakash, G. K., *et al. Synlett* **1997**, 1193)

Note
    For the preparation of (trifluoromethyl)trimethylsilane, see page 206 of this volume.

Surya Prakash, G. K., *et al. Synlett* **1997**, 1193.

4.1.2 Introduction of a Trifluoromethyl Group (Electrophilic, Radical)

1,2,6,7,8,8a-Hexahydro-5-(trifluoromethyl)-3-methoxy-8a-methylnaphthalene,
4,4a,5,6,7,8-Hexahydro-8-(trifluoromethyl)-4a-methyl-3H-naphthalen-2-one

a) CF$_3$I, hv / pyridine, b) 80% CF$_3$CO$_2$H aq.

Synthetic method
    1,2,6,7,8,8a-Hexahydro-5-(trifluoromethyl)-3-methoxy-8a-methylnaphthalene
    Trifluoromethyl iodide (10 mL) was added to a pyridine solution (10 mL) of 2-methoxy-3,4,4a,5,6,7-hexahydro-4a-methylnaphthalene (6.1 g, 34.2 mmol) in a 300-mL quartz tube by using a vacuum line and the vessel was sealed with a grease-free stopcock. The resulting solution was irradiated with a 100W high-pressure Hg lamp for 33 h at room temperature. The cooled reaction mixture was diluted with Et$_2$O and washed with sat. NaHSO$_3$ aq. The separated organic phase was washed with H$_2$O, 1 N HCl, sat. NaHCO$_3$ and brine successively before drying (MgSO$_4$). The filtrate was concentrated to give a crude oil. Purification by column chromatography on silica gel (hexane:AcOEt=20:1) gave 6.6 g of 1,2,6,7,8,8a-hexahydro-5-(trifluoromethyl)-3-methoxy-8a-methylnaphthalene in 78% yield.
Spectral data
    ^{1}H NMR (CDCl$_3$): δ 1.06 (3 H, s), 1.2-1.8 (6 H, m), 2.1-2.5 (4 H, m), 3.64 (3 H, s), 5.63 (1 H, br s); ^{19}F NMR (CDCl$_3$): δ 5.83 (s) from C$_6$H$_5$CF$_3$.

Synthetic method
    4,4a,5,6,7,8-Hexahydro-8-(trifluoromethyl)-4a-methyl-3H-naphthalen-2-one
    A solution of 1,2,6,7,8,8a-hexahydro-5-(trifluoromethyl)-3-methoxy-8a-methylnaphthalene (6.5 g, 27 mmol) was treated with 80% trifluoroacetic acid (30 mL) at 0 °C and the mixture was stirred at room temperature for 5 h. The mixture was poured into ice H$_2$O and extracted with Et$_2$O. The organic phase was washed with H$_2$O, sat. NaHCO$_3$ and brine successively before being dried (MgSO$_4$). The filtrate was concentrated to give a crude oil, which was purified by flash column chromatography on silica gel (hexane:AcOEt=4:1) and subsequent distillation gave 3.75 g of pure 4,4a,5,6,7,8-hexahydro-8-(trifluoromethyl)-4a-methyl-3H-naphthalen-2-one in 61% yield.
Spectral data
    bp. 112 °C/2 mmHg; ^{1}H NMR (CDCl$_3$): δ 1.25 (3 H, s), 1.45-1.95 (6 H, m), 2.1-2.5 (4 H, m), 6.0 (1 H, s); ^{19}F NMR (CDCl$_3$): δ -3.83 (d, J=9 Hz) from C$_6$H$_5$CF$_3$; IR (neat): ν 1680 cm^{-1}.

Taguchi, T., *et al. Chem. Pharm. Bull.* **1991**, *39*, 1035.

1-Chloro-3,3,3-trifluoro-1-phenylpropane, (E)-3,3,3-Trifluoro-1-phenylpropene

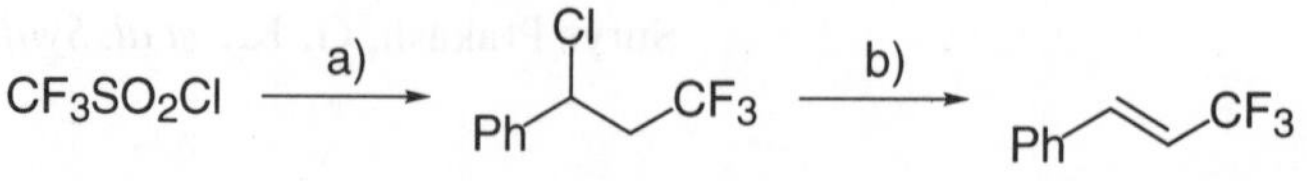

a) PhCH=CH$_2$, RuCl$_2$(PPh$_3$)$_3$ / benzene, b) KOH aq. / MeOH

Synthetic method

   1-Chloro-3,3,3-trifluoro-1-phenylpropane

   A solution containing trifluoromethanesulfonyl chloride (2.0 mmol), styrene (4-10 mmol), and dichlorotris(triphenylphosphine)ruthenium(II) (0.02 mmol) in dry benzene (4 mL) was degassed by a freeze-pump-thaw cycle, sealed in an ampoule, and heated at 120 °C for 16 h.   The reaction mixture was subjected to short column chromatography on Florisil (benzene) to remove the metal complex.   The products were isolated from the reaction mixture by use of gel permeation chromatography and/or column chromatography over silica gel in 87% yield.

Spectral data

   ^{1}H NMR (CDCl$_3$): δ 2.50-3.28 (2 H, m), 5.40 (1 H, t, *J*=7.2 Hz), 7.26 (5 H, s); IR (neat): ν 3030, 1380, 1270, 1140 cm^{-1}.

Synthetic method

   (E)-3,3,3-Trifluoro-1-phenylpropene

   A mixture containing 1-chloro-3,3,3-trifluoro-1-phenylpropane (133 mg, 0.64 mmol), potassium hydroxide (101 mg, 1.08 mmol) and MeOH (6 mL) was stirred at 50 °C for 3 h.   After removal of MeOH under reduced pressure, the organic residue was extracted with 1,1-dichloro-ethane, and the extracts were dried over anhydrous MgSO$_4$.   (E)-3,3,3-Trifluoro-1-phenylpropene was obtained in pure form after removal of the solvent under reduced pressure (68 mg, 62% yield).

Spectral data

   ^{1}H NMR (CDCl$_3$): δ 5.80-6.39 (1 H, m), 6.87-7.30 (1 H, m), 7.30 (5 H, s); IR (neat): ν 1670, 1580, 1500, 1450, 1340 cm^{-1}.

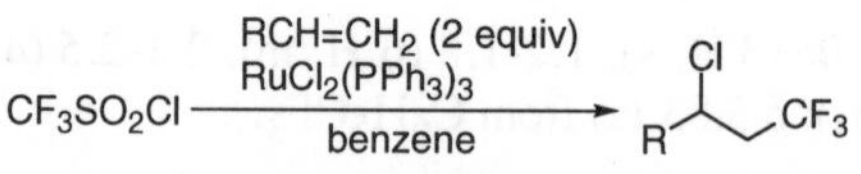

R	Yield (%)	R	Yield (%)
4-Me-C$_6$H$_4$-[a,b]	46	4-Cl-C$_6$H$_4$-	74
3-O$_2$N-C$_6$H$_4$-	79	n-C$_6$H$_{13}$-	66
n-C$_8$H$_{17}$-	72	n-C$_{10}$H$_{21}$-	70
EtO$_2$C	41	EtO$_2$C[a]	70
PhCO$_2$	49		

[a] 5 equiv of alkene were employed.   [b] 16% of olefin was also obtained by further dehydrochlorination.
(Reproduced with permission from Kamigata, N., *et al*. J. Chem. Soc.
Perkin Trans. 1 **1991**, 627)

Note

   The present reaction was carried out with various types of olefins as depicted in the above table, but only one more example of the next dehydrochlorination step was described.

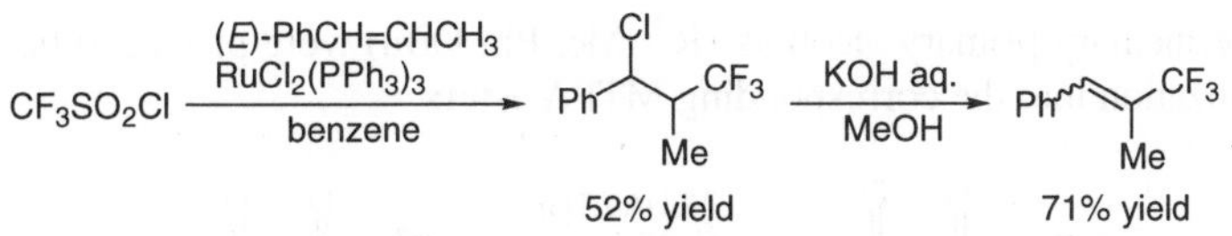

52% yield   71% yield

Kamigata, N., *et al. J. Chem. Soc. Perkin Trans. 1* **1991**, 627.

(*S*)-3-[(*S*)-2-(Trifluoromethyl)propionyl]-4-(prop-2-yl)oxazolidin-2-one

a) LDA / THF, b) CF$_3$I, c) Et$_3$B

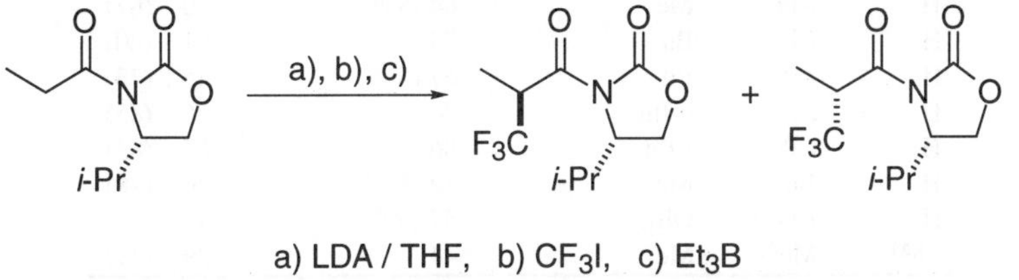

Synthetic method

To a solution of LDA (2.6 mmol) in THF (3 mL) was added a solution of (*S*)-3-propionyl-4-(prop-2-yl)oxazolidin-2-one (370 mg, 2.0 mmol) in THF (3 mL) at -78 °C.   After 1 h of stirring at the same temperature, gaseous trifluoroiodomethane (0.8 mL at -42 °C, 9.9 mmol) was added with a cannula followed by triethylborane (1 *M* in hexane, 2.0 mL, 2.0 mmol) over 1 min.   After stirring at -78 °C for 10 min and at -20 °C for 2 h, the reaction mixture was quenched with sat. NH$_4$Cl and extracted with Et$_2$O.   The combined ethereal extracts were washed with sat. NaHCO$_3$ aq. and brine, dried and filtered.   After evaporation of the solvent, chromatography of the residue (hexane:CH$_2$Cl$_2$=2:1) gave the less polar (*R*)-isomer (55 mg, 11%), more polar (*S*)-isomer (299 mg, 59%) and recovered the starting material (72 mg, 19%).

Spectral data

[α]$_D^{22}$ +63.7 (*c* 5.49, CHCl$_3$); ^{1}H NMR (CDCl$_3$): δ 0.88 (3 H, d, *J*=6.9 Hz), 0.93 (3 H, d, *J*=7.1 Hz), 1.42 (3 H, d, *J*=7.0 Hz), 2.39 (1 H, qqd, *J*=7.1, 6.9, 4.0 Hz), 4.25 (1 H, dd, *J*=9.1, 4.0 Hz), 4.32 (1 H, dd, *J*=9.1, 7.8 Hz), 4.50 (1 H, ddd, *J*=7.8, 4.0, 4.0 Hz), 4.90 (1 H, qq, *J*=8.0, 7.0 Hz); ^{19}F NMR (CDCl$_3$): δ -69.58 (d, *J*=8.0 Hz); IR (KBr): ν 2968, 1782, 1708, 1244, 1174 cm^{-1}.

(*S*)-3-[(*R*)-2-(Trifluoromethyl)propionyl]-4-(prop-2-yl)oxazolidin-2-one

[α]$_D^{23}$ +86.5 (*c* 4.51, CHCl$_3$); mp. 48.6-50.0 °C (hexane); ^{1}H NMR (CDCl$_3$): δ 0.89 (3 H, d, *J*=6.9 Hz), 0.96 (3 H, d, *J*=7.0 Hz), 1.23 (3 H, d, *J*=7.1 Hz), 2.37 (1 H, qqd, *J*=7.0, 6.9, 3.9 Hz), 4.25 (1 H, dd, *J*=9.1, 3.3 Hz), 4.33 (1 H, dd, *J*=9.1, 7.8 Hz), 4.51 (1 H, ddd, *J*=7.8, 3.9, 3.3 Hz), 4.73 (1 H, qq, *J*=7.6, 7.1 Hz); ^{19}F NMR (CDCl$_3$): δ -69.67 (d, *J*=7.6 Hz); IR (KBr): ν 2974, 1789, 1699, 1248, 1174 cm^{-1}.

Note

In spite of the inherent instability of 2-trifluoromethylated carbonyl compounds, products obtained were successfully transformed into the corresponding alcohols, carboxylic acids, and

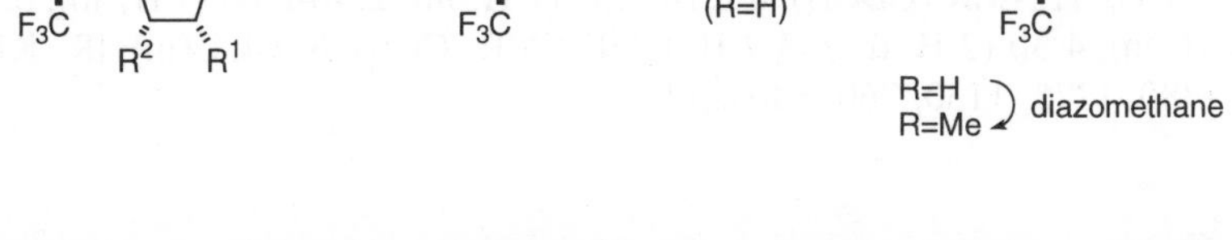

R=H   ⎫ diazomethane
R=Me ⎭

esters.   The intermediary primary alcohols (R^3=Me, Ph, OBn) were proved to be enantiomerically pure after derivatization into the corresponding MTPA esters.

$$R^3 \text{—C(O)—N} \begin{array}{c} \text{(oxazolidinone)} \\ R^2 \quad R^1 \end{array} \xrightarrow[\text{iii) Et}_3\text{B}]{\substack{\text{i) LDA / THF} \\ \text{ii) CF}_3\text{I}}} R^3\text{—CH(CF}_3)\text{—C(O)—N}\begin{array}{c} \text{(oxazolidinone)} \\ R^2 \quad R^1 \end{array}$$

R^1	R^2	R^3	Diastereomeric excess (% de)	Yield (%)[a]	
H	i-Pr	Me	64 (S)[b]	70	(87)
H	i-Pr	Bn	74	61	(80)
H	i-Pr	Ph	67 (S)[b]	45	(73)
H	i-Pr	n-Bu	76	75	(85)
H	i-Pr	t-Bu	86	67	(83)
H	Bn	Me	72 (S)[b]	70	(78)
H	i-Pr[c]	OBn	57 (R)[b]	45	
Ph[a]	Me[c]	Me	44	29	(39)

[a] The conversion yield is shown in parentheses.   [b] Absolute stereochemistry at the newly created carbon atom.   [c] Opposite stereochemistry at the indicated site.
(Reproduced with permission from Iseki, K., et al. Tetrahedron: Asym. **1994**, 5, 961)

Iseki, K., et al. Tetrahedron: Asym. **1994**, 5, 961.

(E)-3-(2,2,2-Trifluoroethyl)-2-(1-iodoethylidene)butan-4-olide

$$\text{Me—C≡C—C(O)—O—CH}_2\text{CH=CH}_2 \xrightarrow{a)} \text{(iodoethylidene butanolide with CF}_3)$$

a) CF$_3$I, Na$_2$S$_2$O$_4$-NaHCO$_3$ / MeCN-H$_2$O

Synthetic method
     To a stirred solution of trifluoromethyl iodide (1.0 mmol) and allyl but-2-ynoate (1.0 mmol) in a mixed solvent of MeCN (3 mL) and H$_2$O (2 mL), was added sodium bicarbonate (100 mg, 1.2 mmol) and sodium dithionite (208 mg, 1.2 mmol).   Stirring was continued at 10-15 °C until TLC monitoring showed complete conversion of allyl but-2-ynoate.   The combined organic layer was washed with brine, and the resulting mixture was diluted with H$_2$O (4 mL) and extracted with Et$_2$O (3x5 mL).   The combined extracts were successively washed with brine (2x5 mL), dried over anhydrous MgSO$_4$ and concentrated under reduced pressure.   The residue was purified by flash chromatography (petroleum ether:AcOEt=9:1) on silica gel to give 3-(2,2,2-trifluoroethyl)-2-(1-iodoethylidene)butan-4-olide in 80% yield as a 95:5 (E):(Z) mixture.
Spectral data
     mp. 45-46 °C; ^{1}H NMR (CDCl$_3$): δ 2.18-2.38 (1 H, m), 2.40-2.60 (1 H, m), 3.10 (3 H, s), 3.28-3.40 (1 H, m), 4.30 (2 H, d, J=5.7 Hz); ^{19}F NMR (CCl$_4$): δ -64.1 (m); IR (KBr): ν 2900, 1750, 1640, 1380, 1275, 1130, 760, 640 cm^{-1}.

(Z)-3-(2,2,2-Trifluoroethyl)-2-(1-iodoethylidene)butan-4-olide

mp. 98-99 °C; ^{1}H NMR (CDCl$_3$): δ 2.20-2.50 (2 H, m), 2.84 (3 H, s), 3.56-3.68 (1 H, m), 4.18 (1 H, d, J=10.3 Hz), 4.27 (1 H, dd, J=10.3, 3.7 Hz); ^{19}F NMR (CCl$_4$): δ -64.3 (m); IR (KBr): ν 2950, 1750, 1630, 1380, 1280, 1140, 760, 640 cm^{-1}.

Note

This reaction was also carried out with the corresponding amide. For the construction of the cyclic material, the tosyl group on amide nitrogen is crucial, and an alkyl substituent at this position led to the formation of the perfluoroalkylated acyclic product at the same time.

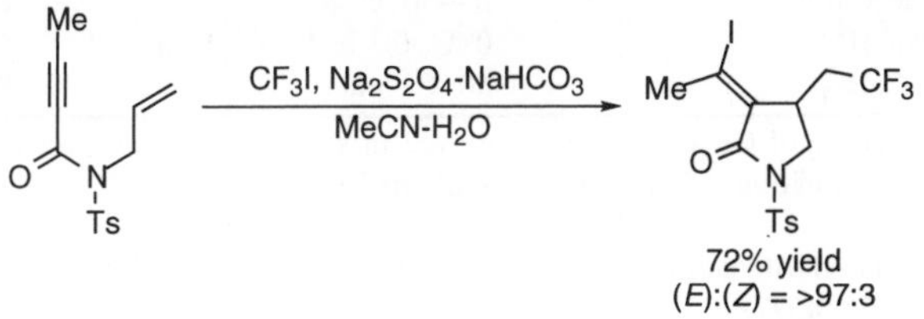

72% yield
(*E*):(*Z*) = >97:3

Wang, Z.; Lu, X.-Y. *Tetrahedron* **1995**, *51*, 2639.

4.1.3 Introduction of a Trifluoromethyl Group (Miscellaneous)

2,2-Dichloro-3,3,3-trifluoro-1-phenylpropanol

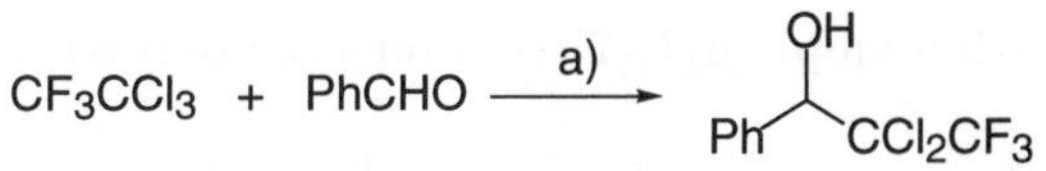

a) Zn / DMF

Synthetic method

To a solution of benzaldehyde (2.12 g, 20.0 mmol) in DMF (20 mL) were added 1,1,1-trichloro-2,2,2-trifluoroethane (4.52 g, 24.1 mmol) and zinc powder (1.44 g, 22.1 mmol), and the resulting mixture was stirred for 3 h at room temperature and for 2 h at 50 °C before treatment with sat. NH$_4$Cl aq. (30 mL). Extraction with Et$_2$O (2x30 mL), drying the ethereal extract over MgSO$_4$, and concentration followed by distillation gave 2,2-dichloro-3,3,3-trifluoro-1-phenylpropanol (4.49 g) in 86% yield as a colorless oil.

Spectral data

bp. 100 °C (bath temperature)/1 mmHg; ^{1}H NMR (CDCl$_3$): δ 2.90 (1 H, d, J=5 Hz), 5.20 (1 H, d, J=5 Hz), 7.25-7.55 (5 H, m); ^{19}F NMR (CDCl$_3$): δ -74.4 (s); IR (neat): ν 3460, 1248, 1188, 1061, 874, 766, 712, 700, 666, 612 cm^{-1}.

Note

For related work, see i) Solladié-Cavallo, A.; Quazzotti, S. *J. Fluorine Chem.* **1990**, *46*, 221, ii) Solladié-Cavallo, A.; Quazzotti, S. *Synthesis* **1991**, 177, iii) Solladié-Cavallo, A., *et al.* *J. Org. Chem.* **1992**, *57*, 174.

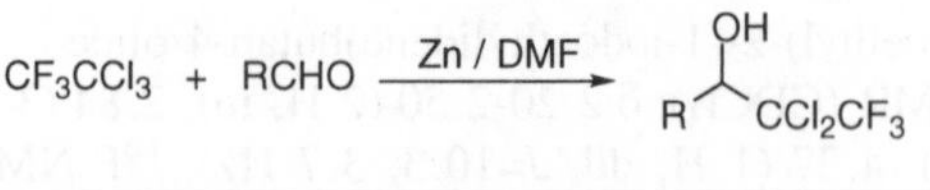

$$CF_3CCl_3 + RCHO \xrightarrow{Zn/DMF} \underset{R}{\overset{OH}{\underset{|}{\bigwedge}}}CCl_2CF_3$$

R	Conditions	Yield (%)
3,4-(OCH$_2$O)-C$_6$H$_3$-	0 °C, 0.2 h; 50 °C, 2 h	80
4-Cl-C$_6$H$_4$-	0 °C, 0.2 h; 50 °C, 2 h	87[a]
3,4-Cl$_2$-C$_6$H$_3$-	rt, 0.5 h; 50 °C, 4.5 h	96
PhCH=CH-	rt, 1 h; 50 °C, 17 h	82
Me$_2$C=CH-	60 °C, 12 h	22[a]
Me$_2$C=CH-	rt→50 °C, 8 h	83[a,b]
PhCHMe-	0 °C, 0.3 h; 50 °C, 12 h	60
c-C$_6$H$_{11}$	50 °C, 5 h	87[c]

[a] 1.5 equiv of CF$_3$CCl$_3$ were employed in spite of 1.2 equiv usage for the standard condition.   [b] Under ultrasonic irradiation.   [c] 5 mol% of CuCl was added.

(Reproduced with permission from Fujita, M.; Hiyama, T. *Bull. Chem. Soc. Jpn.* **1987**, *60*, 4377)

Fujita, M.; Hiyama, T. *Bull. Chem. Soc. Jpn.* **1987**, *60*, 4377.

4.2 Conversion to a Trifluoromethyl Group

2-(4-Bromophenyl)-1-(trifluoromethoxy)ethane

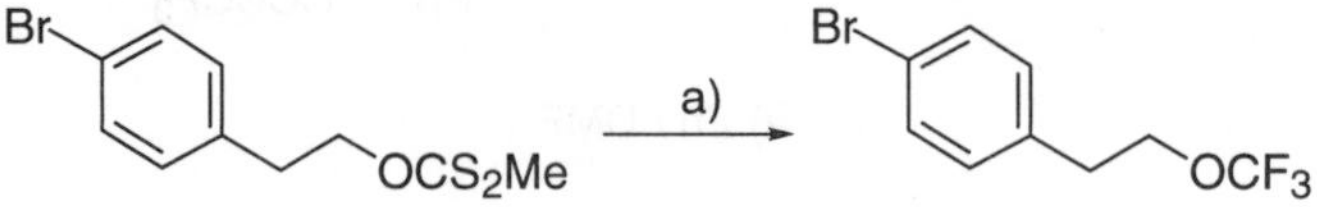

a) 70% pyridine-(HF)$_n$, DBH / CH$_2$Cl$_2$

Synthetic method

An oven-dried 500-mL polypropylene bottle (dried at 60 °C for 12 h and used immediately) equipped with a rubber septum, a Teflon-coated magnetic stirring bar, and an argon inlet, was flushed with argon and charged with 17.2 g of 1,3-dibromo-5,5-dimethylhydantoin (DBH, 60 mmol) and 60 mL of dry CH$_2$Cl$_2$.   To this suspension at -78 °C was added 20 mL of 70% pyridine-(HF)$_n$ over 5 min through a disposable 20-mL volume polypropylene-polyethylene syringe under an argon atmosphere.   The resulting suspension was stirred vigorously, and a solution of *S*-methyl *O*-2-(4-bromophenyl)ethyl dithiocarbonate (5.82 g, 20 mmol) in 20 mL of CH$_2$Cl$_2$ was added dropwise at -78 °C *via* a cannula applying an argon positive pressure.   After the addition was completed, the dry ice-acetone bath was replaced by an ice-salt bath for keeping the temperature between -3 to 0 °C (higher temperature accelerates the aromatic bromination).   The resulting red-brown reaction mixture was stirred at the same temperature for 30 min, diluted carefully with 100 mL of Et$_2$O at 0 °C, and quenched by a dropwise addition of an ice-cold NaHSO$_3$/NaHCO$_3$/NaOH (pH=10) aq. at 0 °C until the red-brownish color of the mixture disappeared.   The pH of the mixture was adjusted to 10 by careful addition of an ice-cold 30% NaOH aqueous solution (*ca.* 100 mL as checked by a pH test paper).   The whole was diluted with

200 mL of Et$_2$O, and the organic phase was separated.   The aqueous phase was extracted with
Et$_2$O (4x100 mL), and the combined organic phase was dried over anhydrous MgSO$_4$, filtered, and
concentrated under reduced pressure.   Pyridine in the residue was removed by the toluene
azeotrope under reduced pressure using 100 mL of toluene 2 times.   The resulting residue was
bulb-to-bulb distilled to afford 4.36 g (81%) of 2-(4-bromophenyl)-1-(trifluoromethoxy)ethane as a
colorless oil with >98% purity.

Spectral data

bp. 120 °C (bath temperature)/6 mmHg; Rf 0.47 (hexane); ^{1}H NMR (CDCl$_3$) δ 2.95 (2 H, t,
J=6.8 Hz), 4.13 (2 H, t, J=6.8 Hz), 7.10 (2 H, d, J=8.1 Hz), 7.44 (2 H, d, J=8.1 Hz); ^{13}C NMR
(CDCl$_3$) δ 34.6, 67.3 (q, J=3.5 Hz), 120.8, 121.5 (q, J=254.6 Hz), 130.6, 131.7, 135.6.

Caution

When the reaction mixture was quenched with NaHSO$_3$/NaHCO$_3$/NaOH (pH=10) aq., a
vigorous evolution of HF was observed.

Note

It is very important to keep the ratio of 70% pyridine-(HF)$_n$ to dithiocarbonate at 1.0 L/mol to
prevent the undesired aromatic bromination.   Thus, 70% pyridine-(HF)$_n$/dithiocarbonate=1.5
L/mol led to the formation of a mixture of 2-(2,4-dibromophenyl)- and 2-(4-bromophenyl)ethyl
trifluoromethyl ether (47% and 43%, respectively) and 0.5 L/mol afforded the desired material in
only 20% yield.

An aqueous solution of NaHSO$_3$/NaHCO$_3$/NaOH (pH=10) for quenching the reaction was
prepared by mixing a solution of NaOH (10 g) and NaHSO$_3$ (15 g) in 100 mL H$_2$O and sat.
NaHCO$_3$ aq. (100 mL).

See also the following reports related to the present transformation: i) Hiyama, T., *et al.*
Tetrahedron Lett. **1992**, *33*, 4173, ii) Hiyama, T., *et al.* *Chem. Lett.* **1997**, 827, iii) Furuta, S.;
Hiyama, T. *Synlett* **1996**, 1199.

Hiyama, T., *et al. Chem. Commun.* **1997**, 309.

Ethyl (2-trifluoromethyl)propionate

a) KF / DMSO

Synthetic method

Diethyl 2-(bromodifluoromethyl)-2-methylmalonate (0.02 mol) was dissolved in dry DMSO
(30 mL) in a three-neck round-bottomed flask fitted with a short path distilling unit and thermometer.   A two-fold excess of KF (0.04 mol) was added and heating of the heterogeneous mixture
was commenced with stirring.   In all, three thermometers were used to monitor the reaction and
insure the desired rate and extent of heating.   Temperatures of the oil bath, a DMSO solution, and
distilling vapors were checked.   The DMSO solution was quickly heated to 150 °C in the first 0.5
h and then maintained at approximately 170 °C for 1-1.5 h during which time gas evolved, distillate
was collected, and the DMSO solution developed a dark color.

Distillation from the above mixture afforded the crude product, and subsequent distillation
removed low boiling dimethyl sulfide and acetaldehyde, resulting in a product 70-90% pure.
Distillation allowed analytically pure ethyl (2-trifluoromethyl)propionate to be obtained in 44%

yield.
Spectral data

bp. 112 °C; ^1H NMR (CDCl$_3$): δ 1.3 (3 H, t), 1.4 (3 H, d), 3.2 (1 H, sept), 4.2 (2 H, q); ^{19}F NMR (CDCl$_3$): δ -71 (d).

Note

For the preparation of the starting bromodifluoromethylated material, see page 85 of this volume.

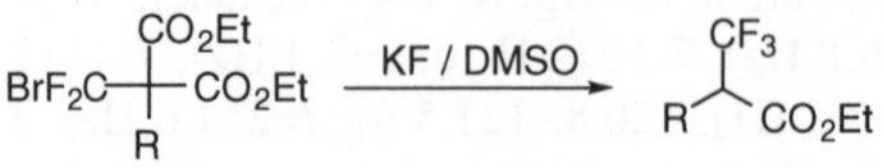

R	Yield (%)	R	Yield (%)
Et	44	*n*-Pr	40
n-Bu	34	Allyl	44
CH$_2$CH$_2$CN	35	CH$_2$CO$_2$Et	37
Ph	42	Bn	61

(Reproduced with permission from Purrington, S. T., *et al. J. Org. Chem.* **1984**, *49*, 3702)

Purrington, S. T., *et al. J. Org. Chem.* **1984**, *49*, 3702.

4-(2,2,2-Trifluoroethyl)nitrobenzene

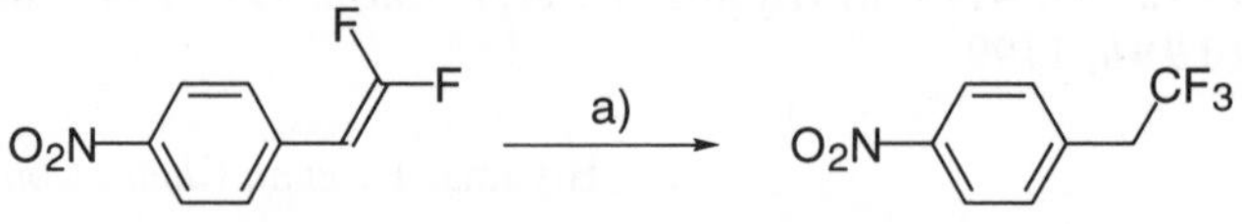

a) KF, H$_2$O / DMF

Synthetic method

A 50-mL, two-necked, round-bottomed flask equipped with a septum, a Teflon-coated magnetic stir bar, and a nitrogen inlet which was connected to a mineral oil bubbler was charged with 1.1 g (6 mmol) of 4-(2,2-difluorovinyl)nitrobenzene, 0.6 g (10 mmol) of potassium fluoride, 10 mL of DMF, and 0.5 mL of H$_2$O. The reaction was stirred at room temperature for 10 h and the NMR showed that the reaction was finished. The reaction mixture was filtered through a short silica gel column. The filtrate was concentrated *via* rotary evaporation with some silica gel; then the dry silica gel and product were introduced onto a silica gel column and eluted (hexane:CH$_2$Cl$_2$= 10:2) to give 1.1 g of 4-(2,2,2-trifluoroethyl)nitrobenzene in 92% yield as a white solid.

Spectral data

mp. 67-68 °C; ^1H NMR (CDCl$_3$): δ 3.5 (2 H, q, *J*=10.5 Hz), 7.5 (2 H, q, *J*=8.5 Hz), 8.2 (2 H, dm, *J*=8.8 Hz); ^{13}C NMR (CDCl$_3$): δ 40.1 (q, *J*=30.3 Hz), 124.0, 125.6 (q, *J*=277.0 Hz), 131.6, 137.8 (m), 148.2; ^{19}F NMR (CDCl$_3$): δ -65.9 (t, *J*=10.2 Hz); IR (CCl$_4$): ν 3078.6, 2949.3, 2862.5, 1923.1, 1604.8, 1523.8, 1498.8, 1421.6, 1359.9, 1267.3, 1199.8, 1149.6, 1118.8, 1076.3 cm^{-1}.

Note

For the preparation of the starting difluorinated olefins, see pages 83 or 84 of this volume.

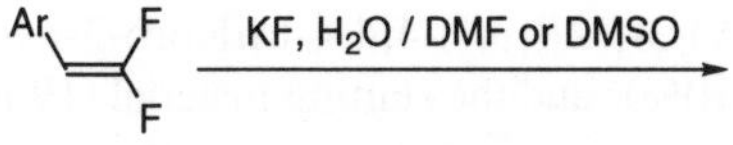

Ar	Yield (%)	Ar	Yield (%)
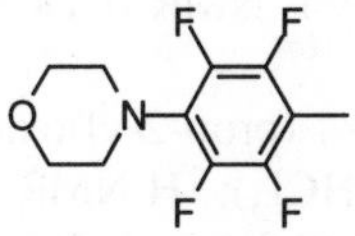	94	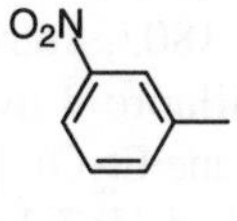	95
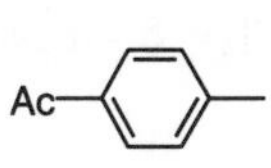	90	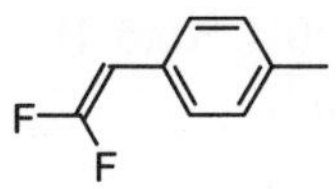	83

(Reproduced with permission from Nguyen, B. V.; Burton, D. J. *J. Org. Chem.* **1997**, 62, 7758)

Nguyen, B. V.; Burton, D. J. *J. Org. Chem.* **1997**, 62, 7758.

4.3 Carbon-Carbon Bond-forming Reactions

4.3.1 Aldol Type Reactions

(S)-3-[(2R,3S)-4,4,4-Trifluoro-3-hydroxy-2-methylbutyryl]-4-(prop-2-yl)-oxazolidin-2-one

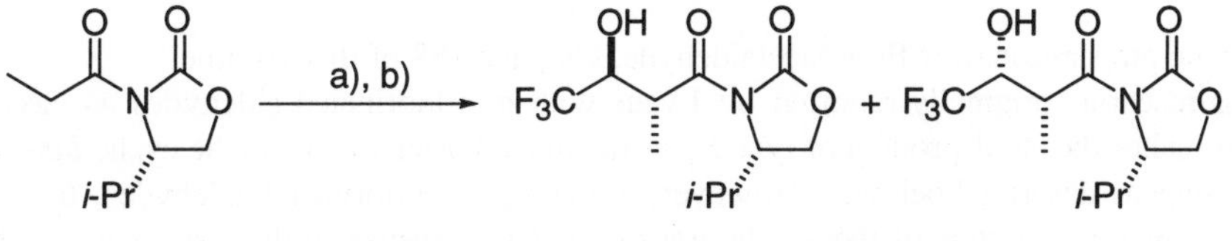

a) *n*-Bu₂BOTf, Et₃N / CH₂Cl₂,   b) CF₃CHO

Synthetic method

To a solution of (S)-3-propionyl-4-(prop-2-yl)oxazolidin-2-one (371 mg, 2.0 mmol) in 3 mL of CH₂Cl₂ was added dibutylboron trifluoromethanesulfonate (2.3 mmol) at -78 °C over 2 min. After 10 min at the same temperature, Et₃N (2.7 mmol) was added over 10 min and the reaction mixture was allowed to warm to 0 °C.   After 1 h at 0 °C, the solution was cooled to -78 °C and gaseous trifluoroacetaldehyde (6 mmol) was added with a cannula at -78 °C.   After 30 min at -78 °C, the reaction mixture was brought to and left at 0 °C for 2 h and quenched with pH 7.0 phosphate buffer (0.1 M, 4 mL) and MeOH (6 mL) followed by the addition of 30% H₂O₂-MeOH (3 mL-9 mL).   After 1 h at 0 °C, the mixture was concentrated *in vacuo*.   The residue was diluted with 10% NaHCO₃ aq. and extracted with CH₂Cl₂.   The combined extracts were washed with brine, dried and filtered.   After evaporation of the solvent, chromatography of the residue (hexane: AcOEt=5:1)   gave   (S)-3-[(2R,3S)-4,4,4-trifluoro-3-hydroxy-2-methylbutyryl]-4-(prop-2-yl)oxa-

zolidin-2-one (295 mg, 52%), (S)-3-[(2R,3R)-4,4,4-trifluoro-3-hydroxy-2-methylbutyryl]-4-(prop-2-yl)oxazolidin-2-one (58 mg, 10%), and the starting material (19 mg, 5%).
Spectral data

$[\alpha]_D^{24}$ +49.4 (c 0.96, CHCl$_3$); mp. 99.0-99.5 °C (hexane:Et$_2$O); ^{1}H NMR (CDCl$_3$): δ 0.88 (3 H, d, J=6.9 Hz), 0.93 (3 H, d, J=7.0 Hz), 1.41 (3 H, d, J=7.0 Hz), 2.03-2.46 (1 H, m), 3.92-4.10 (1 H, m), 4.22-4.49 (4 H, m), 4.57 (1 H, d, J=9.9 Hz); ^{19}F NMR (CDCl$_3$): δ -77.57 (d, J=7.9 Hz); IR (KBr): ν 3422, 1803, 1680 cm^{-1}.

(S)-3-[(2R,3R)-4,4,4-trifluoro-3-hydroxy-2-methylbutyryl]-4-(prop-2-yl)oxazolidin-2-one

mp. 99.2-100.4 °C (hexane:Et$_2$O); $[\alpha]_D^{26}$ +62.6 (c 0.87, CHCl$_3$); ^{1}H NMR (CDCl$_3$): δ 0.88 (3 H, d, J=7.0 Hz), 0.93 (3 H, d, J=7.1 Hz), 1.32 (3 H, dd, J=7.0, 0.9 Hz), 2.30-2.46 (1 H, m), 2.95 (1 H, d, J=5.8 Hz), 4.20-4.50 (5 H, m); ^{19}F NMR (CDCl$_3$): δ -76.90 (d, J=7.3 Hz); IR (KBr): ν 3430, 1794, 1697 cm^{-1}.

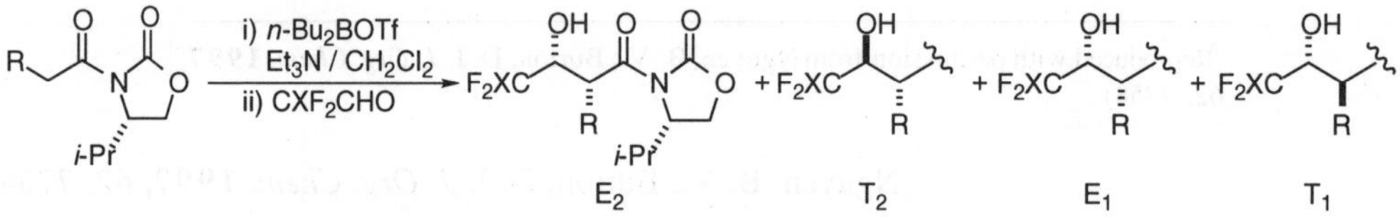

R	X	Yield (%)[a]		E_2:T_2:E_1:T_1			
Me	F	62	(65)	15 :	85 :	0 :	0
Bn	F	64	(75)	30 :	70 :	0 :	0
n-Bu	F	60	(64)	19 :	81 :	0 :	0
Me	F	80[b]		22 :	78 :	0 :	0
Me	F	83[c]	(88)	54 :	46 :	0 :	0
Me	Ph(CH$_2$)$_3$	33	(56)	12 :	82 :	6 :	0

[a] The conversion yield is shown in parentheses.   [b] Oxazolidinone with a benzyl group instead of an isopropyl moiety was employed.   [c] TiCl$_4$ was added.
(Reproduced with permission from Iseki, K., *et al. Tetrahedron* **1996**, *52*, 71)

Note

For the preparation of trifluoroacetaldehyde, see page 188 of this volume.

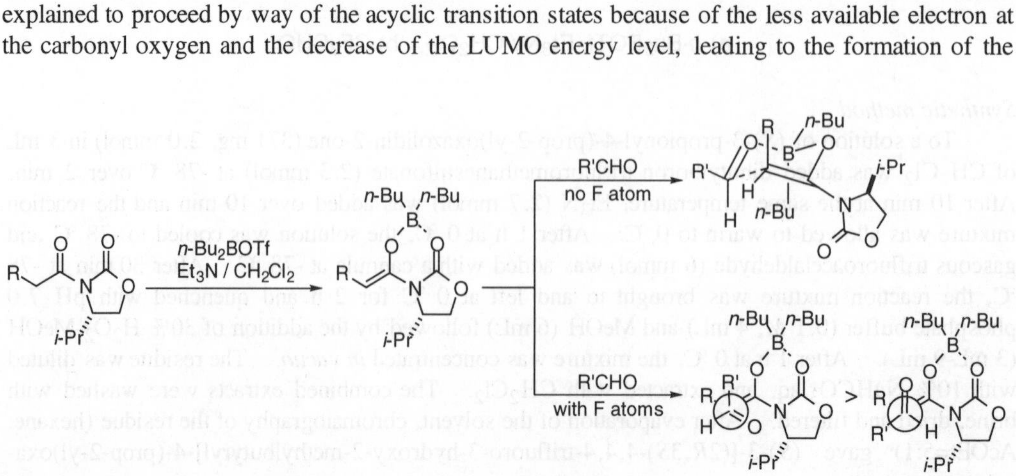

This reaction, originally reported by Evans with nonfluorinated aldehydes as electrophiles, usually furnishes the aldol product of type E_1 as the main diastereomer *via* the cyclic six-membered transition state as described below.   However, in the case of fluorinated aldehydes, the reaction is explained to proceed by way of the acyclic transition states because of the less available electron at the carbonyl oxygen and the decrease of the LUMO energy level, leading to the formation of the "abnormal" stereoisomer as the main product. Very high stereoselectivity was obtained by hexafluoroacetone as the electrophile, again *via* the acyclic transition state.

See also the following work by Iseki *et al.* on the aldol reaction of fluorinated carbonyl materials with the boryl enolate derived from Oppolzer's sultam: Iseki, K., *et al. Chem. Pharm. Bull.* **1996**, *44*, 2003.

Iseki, K., *et al. Tetrahedron* **1996**, *52*, 71.

(2S, 3S)-2-Amino-3-(trifluoromethyl)-3-hydroxydecanoic acid

a) $CF_3C(O)C_7H_{15}^{-n}$, DBU / MeCN, b) 5% AcOH aq.,
c) HCl / MeOH, d) Dowex-H⁺ / NH₄OH

Synthetic method

To a solution of 500 mg (1 mmol) of a complex prepared from (S)-o-[(N-benzylprolyl)-amino]benzophenone (BPB), glycine, nickel(II) nitrate hexahydrate, and sodium hydroxide in MeOH, in 5 mL of MeCN was added 450 mg (3 mmol) of 1,8-diazabicyclo[5.4.0]undec-7-ene. The mixture was stirred for 1-2 min to a homogeneous solution, and then 392 mg (2 mmol) of neat 1,1,1-trifluorononan-2-one was added with stirring at ambient temperature. The reaction mixture was vigorously stirred for 15 min then quenched with 5% aqueous acetic acid. The oily product was purified on silica gel (CHCl₃:acetone=7:1). Two main fractions isolated in the order of their emergence from the column yielded the minor diastereomer, followed by the major isomer. The isolated yield of the latter was 70% and the diastereomeric ratio was found to be more than 99:1.

Spectral data

$[\alpha]_D^{25}$ +2000.0 (c 0.04, CHCl₃); mp. 99-100 °C; ¹H NMR (CDCl₃): δ 0.89 (3 H, t, J=6.9 Hz), 1.05-1.35 (12 H, m), 1.55-3.50 (7 H, m), 3.39 (1 H, d, J=12.5 Hz), 4.26 (1 H, d, J=12.5 Hz), 4.24 (1 H, s), 4.28 (1 H, s), 6.55-7.60 (11 H, m), 8.06 (2 H, m), 8.45 (1 H, m); ¹⁹F NMR (CDCl₃): δ -72.94 (s).

The complex thus obtained in MeOH was added in portions to diluted HCl aq. (1 mL of conc HCl and 2 mL of H₂O for 1 g of the complex). The mixture was refluxed for 10-20 min then cooled to room temperature; the precipitated crystalline, BPB hydrochloride, was filtered, thoroughly washed with H₂O and air dried (the recovery was 90-92%). The filtrate was evaporated to dryness in vacuum. The residue was dissolved in H₂O, neutralized with conc NH₃ until pH 7 and extracted three times with CHCl₃. From the combined organic layers an additional amount (5%) of BPB was isolated. From the H₂O layer the crude (2S,3S)-2-amino-3-(trifluoro-methyl)-3-hydroxydecanoic acid was isolated using a cation exchange technique (10 mL of Dowex-H⁺ with the capacity 2 mmol/mL was used per 1 g of the decomposed complex). The crude acid was dissolved in a minimum volume of boiling H₂O and upon addition of 4-5 volume of EtOH the amino acid crystallized immediately. This mixture was kept overnight at room temperature and

(2*S*,3*S*)-2-amino-3-(trifluoromethyl)-3-hydroxydecanoic acid was filtered and dried.   The yield was 49% and the chiral auxiliary was recovered almost in a quantitative yield.

Spectral data

[α]$_{578}^{20}$ -35.0 (*c* 0.79, acetone); mp. 174-178 ℃; ^{1}H NMR ((CD$_3$)$_2$CO): δ 0.87 (3 H, t, *J*=7.2 Hz), 1.17-1.73 (12 H, m), 4.30 (1 H, s); ^{19}F NMR ((CD$_3$)$_2$SO): δ -75.10 (s).

Note

For the preparation of the starting trifluoromethylated ketone, see page 158 of this volume.

Various CF$_3$-containing β-hydroxy-α-amino acids were synthesized by this procedure as depicted below.   See also the following articles closely related to this topic: i) Soloshonok, V. A., *et al*. *Tetrahedron Lett*. **1997**, *38*, 4671.   ii) Soloshonok, V. A., *et al*. *J. Chem. Soc. Perkin Trans. 1* **1993**, 3143.

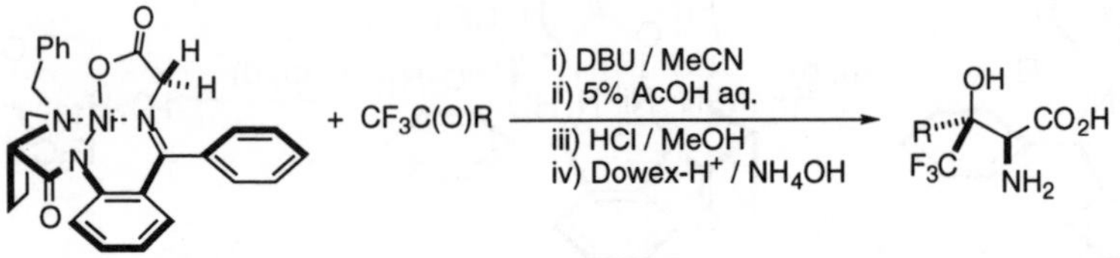

R	Reaction time[a] (min)	Yield (%)	Diastereomeric excess (% de)
Me	15	75	95
	1	73	>98
n-Bu	10	71	98
n-C$_7$H$_{15}$	15	71	97
	1	70	>98
n-C$_8$H$_{17}$	1	71	97
Ph(CH$_2$)$_3$	1	87	96
PhC≡C[b]	30	56	90

[a] For the first step.   [b] Et$_3$N was used instead of DBU.
(Reproduced with permission from Soloshonok, V. A., *et al*.
Tetrahedron **1996**, *52*, 12433)

Soloshonok, V. A., *et al*. *Tetrahedron* **1996**, *52*, 12433.

Ethyl *syn*-2-dibenzylamino-4,4,4-trifluoro-3-hydroxybutyrate, Ethyl *syn*-2-amino-4,4,4-trifluoro-3-hydroxybutyrate

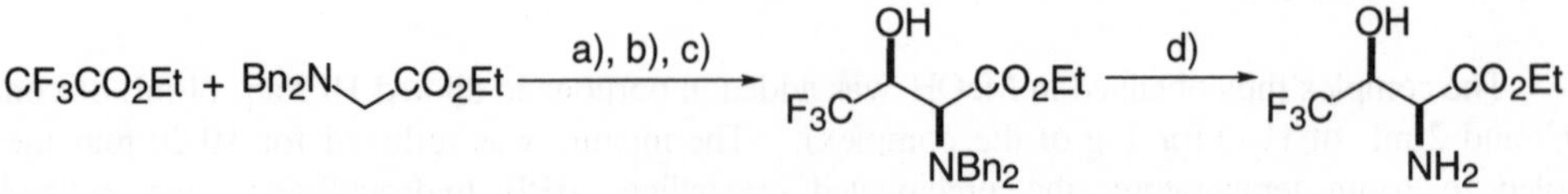

a) NaH / THF, b) AcOH, c) NaBH$_4$, d) H$_2$, 10% Pd/C / EtOH

Synthetic method

Ethyl *syn*-2-amino-4,4,4-trifluoro-3-hydroxybutyrate

To a solution of ethyl dibenzylaminoacetate (2 g, 7.06 mmol) and ethyl trifluoroacetate (1.85 mL, 15.53 mmol) in THF (15 mL), sodium hydride (50% suspension in mineral oil; 1.02 g, 21.18 mmol) was added.   The resulting suspension was refluxed for 5 h (after this time it turned deep red), cooled to 0 °C and treated with acetic acid (1.42 mL, 24.71 mmol).   Sodium borohydride (668 mg, 17.66 mmol) was then added and the suspension was stirred overnight at room temperature, treated with 1 N HCl to pH 5, stirred for 10 min, and treated with 1 N KOH to pH 10. Extraction with Et_2O (3x150 mL), after drying of the extract with Na_2SO_4 and evaporation to dryness, gave 3.3 g of a crude product.   This was purified by flash chromatography (hexane: Et_2O) to give pure ethyl *syn*-2-dibenzylamino-4,4,4-trifluoro-3-hydroxybutyrate as a white solid which was crystallized from hexane:Et_2O.   The yield was 65%.

Spectral data

mp. 78-79 °C (hexane:Et_2O); ^{1}H NMR ($CDCl_3$): δ 1.40 (3 H, t, J=7.1 Hz), 3.44 (2 H, d, J=13 Hz), 3.52 (1 H, d, J=9.4 Hz), 3.96 (2 H, d, J=13 Hz), 4.33, 4.34 (ABX_3 system, 2 H, J=7.1 Hz), 4.35 (1 H, br s), 4.10-4.65 (1 H, m), 7.30 (10 H, s); IR ($CHCl_3$): ν 1730, 1170, 1145 cm^{-1}.

Synthetic method

Ethyl *syn*-2-amino-4,4,4-trifluoro-3-hydroxybutyrate

A solution of ethyl *syn*-2-dibenzylamino-4,4,4-trifluoro-3-hydroxybutyrate (100 mg, 0.262 mmol) in absolute EtOH (10 mL) was hydrogenated over 10% palladium on carbon (20 mg) for 3 h at room temperature.   Filtration of the catalyst and evaporation of the filtrate to dryness gave 48 mg of white solid corresponding to pure ethyl *syn*-2-amino-4,4,4-trifluoro-3-hydroxybutyrate in 91% yield.

Spectral data

mp. 88-89 °C; ^{1}H NMR ($CDCl_3$): δ 1.30 (3 H, t, J=7 Hz), 3.87 (1 H, d, J=1.8 Hz), 4.05-4.52 (3 H, m); IR ($CHCl_3$): ν 3360, 3280, 1730, 1600, 1465 cm^{-1}.

Note

See the following articles on the chiral fluorinated threonine synthesis: 4,4,4-F_3 derivatives, i) Seebach, D., *et al. Helv. Chim. Acta* **1987**, *70*, 237, ii) Guanti, G., *et al. Tetrahedron* **1988**, *44*, 5553, iii) Kitazume, T., *et al. Tetrahedron: Asym.* **1991**, *2*, 235.   4,4-F_2 derivatives, i) Kitazume, T., *et al. Bioorg. Med. Chem. Lett.* **1991**, *1*, 271, ii) Fujisawa, T., *et al. Tetrahedron: Asym.* **1993**, *4*, 835.

Scolastico, C., *et al. Synthesis* **1985**, 850.

(3S^*,4S^*)-3-Benzyloxy-4-trifluoromethyl-N-(4-methoxyphenyl)azetidin-2-one

a) BnOCH$_2$C(O)Cl, Et$_3$N / CH$_2$Cl$_2$

Synthetic method

A solution of N-(2,2,2-trifluoroethylidene)-4-methoxyaniline (2.5 g, 12.3 mmol) and 2-(benzyloxy)acetyl chloride (7.8 mL, 49.2 mmol) in dry CH$_2$Cl$_2$ (30 mL) was treated with triethyl-amine (7.8 mL, 61.6 mmol).   The resulting mixture was then stirred overnight at 40 °C.   The solution was poured in H$_2$O (20 mL) and extracted with AcOEt (2x50 mL).   The organic phase was washed with brine and dried over MgSO$_4$.   After evaporation of the solvent, the crude product was purified by crystallization in cold EtOH to give 2.1 g of (3S^*,4S^*)-3-benzyloxy-4-

trifluoromethyl-*N*-(4-methoxyphenyl)azetidin-2-one in 50% yield as a white solid.
Spectral data
    mp. 132 °C (EtOH); ^{1}H NMR (CDCl$_3$): δ 3.8 (3 H, s), 4.6 (1 H, quint, *J*=5.3 Hz), 4.8 (2 H, d, *J*=14.0 Hz), 5.0 (1 H, d, *J*=5.3 Hz), 6.8 (2 H, d, *J*=10.6 Hz), 7.5 (7 H, m); ^{13}C NMR (CDCl$_3$): δ 55.6, 57.8 (q, *J*=32.8 Hz), 73.9, 80.4, 114.6, 119.6, 124.8 (q, *J*=278.0 Hz), 128.0, 128.4, 128.7, 129.5, 136.3, 157.4, 164.0; ^{19}F NMR (CDCl$_3$): δ -68.6 (d, *J*=5.5 Hz).
Note
    For the preparation of the starting imine, see page 175 of this volume.
    Further elaboration of the resultant azetidinone led to the formation of the *syn*-trifluoro-methylated isoserinate in approximately 50% total yield *via* the procedure described below.   This β-lactam was also employed for the construction of the 3'-CF$_3$-taxoids from C10-modified baccatin III.   See Ojima, I., *et al. Bioorg. Med. Chem. Lett.* **1997**, *7*, 133, and in *Biomedical Frontiers of Fluorine Chemistry (ACS Symposium Series No. 639)*; Ojima, I.; McCarthy, J. R.; Welch, J. T., Eds.; ACS: New York, 1996, p. 228.
    For the preparation of a similar β-lactam with a CF$_3$ moiety, see Guanti, G., *et al. Synthesis* **1985**, 609.   The corresponding difluoro- and chlorodifluoromethylated counterparts have also been reported recently.   See Bégué, J.-P., *et al. J. Org. Chem.* **1997**, *62*, 8826.

Bégué, J.-P., *et al. Synlett* **1996**, 399.

4.3.2 Michael Reactions

Ethyl *trans*-3-(trifluoromethyl)pyroglutamate

a) Ph$_2$C=N-CH$_2$CO$_2$Et, cat. (*n*-Bu)$_4$NHSO$_4$ / CH$_2$Cl$_2$-NaOH aq.,
b) 15% Citric acid / THF

Synthetic method
    10 g of ethyl *N*-(diphenylmethylene)glycinate (37.4 mmol), 1.5 g of tetra-*n*-butylammonium hydrogen sulfate, 60 mL of 10% NaOH and 60 mL of CH$_2$Cl$_2$ were stirred for 15 min at 0 °C. Thereafter, 6.3 g of ethyl (*E*)-4,4,4-trifluorobut-2-enoate (5.6 mL, 37.4 mmol) was added and the whole two-phase system was vigorously stirred at 0 °C for 2 h.   The reaction mixture was diluted with H$_2$O/CH$_2$Cl$_2$ (500 mL/250 mL) and the aqueous layer was extracted three times with 150 mL of CH$_2$Cl$_2$.   The combined organic layers were washed with H$_2$O (150 mL) and brine (150 mL). After drying over MgSO$_4$, the solvent was evaporated *in vacuo* at 30 °C.   The resulting, slightly yellow oil (*ca.* 15 g) was dissolved in 100 mL of THF and stirred for 7 days with 70 mL of 15% citric acid at room temperature.   THF was evaporated *in vacuo* (30 °C).   The resulting solid was

crystallized (hexane:EtOH=9:1) to give 8.0 g of ethyl *trans*-3-(trifluoromethyl)pyroglutamate in 95% yield as colorless needles.
Spectral data
        mp. 96-97 °C; ^{1}H NMR (CDCl$_3$): δ 1.35 (3 H, t, *J*=7.1 Hz), 2.57 (1 H, dd, *J*=17.8, 5.3 Hz), 2.70 (1 H, dd, *J*=17.8, 10.0 Hz), 3.44 (1 H, m), 4.31 (2 H, q, *J*=7.1 Hz), 4.34 (1 H, d, *J*=3.2 Hz), 6.74 (1 H, br s).

Prati, F., *et al. Tetrahedron:Asym.* **1996**, *7*, 3309.

(*E*)-4,4,4-Trifluorobut-2-enoic acid,
3-Amino-4,4,4-trifluorobutyric acid

a) 1 *N* NaOH aq. / THF, b) NH$_3$

Synthetic method
        (*E*)-4,4,4-Trifluorobut-2-enoic acid
        To a stirred solution of ethyl (*E*)-4,4,4-trifluorobut-2-enoate (120 g, 0.71 mol) in THF (1.4 L) was added 1 *N* NaOH aq. (715 mL, 715 mmol) and the resulting mixture was stirred at ambient temperature for 1.5 h.   The pH was adjusted to 2 with 1 *N* HCl aq. and THF was removed under vacuum.   The resulting aqueous solution was saturated with NaCl and extracted with Et$_2$O (4x1 L).   The organic phase was washed with sat. NaCl aq. and dried, and the solvent was removed to afford (*E*)-4,4,4-trifluorobut-2-enoic acid as a white crystalline solid (93 g) in 93% yield.
Spectral data
        mp. 54-55 °C; ^{1}H NMR (CDCl$_3$): δ 6.50 (1 H, dq, *J*=15.5 1.5 Hz), 6.88 (1 H, dq, *J*=15.5, 6.1 Hz), 11.85 (1 H, s); ^{19}F NMR (CDCl$_3$): δ -66.5 (dd, *J*=6.1, 1.5 Hz) ; IR (CHCl$_3$): ν 2700, 2605, 2530, 1743, 1717, 1683, 1665, 1420, 1307, 1280, 1145, 980 cm^{-1}.

Synthetic method
        3-Amino-4,4,4-trifluorobutyric acid
        A solution of (*E*)-4,4,4-trifluorobut-2-enoic acid (14 g, 0.1 mmol) and liquid NH$_3$ (35 mL) in a glass-lined steel cylinder was heated at 100 °C for 20 h.   Ammonia was allowed to evaporate, EtOH (50 mL) was added, and the solution was concentrated to dryness.   The residue was redissolved in EtOH (50 mL), concentrated to dryness, then triturated with CH$_2$Cl$_2$ (50 mL), and the solvent was removed under vacuum.   The resulting cream-colored solid was dried over P$_2$O$_5$ to afford 3-amino-4,4,4-trifluorobutyric acid (15.1 g) in quantitative yield.
Spectral data
        mp. 180-182 °C; ^{1}H NMR (DMSO-d_6): δ 2.26 (1 H, dd, *J*=15.5, 9.0 Hz), 2.53 (1 H, dd, *J*=15.5, 4.5 Hz), 3.58 (1 H, dqd, *J*=9.0, 8.7, 4.5 Hz), 5.15 (3 H, br s); ^{19}F NMR (DMSO-d_6): δ -76.63 (d, *J*=8.7 Hz); IR (KBr): ν 2980-2620, 2220, 1670, 1627, 1595, 1390, 1365, 1250, 1185, 1120 cm^{-1}.

Roberts, J. L., *et al. J. Org. Chem.* **1984**, *49*, 1430.

2-(1,1,1-Trifluoro-3-nitroprop-2-yl)-1-methylpyrrole

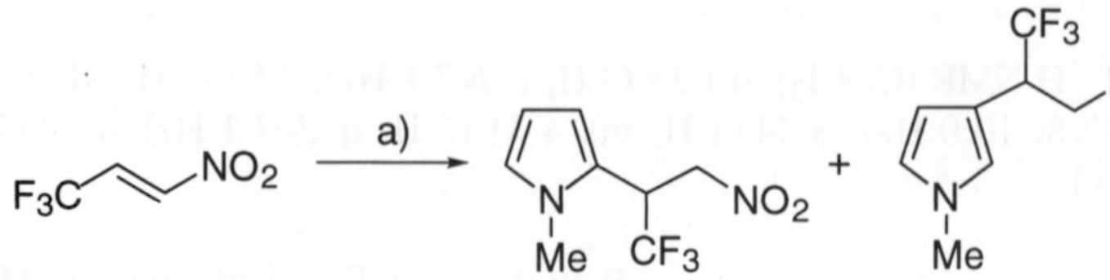

a) 1-Methylpyrrole / CHCl$_3$ (or CH$_2$Cl$_2$)

Synthetic method

To a solution of 1.0 g (10 mmol) of 1-methylpyrrole in CHCl$_3$ (10 mL) or CH$_2$Cl$_2$ (10 mL), 1.4 g (10 mmol) of 3,3,3-trifluoro-1-nitropropene was added dropwise at -78 °C under stirring. Then the mixture was allowed to warm to room temperature within 5 h and the stirring was continued at this temperature for a further 12 h.   After evaporation of the solvent, the crude product was purified by column chromatography (hexane:AcOEt=3:1) giving 1.20 g of 2-(1,1,1-trifluoro-3-nitroprop-2-yl)-1-methylpyrrole in 50% yield and 100 mg of 3-(1,1,1-trifluoro-3-nitroprop-2-yl)-1-methylpyrrole in 4% yield as a colorless oil.

Spectral data

Rf 0.5 (hexane:AcOEt=3:1); ^{1}H NMR (DMSO-d_6): δ 3.63 (3 H, s), 4.78 (1 H, dqd, J=8.9, 8.6, 6.0 Hz), 5.15 (1 H, dd, J=14.4, 8.9 Hz), 5.31 (1 H, dd, J=14.4, 6.0 Hz), 6.04 (1 H, dd, J=4.0, 2.7 Hz), 6.29 (1 H, dd, J=4.0, 1.8 Hz), 6.79 (1 H, dd, J=2.7, 1.8 Hz); ^{13}C NMR (DMSO-d_6): δ 33.5, 38.9 (q, J=29.3 Hz), 73.4 (q, J=2.9 Hz), 107.2, 108.7, 121.6 (q, J=2.0 Hz), 124.3, 124.8 (q, J=280.4 Hz); ^{19}F NMR (DMSO-d_6): δ -68.9 (d).

Note

For the preparation of the starting nitropropene, see page 161 of this volume.

Diels-Alder reaction of 3,3,3-trifluoro-1-nitropropene was carried out with dienes such as cyclopentadiene, cyclohexa-1,3-diene, or furan, furnishing the corresponding adducts in 54%, 38%, or 59% yield, but as an equal diastereomer mixture.

Miethchen, R., *et al. J. Fluorine Chem.* **1997**, *81*, 205.

2-(Trifluoromethyl)butan-4-olide

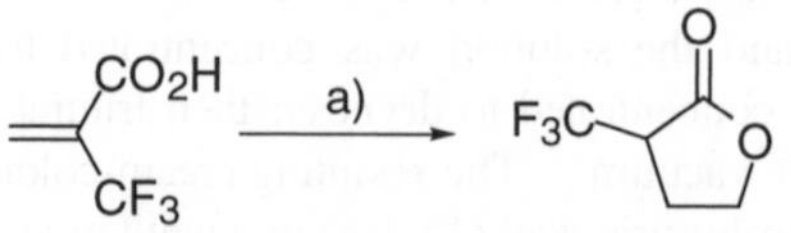

a) Ph$_2$C=O / MeOH

Synthetic method

A solution of 2-(trifluoromethyl)acrylic acid (1.04 g, 7.5 mmol) and benzophenone (1.30 g, 7.2 mmol) in MeOH (30 mL) was placed in a Schlenk tube and degassed using a stream of nitrogen. The mixture was irradiated for 5 h, after which the solvent was removed and distillation gave 2-(trifluoromethyl)butan-4-olide as a colorless oil (0.58 g, 64% yield).

Spectral data

bp. 50 °C/0.2 mmHg; ^{1}H NMR (CDCl$_3$): δ 2.3-2.6 (2 H, m), 3.2-3.4 (1 H, m), 4.1-4.4 (2

H, m); ^{13}C NMR (CDCl$_3$): δ 23.9, 44.4 (q, J=30.4 Hz), 66.9, 119.4 (q, J=277.7 Hz), 170.1; ^{19}F NMR (CDCl$_3$): δ -69.5 (s); IR (neat) v 1780 cm^{-1}.

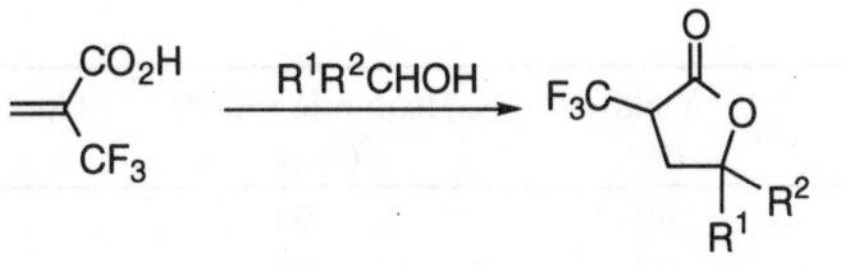

R^1	R^2	Yield (%)	Diastereomeric ratio		
Me	H	77	2	:	1
Ph	H	93	2	:	1[a]
Me	Me	79			
Ph	Me	46	2	:	1[a]

[a] The main product possesses the *cis* Ph moiety with respect to the CF$_3$ group.

Note

For other examples of Michael addition to 2-(trifluoromethyl)acrylic acid or its derivatives, see the followings: i) Fuchikami, T., *et al. Tetrahedron Lett.* **1986**, *27*, 3173, ii) Ojima, I., *et al. J. Org. Chem.* **1989**, *54*, 4511, iii) Ojima, I.; Jameison, F. A. *Bioorg. Med. Chem. Lett.* **1991**, *1*, 581, iv) Yamazaki, T., *et al. J. Org. Chem.* **1994**, *59*, 5100.

O'Hagan, D., *et al. J. Chem. Soc., Perkin Trans. 1* **1995**, 147.

Ethyl (3R,R_S)-3-(trifluoromethyl)-4-(4-methylphenylsulfinyl)butanoate

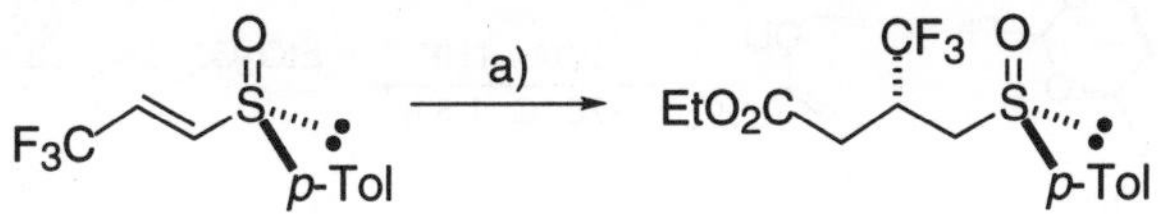

a) CH$_3$CO$_2$Et, LDA / THF

Synthetic method

To an LDA solution in THF (16.1 mmol in 50 mL) was added 16.3 mmol of ethyl acetate and the whole was stirred for 0.5 h at -78 °C. To the enolate formed, 14.4 mmol of (R_S,E)-3,3,3-trifluoro-1-(4-methylphenylsulfinyl)propene was added in 10 mL of THF and stirring was continued for 1.5 h at that temperature, followed by 0.5 h at 0 °C. The reaction was quenched with sat. NH$_4$Cl aq., the organic material was extracted with CH$_2$Cl$_2$, washed with H$_2$O and sat. NaCl aq. successively, dried over anhydrous MgSO$_4$, and evaporated to afford, after chromatographic separation, ethyl (3R,R_S)-3-(trifluoromethyl)-4-(4-methylphenylsulfinyl)butanoate in 83% yield.

Spectral data

[α]$_D^{21}$ +183.13 (*c* 1.3, CHCl$_3$); ^{1}H NMR (CDCl$_3$): δ 1.28 (3 H, t, J=7.14 Hz), 2.43 (3 H, s), 2.67 (2 H, d, J=6.33 Hz), 2.91 (1 H, dd, J=13.71, 7.32 Hz), 2.99 (1 H, dd, J=13.69, 5.72 Hz), 3.39 (1 H, qdtd, J=8.57, 7.30, 6.33, 5.78 Hz), 4.19 (2 H, q, J=7.16 Hz), 7.3-7.6 (4 H, m); ^{13}C NMR (CDCl$_3$): δ 14.10, 21.46, 33.23 (q, J=2.5 Hz), 36.32 (q, J=28.0 Hz), 56.16 (q, J=1.4 Hz), 61.38, 124.03, 126.51 (q, J=280.1 Hz), 130.25, 140.45, 142.34, 169.82; ^{19}F NMR (CDCl$_3$): δ 7.47 (d, J=8.30 Hz) from CF$_3$CO$_2$H; IR (neat): v 3060, 3000, 2950, 2880, 1740 cm^{-1}.

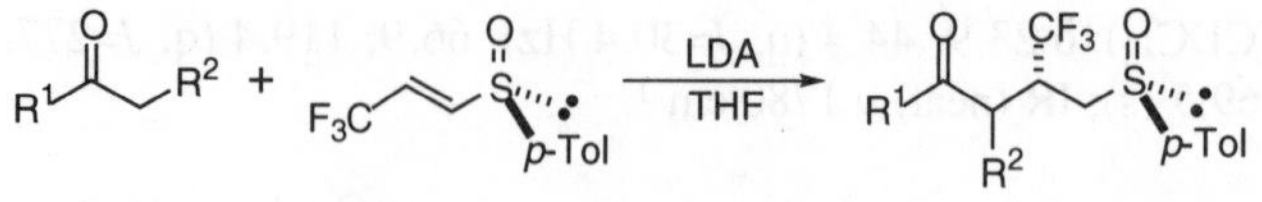

R[1]	R[2]	Yield (%)	Isomeric ratio[a] (% de)	Configuration[b]
Ph	H	99	94	R
Ph[c]	H	92	>98	S
t-Bu	H	96	>98	R
Et	Me	86	>98[d]	
			>98[d]	
OEt	CO$_2$Et	95	85	R

a) Between CF$_3$-CH-CH$_2$-S(O)-p-Tol.   b) At CF$_3$-C.   c) (Z)-vinylic sulfoxide was employed.   d) These are the diastereomers resulting from CF$_3$-CH-CH-CH$_3$- in a ratio of 73:27.

(Reproduced with permission from Yamazaki, T., *et al. J. Chem. Soc., Chem. Commun.* **1987**, 1340)

Note

For the preparation of the starting vinylic sulfoxide, see page 161 of this volume.

The data in the above table were extracted from the following report: Yamazaki, T., *et al. J. Chem. Soc., Chem. Commun.* **1987**, 1340.

A similar type of materials was obtained in a virtually optically pure form by Michael addition of the corresponding chiral sulfonamide depicted below. See Eguchi, S., *et al. Tetrahedron* **1997**, *53*, 823. See also the following on the Diels-Alder reaction of this chiral sulfonamide as the dienophile: Eguchi, S., *et al. Synlett* **1996**, 1106.

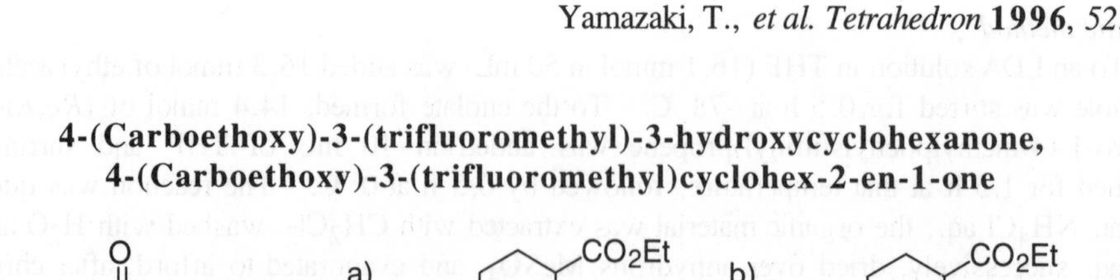

Yamazaki, T., *et al. Tetrahedron* **1996**, *52*, 199.

4-(Carboethoxy)-3-(trifluoromethyl)-3-hydroxycyclohexanone,
4-(Carboethoxy)-3-(trifluoromethyl)cyclohex-2-en-1-one

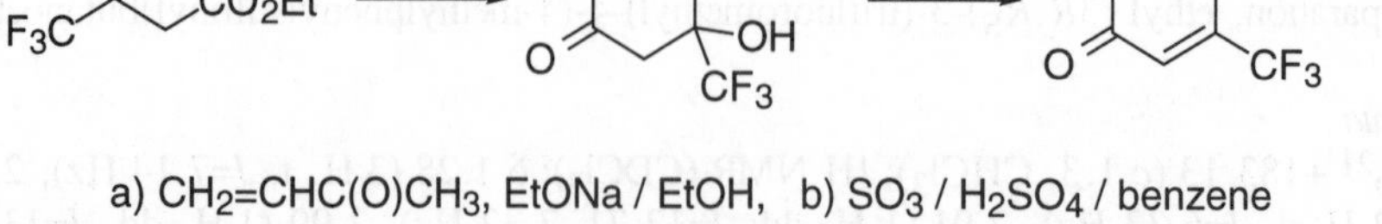

a) CH$_2$=CHC(O)CH$_3$, EtONa / EtOH,   b) SO$_3$ / H$_2$SO$_4$ / benzene

Synthetic method

4-(Carboethoxy)-3-(trifluoromethyl)-3-hydroxycyclohexanone

Sodium (1.15 g, 50 mmol) was dissolved in absolute EtOH (50 mL).   Ethyl 4,4,4-trifluoro-3-oxobutyrate (9.2 g, 50 mmol) in absolute EtOH (50 mL) was added.   The mixture was cooled at

0 °C.   But-3-en-2-one (7 g, 100 mmol) in absolute EtOH (50 mL) was added with stirring.   After additional stirring for 2 h, the solvent was evacuated under vacuum. Et_2O (100 mL) was added. The organic layer was washed successively with diluted HCl, sat. $NaHCO_3$ aq., and brine.   The organic layer was dried on $MgSO_4$. Et_2O was evacuated under vacuum.   The residue was stirred for 15 h with benzene (50 mL) and 20% H_2SO_4 (20 mL).   After extraction with Et_2O, washing with brine and drying on $MgSO_4$, the solvent was evacuated under vacuum.   The two isomers of 4-(carboethoxy)-3-(trifluoromethyl)-3-hydroxycyclohexanone (6.8 g, 27 mmol) were obtained as a non-separated mixture by silica gel column chromatography ($CHCl_3$) in 54% yield.
Spectral data
   ^1H NMR: δ 1.17-1.5 (3 H, m), 1.7-2.7 (7 H, m), 4-4.5 (2 H, m); ^{19}F NMR: δ -77 and -83.

Synthetic method
   4-(Carboethoxy)-3-(trifluoromethyl)-cyclohex-2-enone
   4-(Carboethoxy)-3-(trifluoromethyl)-3-hydroxycyclohexanone (6.8 g, 27 mmol), benzene (20 mL) and Nordhausen acid (30 mL; 20% SO_3/H_2SO_4) were stirred for a day.   The mixture was poured in ice.   After extraction with Et_2O, washing with sat. $NaHCO_3$ aq., then brine, the organic layer was dried on $MgSO_4$.   4-(Carboethoxy)-3-(trifluoromethyl)cyclohex-2-en-1-one was obtained by silica gel column chromatography ($CHCl_3$) in 40% yield.
Spectral data
   ^1H NMR: δ 1.3 (3 H, t, *J*=7.5 Hz), 2.36-2.63 (4 H, m), 3.57 (1 H, br t, *J*=4 Hz), 4.27 (2 H, q, *J*=7.5 Hz), 6.53 (1 H, br s); ^{19}F NMR: δ -68 (s).

Tordeux, M.; Wakselman, C. *Synth. Commun.* **1991**, *21*, 1243.

Ethyl *anti*-3-(trifluoromethyl)-4-methyl-5-oxo-5-phenylpentanoate

a) LDA / THF

Synthetic method
   A dry two-necked flask equipped with a rubber septum was placed under a nitrogen atmosphere and charged with 2.4 mmol of LDA in freshly distilled THF (5 mL).   To this solution at -78 °C was added propiophenone (2.4 mmol), and the whole was stirred for 30 min.   The reaction mixture was then treated with ethyl (*E*)-4,4,4-trifluorobut-2-enoate (2.0 mmol) and further stirred at -78 °C for 0.5 h.   The reaction was quenched with 3 *N* HCl aq. and diluted with Et_2O, and the resulting organic layer was separated.   The aqueous layer was extracted twice with Et_2O and the combined ethereal layers were washed with brine, dried over $MgSO_4$, and evaporated.   The resulting crude material was purified by silica gel chromatography to give ethyl *anti*-3-(trifluoromethyl)-4-methyl-5-oxo-5-phenylpentanoate in 98% yield as a single stereoisomer.
Spectral data
   Rf 0.30 (hexane:AcOEt=10:1); ^1H NMR ($CDCl_3$): δ 1.24 (3 H, dq, *J*=7.04, 0.78 Hz), 1.24 (3 H, t, *J*=7.14 Hz), 2.57 (1 H, dd, *J*=16.77, 6.94 Hz), 2.67 (1 H, dd, *J*=16.77, 5.54 Hz), 3.44 (1 H, qddd, *J*= 9.40, 6.95, 5.61, 4.88 Hz), 3.94 (1 H, qd, *J*=7.00, 4.85 Hz), 4.13 (1 H, dq, *J*=10.68, 7.07 Hz), 4.17 (1 H, dq, *J*=10.82, 7.08 Hz), 7.49-7.94 (5 H, m); ^{13}C NMR ($CDCl_3$): δ 12.88, 14.05, 29.80 (q, *J*=2.5 Hz), 37.88 (q, *J*=1.9 Hz), 40.68 (q, *J*=26.0 Hz), 61.13, 127.50 (q,

J=280.8 Hz), 128.37, 128.89, 133.48, 135.35, 170.58, 200.75; ^{19}F NMR (CDCl$_3$): δ 8.2 (d, J= 9.2 Hz) from CF$_3$CO$_2$H; IR (neat): ν 3050, 3000, 2925, 1740, 1690, 1600, 1580 cm^{-1}.

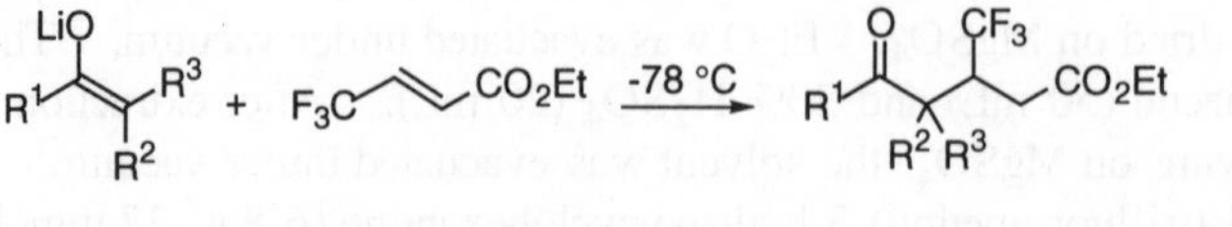

R^1	R^2	R^3	Time (h)	Yield (%)	Selectivity (% de)
Ph[a)]	H	H	1.0	33	
Et[b)]	Me	H	0.5	97	96 (anti)
Et[b)]	H	Me	0.5	97	>98 (anti)
EtO	H	H	1.5	87	
EtO	Me	H	1.5	54	78 (anti)
EtO	Me	Me	1.0	98	
Me$_2$N	H	H	1.5	86	
Me$_2$N	H	Me	1.5	89	70 (syn)
-(CH$_2$)$_3$-[a)]		H	4.0	93	20
-O-(CH$_2$)$_3$-		H	2.0	>98	74 (anti)
Ox[c)]	H	Me	1.5	88	>98 (anti)
Ox[c)]	H	i-Pr	1.5	97	97 (anti)
Ox[c)]	H	BnO	1.5	36	>98 (anti)

[a)] Reaction was carried out at -78 °C to rt.   [b)] Prepared from the corresponding silyl enol ether ((E):(Z)=10:90 and (E):(Z)=76:24, respectively).   [c)] (S)-4-(prop-2-yl)oxazolidinon-3-yl.

(Reproduced with permission from Yamazaki, T., et al. J. Org. Chem. **1995**, 60, 4363)

Yamazaki, T., et al. J. Org. Chem. **1995**, 60, 4363.

4.3.3 Ene Reactions

(2S*, 3S*)-1,1,1-Trifluoro-3-methylnon-4-en-2-ol

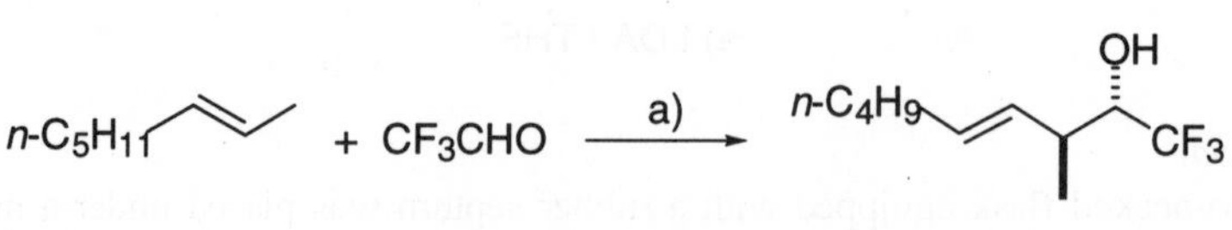

a) MeAlCl$_2$ / CH$_2$Cl$_2$

Synthetic method

In an atmosphere of argon, trifluoroacetaldehyde (3.3 mL) and methylaluminum dichloride (20 wt% in hexane, 5 mL) were added to a solution of (E)-2-octene (3.0 g, 27 mmol) in dry CH$_2$Cl$_2$ (10 mL) at -78 °C and the mixture was stirred at this temperature for 1.5 h.   The solution was poured into a mixture of 10% HCl and ice, and extracted with CH$_2$Cl$_2$.   The CH$_2$Cl$_2$ layer was washed with H$_2$O and dried over MgSO$_4$.   After evaporation of the solvent, the residue was separated by silica gel column chromatography (hexane:CH$_2$Cl$_2$=4:1 to 1:1) to give 4.28 g of anti-1,1,1-trifluoro-3-methylnon-4-en-2-ol in 76% yield as a colorless oil.

Spectral data

^{1}H NMR (CDCl$_3$): δ 0.89 (3 H, t, J=6.2 Hz), 1.16 (3 H, d, J=7.5 Hz), 1.21-1.45 (4 H, m), 1.77-2.23 (2 H, m), 2.48 (1 H, s), 2.67 (1 H, ddq, J=7.5, 7.2, 5.4 Hz), 3.76 (1 H, dq, J=6.8,

5.4 Hz), 5.40 (1 H, dd, J=15.4, 7.2 Hz), 5.53-5.87 (1 H, m); ^{19}F NMR (CDCl$_3$): δ 11.82 (d, J=6.8 Hz) from C$_6$H$_5$CF$_3$.

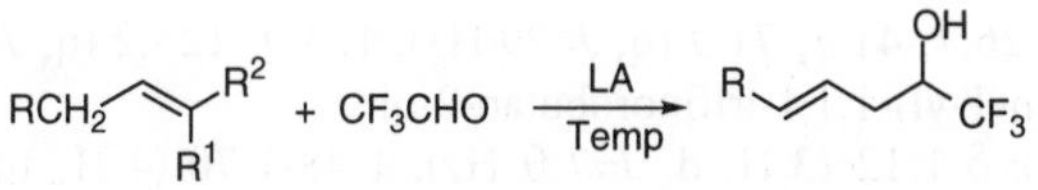

R	R^1	R^2	LA	Temp. (°C)	Yield (%)
n-C$_5$H$_{11}$	H	H	FeCl$_3$	80	42
			none	120	0
			AlCl$_3$	-78	36
-(CH$_2$)$_3$-		H	AlCl$_3$	-78	9
			MeAlCl$_2$	-78	32
Ph	H	H	AlCl$_3$	-78	20
			MeAlCl$_2$	-78	38
n-Bu	Me	H	FeCl$_3$	80	22
n-C$_7$H$_{15}$	H	H	MeAlCl$_2$	-78	27
			AlCl$_3$	-78	77

(Reproduced with permission from Kumadaki, I., *et al. Chem. Pharm. Bull.* **1991**, 39, 1707)

Note

For the preparation of trifluoroacetaldehyde, see page 188 of this volume.

See the following articles on the Lewis acid-induced ene cyclization using trifluoromethylated substrates: i) Bégué, J.-P., *et al. J. Chem. Soc., Perkin Trans. 1* **1991**, 1397, ii) Bégué, J.-P., *et al. J. Org. Chem.* **1991**, 56, 5800.

Kumadaki, I., *et al. Chem. Pharm. Bull.* **1991**, 39, 1707.

syn-3-(Cyclohexen-1-yl)-1,1,1-trifluorobutan-2-ol

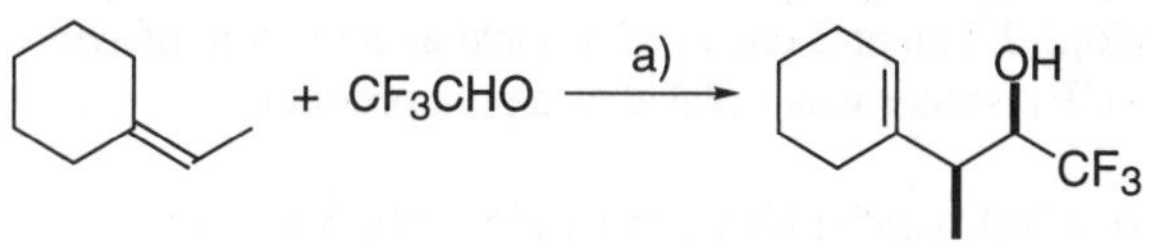

a) (*R*)-BINOL-TiCl$_2$, MS 4 Å / CH$_2$Cl$_2$

Synthetic method

To a suspension of MS 4Å (activated powder, 200 mg) in CH$_2$Cl$_2$ (2 mL) was added the (*R*)-BINOL-derived titanium complex (0.10 mmol) at room temperature. After stirring for 5 min, ethylidenecyclohexane (1.0 mmol) and freshly dehydrated and distilled trifluoroacetaldehyde (*ca.* 2 equiv) in CH$_2$Cl$_2$ (0.5 mL) were added to the mixture at 0 °C. After stirring for 30 min at that temperature, Et$_2$O (2 mL) and sat. NaHCO$_3$ aq. (2 mL) were added to the mixture. MS 4 Å was filtered off through a pad of Celite and the filtrate was extracted three times with Et$_2$O. The combined organic layer was washed with brine, dried over MgSO$_4$, and evaporated under reduced pressure. Chromatographic separation by silica gel gave 3-(cyclohexen-1-yl)-1,1,1-trifluoro-butan-2-ol in 94% yield in a *syn:anti* ration of 98:2, and the *syn* isomer was found to be 96% ee after derivatization to the corresponding MTPA ester.

Spectral data

$[\alpha]_D^{24}$ +13.5 (*c* 1.0, CHCl$_3$); ^{1}H NMR (CDCl$_3$): δ 1.12 (3 H, d, *J*=7.0 Hz), 1.48-1.70 (4 H, m), 1.84-2.08 (4 H, m), 2.45 (1 H, m), 3.95 (1 H, m), 5.58 (1 H, br s); ^{13}C NMR (CDCl$_3$): δ 13.2, 22.5, 23.0, 25.4, 26.7, 41.3, 71.9 (q, *J*=29 Hz), 123.9, 125.2 (q, *J*=281 Hz).

anti-3-(Cyclohexen-1-yl)-1,1,1-trifluorobutan-2-ol

^{1}H NMR (CDCl$_3$): δ 1.12 (3 H, d, *J*=7.0 Hz), 1.48-1.70 (4 H, m), 1.84-2.08 (4 H, m), 2.45 (1 H, m), 3.72 (1 H, m), 5.66 (1 H).

Note

For the preparation of trifluoroacetaldehyde, see page 188 of this volume.

Mikami, K., *et al. Tetrahedron* **1996**, *52*, 85.

4.3.4 Diels-Alder Reactions and 1,3-Dipolar Cycloadditions

5-(Trifluoromethyl)-5-(phenylsulfenyl)bicyclo[2.2.1]hept-2-ene

a) Cyclopentadiene / benzene

Synthetic method

A Carius tube was charged with 0.50 g (2.4 mmol) of 3,3,3-trifluoro-2-(phenylsulfenyl)-propene, 5 mL of cyclopentadiene, hydroquinone and 1 mL of dry benzene. The tube was sealed under vacuum and heated at 110 °C for 24 h. After evaporation, the residue was purified by column chromatography on silica gel (petroleum ether) to provide 0.48 g of 5-(trifluoromethyl)-5-(phenylsulfenyl)-bicyclo[2.2.1]hept-2-ene in 72% yield as a mixture of *endo/exo* isomers in a ratio of 1:1 from which *exo* CF$_3$ isomer was isolated as a pure product.

Spectral data

IR (neat): ν 2900, 1500, 1285-1120, 1250-1015, 700, 600 cm^{-1}.

5-(Trifluoromethyl)-5-(phenylsulfenyl)bicyclo[2.2.1]hept-2-ene (*exo* CF$_3$ form)

Rf 0.30 (petroleum ether); ^{1}H NMR (CDCl$_3$): δ 1.27 (1 H, dd, *J*=13.4, 2.9 Hz), 1.57 (1 H, ddd, *J*=9.3, 2.9, 1.5 Hz), 1.89 (1 H, dd, *J*=9.3, 1.5 Hz), 2.06 (1 H, dd, *J*=13.3, 3.5 Hz), 2.97 (1 H, br s), 3.22 (1 H, br s), 6.31 (1 H, dd, *J*=5.6, 3.0 Hz), 6.39 (1 H, dd, *J*=5.4, 3.4 Hz), 7.25-7.55 (5 H, m); ^{13}C NMR (CDCl$_3$): δ 34.38 (td, *J*=139.0 Hz), 41.97 (dd, *J*=147.5, 6.4 Hz), 47.83 (tdd, *J*=134.4, 7.0, 4.0 Hz), 47.90 (q, *J*=26.1 Hz), 48.77 (dd, *J*=158.7, 8.5 Hz), 128.15 (q, *J*=286.0 Hz), 128.70 (dd, *J*=161.0, 6.9 Hz), 128.94 (dt, *J*=161.1, 7.9 Hz), 136.09 (dd, *J*=164.9 Hz), 137.07 (d, *J*=173.4 Hz), 137.15 (d, *J*=163.2 Hz), 140.20; ^{19}F NMR (CDCl$_3$): δ -66.08 (s).

5-(Trifluoromethyl)-5-(phenylsulfenyl)bicyclo[2.2.1]hept-2-ene (*endo* CF$_3$ form)

Rf 0.40 (petroleum ether); ^{1}H NMR (CDCl$_3$): δ 1.56 (1 H, d, *J*=8.1 Hz), 1.59 (1 H, d, *J*=12.4 Hz), 1.89 (1 H, d, *J*=12.4 Hz), 2.46 (1 H, d, *J*=8.1 Hz), 2.88 (1 H, br s), 3.00 (1 H, br s), 5.93 (1 H, dd, *J*=5.5, 2.7 Hz), 6.25 (1 H, dd, *J*=5.8, 3.0 Hz), 7.26-7.64 (5 H, m); ^{13}C NMR (CDCl$_3$): δ 36.62 (td, *J*=136.5, 6.7 Hz), 43.31 (dd, *J*=148.2, 6.8 Hz), 47.87 (q, *J*=26.1 Hz), 49.38 (tdd, *J*=151.0, 16.0, 8.5 Hz), 49.68 (d, *J*=135.4 Hz), 127.68 (q, *J*=280.1 Hz), 128.70 (dd, *J*=161.1, 6.3 Hz), 129.08 (dt, *J*=161.2, 7.9 Hz), 129.52 (dd, *J*=161.0, 7.5 Hz), 137.04 (d,

J=163.6 Hz), 139.21 (d, J=164.2 Hz), 167.07; ^{19}F NMR (CDCl$_3$): δ -63.57 (s).
Note

For the preparation of the starting vinylic sulfide, see page 187 of this volume.

For the representative Diels-Alder reaction using trifluorinated substrates, see the following: i) Ojima, I., *et al. J. Org. Chem.* **1982**, *47*, 2051, ii) Gaede, B.; Balthazor, T. M. *J. Org. Chem.* **1983**, *48*, 276, iii) Kobayashi, Y., *et al. Chem. Pharm. Bull.* **1985**, *33*, 3670, iv) Sicsic, S., *et al. Synthesis* **1987**, 155, v) Schuler, B.; Sundermeyer, W. *Tetrahedron Lett.* **1989**, *30*, 4111, vi) Bonnet-Delpon, D., *et al. Synlett* **1992**, 146, iii) Buback, M., *et al. Liebigs Ann.* **1996**, 1151.

Viehe, H. G., *et al. Tetrahedron* **1997**, *53*, 6861.

4-*exo*- and 4-*endo*-(Trifluoromethyl)bicyclo[2.2.1]hept-6-ene-4-carboxylic acid

a) Cyclopentadiene / CH$_2$Cl$_2$, b) I$_2$, KI / H$_2$O, c) Zn / EtOH

Synthetic method

To a solution of 2-(trifluoromethyl)acrylic acid (0.5 g, 3.5 mmol) in CH$_2$Cl$_2$ (5 mL) was added cyclopentadiene (0.28 g, 4.2 mmol) at 0 °C, and the mixture was stirred at room temperature for 2 h.   Evaporation of the solvent gave a diastereomeric mixture of the adduct (0.59 g, 81% yield, *endo*-CO$_2$H:*exo*-CO$_2$H=2:1).

The mixture (200 mg) was dissolved in sat. NaHCO$_3$ aq. (5 mL) and treated with an excess I$_2$-KI solution (I$_2$ 245 mg, KI 1.4 g in H$_2$O 10 mL).   After the mixture was stirred at room temperature for 2 h, the solution was extracted with Et$_2$O.   The combined Et$_2$O layers were washed with 5% Na$_2$S$_2$O$_2$ aq, H$_2$O, and brine before being dried (MgSO$_4$).   Removal of the solvent gave iodo lactone.

The aqueous layer was made acidic with 1 *N* HCl and extracted with Et$_2$O.   The combined organic layer was dried (MgSO$_4$).   Concentration of the filtrate gave 60 mg of 4-*exo*-(trifluoro-methyl)bicyclo[2.2.1]hept-6-ene-4-carboxylic acid.

A solution of the above iodo lactone (1.5 g, 4.5 mmol) in EtOH (10 mL) was heated to reflux in the presence of zinc powder (792 mg), and the mixture was stirred at the same temperature for 1 h.   The cooled mixture was filtered and concentrated to give a crude material, which was dissolved in sat. NaHCO$_3$ aq. and washed with Et$_2$O.   The aqueous layer was acidified with 1 *N* HCl and extracted with Et$_2$O.   Removal of the solvent gave 713 mg of pure crystals of 4-*endo*-(trifluoro-methyl)bicyclo[2.2.1]hept-6-ene-4-carboxylic acid in 77% yield.

Spectral data

4-*exo*-(Trifluoro-methyl)bicyclo[2.2.1]hept-6-ene-4-carboxylic acid

mp. 87-88 °C; ^{1}H NMR (CDCl$_3$): δ 1.42 (1 H, d, J=8.8 Hz), 1.48-1.54 (2 H, m), 2.6 (1 H, dd, J=13.6, 3.5 Hz), 3.01 (1 H, br s), 3.49-3.51 (1 H, br s), 6.07 (1 H, m), 6.33 (1 H, dd, J=5.5,

3.0 Hz); ^{19}F NMR (CDCl$_3$): δ 1.53 (s) from PhCF$_3$.

4-*endo*-(Trifluoro-methyl)bicyclo[2.2.1]hept-6-ene-4-carboxylic acid

mp. 75 °C; ^{1}H NMR (CDCl$_3$): δ 1.5 (1 H, d, *J*=9.2 Hz), 1.8 (1 H, d, *J*=9.2 Hz), 2.04 and 2.06 (2 H, AB q, *J*=12.9 Hz), 3.0 (1 H, br s), 3.39 (1 H, br s), 6.11 (1 H, dd, *J*=5.7, 3.0 Hz), 6.34 (1 H, dd, *J*=5.7, 3.0 Hz); ^{19}F NMR (CDCl$_3$): δ 2.6 (s) from PhCF$_3$; IR (KBr): v 3100, 1715 cm^{-1}.

Iodo lactone

mp. 65-66 °C; ^{1}H NMR (CDCl$_3$): δ 2.05-2.12 (2 H, m), 2.20 (1 H, dd, *J*=14, 4 Hz), 2.45 (1 H, d, *J*=11.8 Hz), 2.8 (1 H, br s), 3.39 (1 H, d, *J*=5.2 Hz), 3.88 (1 H, d, *J*=2.7 Hz), 5.16 (1 H, d, *J*=5.2 Hz); ^{19}F NMR (CDCl$_3$): δ 8.0 (s) from PhCF$_3$; IR (CCl$_4$): v 1800 cm^{-1}.

Note

When the Diels-Alder reaction was carried out with 2-(trifluoromethyl)acrylate possessing D-pantolactone-derived auxiliary and cyclopentadiene in the presence of a catalytic amount of TiCl$_4$ (13 mol%) at -23 °C, the product was obtained in 95% total yield in a ratio of *endo:exo*=8:1 with 98% diastereomeric excess.

See also the following report on the Ene reaction of the above pantolactone-derived ester with alkenes: Taguchi, T., *et al. J. Chem. Soc., Chem. Commun.* **1991**, 721.

Taguchi, T., *et al. J. Org. Chem.* **1991**, *56*, 1718.

Methyl (1R^*,2R^*,10bR^*)-2-(trifluoromethyl)-1,5,6,10b-tetrahydro-2H-isoxazolo[3,2-a]isoquinoline-1-carboxylate

a) 3,4-Dihydroisoquinoline *N*-oxide / toluene

Synthetic method

A solution of 3,4-dihydroisoquinoline *N*-oxide (0.67 g, 4.53 mmol) and methyl (*E*)-4,4,4-trifluorobut-2-enoate (0.58 g, 3.78 mmol) in 20 mL of toluene was stirred at 20 °C for 20 h and the solvent was then removed. The resulting oil was chromatographed on silica gel (hexane:AcOEt= 10:1) to give 0.18 g (53% yield) of methyl (1R^*,2R^*,10bR^*)-2-(trifluoromethyl)-1,5,6,10b-tetra-hydro-2H-isoxazolo[3,2-a]isoquinoline-1-carboxylate and 0.03 g (18% yield) of the corresponding (1R^*,2R^*,10bS^*) isomer. Only the former diastereomer was formed when the reaction was carried out at 80 °C for 20 h (70% yield).

Spectral data

mp. 72-73 °C (hexane:AcOEt); ^{1}H NMR (CDCl$_3$): δ 2.84 (1 H, dt, *J*=16.6, 4.3 Hz), 3.03 (1 H, ddd, *J*=16.6, 10.8, 5.4 Hz), 3.28 (3 H, s), 3.31 (1 H, m), 3.58 (1 H, ddd, *J*=10.8, 10.8, 4.3 Hz), 3.88 (1 H, dd, *J*=9.8, 6.1 Hz), 4.98 (1 H, d, *J*=9.8 Hz), 5.13 (1 H, dq, *J*=6.1, 6.1 Hz), 7.20 (4 H, m); IR (KBr): v 1740, 1200-1140 cm^{-1}.

Methyl (1R^*,2R^*,10bS^*)-2-(trifluoromethyl)-1,5,6,10b-tetra-hydro-2H-isoxazolo[3,2-a]iso-quinoline-1-carboxylate

mp. 121-122 °C (hexane:AcOEt); ^{1}H NMR (CDCl$_3$): δ 2.97 (1 H, t, *J*=5.4 Hz), 3.26 (1 H, m), 3.35 (1 H, m), 3.61 (1 H, dd, *J*=8.6, 8.6 Hz), 3.86 (3 H, s), 4.78 (1 H, dq, *J*=8.6, 6.5 Hz),

4.93 (1 H, dm, *J*=8.6 Hz), 6.96 (1 H, m), 7.22 (3 H, m); IR (KBr): ν 1740, 1200-1140 cm⁻¹.

Note

See the following recent articles as representative examples of similar types of 1,3-dipolar cycloadditions with trifluoromethylated olefins: i) Bravo, P., *et al. J. Chem. Res. (S)* **1992**, 40, ii) Tanaka, K., *et al. Bull. Chem. Soc. Jpn.* **1993**, *66*, 263, iii) Bégué, J.-P., *et al. Tetrahedron Lett.* **1993**, *34*, 3279, iv) Eguchi, S., *et al. J. Chem. Soc., Perkin Trans. 1* **1995**, 2761, v) Bruché, L., *et al J. Chem. Res. (S)* **1996**, 198, vi) Bouillon, J.-P., *et al. J. Chem. Soc., Perkin Trans. 1* **1996**, 1853, vii) Viehe, H. G., *et al. Tetrahedron* **1996**, *52*, 4383, viii) Viehe, H. G., *et al. Tetrahedron Lett.* **1996**, *37*, 5515.

Tanaka, K., *et al. Bull. Chem. Soc. Jpn.* **1996**, *69*, 2243.

4-Ethyl-3-(trifluoromethyl)-2-methyl-5-phenylpyrrolidine-2,4-dicarboxylate

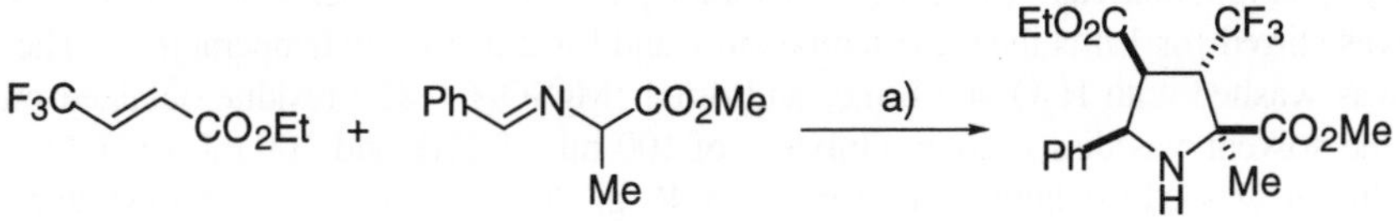

a) AgOAc, Et₃N / toluene

Synthetic method

To a mixture of methyl *N*-benzylidene alaninate (1 g, 5.2 mmol), silver acetate (1.1 g, 6.5 mmol), triethylamine (0.8 mL, 6 mmol) in anhydrous toluene under an argon atmosphere was added ethyl (*E*)-4,4,4-trifluorobut-2-enoate (1.0 g, 6 mmol) in toluene. The suspension was stirred at room temperature for 18 h and then the solvent was removed under reduced pressure to give a colorless solid (a 96:4 mixture of isomers was observed by ¹⁹F NMR). Recrystallization gave 1.65 g of pure 4-ethyl-3-(trifluoromethyl)-2-methyl-5-phenylpyrrolidine-2,4-dicarboxylate in 88% yield.

Spectral data

mp. 43-45 °C (pentane); ¹H NMR (CDCl₃): δ 0.75 (3 H, t, *J*=7.1 Hz), 1.6 (3 H, q, *J*=2 Hz), 2.8 (1 H, br s), 3.5 (1 H, dd, *J*=8.5, 7.4 Hz), 3.6 (2 H, m), 3.85 (3 H, s), 3.9 (1 H, qd, *J*=9.8, 7.4 Hz), 4.8 (1 H, d, *J*=8.5 Hz), 7.3 (5 H, m); ¹³C NMR (CDCl₃): δ 13.2, 19.8, 50.5, 51.6 (q, *J*=27 Hz), 52.7, 60.7, 62.7, 65.8, 125.8 (q, *J*=283 Hz), 127.2, 127.9, 137.7, 172.8, 173.4; ¹⁹F NMR (CDCl₃): δ -67.1 (dq, *J*=9.8, 2 Hz); IR (neat): ν 3350, 1734 cm⁻¹.

Bonnet-Delpon, D., *et al. Bull. Soc. Chim. Fr.* **1995**, *132*, 402.

4.3.5 Wittig Type Reactions

(E)-4,4,4-Trifluoro-3-methylbut-2-enoic acid

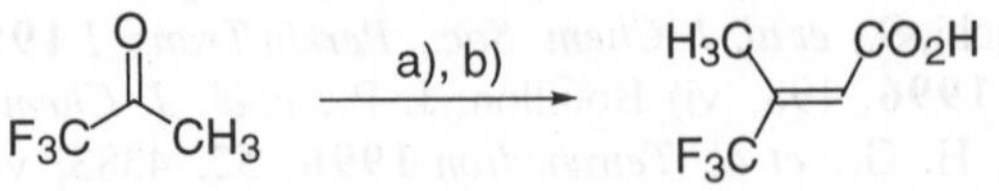

a) (EtO)$_2$P(O)CH$_2$CO$_2$Et, NaH / benzene,   b) LiOH / THF-H$_2$O

Synthetic method

A solution of 1,1,1-trifluoroacetone (10 mL, 108 mmol) in 50 mL of benzene was added to a stirred suspension of the sodium salt of triethyl phosphonoacetate [from NaH (3.90 g, 165 mmol) and triethyl phosphonoacetate (33 mL, 165 mmol)] in 150 mL of benzene at 0 °C.   Then the mixture was stirred for 1 h at the same temperature and for 2 h at room temperature.   The reaction mixture was washed with H$_2$O and brine, and dried (MgSO$_4$).   The residue obtained on evaporation of the solvent was dissolved in a mixture of 100 mL of THF and 100 mL of H$_2$O.   To this stirred solution was added lithium hydroxide (8.46 g, 202 mmol) at room temperature.   After stirring had continued for 7 h at the same temperature, the reaction mixture was diluted with Et$_2$O and extracted with sat. NaHCO$_3$ aq.   The aqueous layer was acidified by the addition of 10% HCl and extracted with Et$_2$O.   The combined extracts were dried (Na$_2$SO$_4$).   Distillation of the residue on evaporation of the solvent gave 5.66 g of (E)-4,4,4-trifluoro-3-methylbut-2-enoic acid in 34% yield as a colorless oil.

Spectral data

bp. 80-81 °C/21 mmHg; ^{1}H NMR (CCl$_4$): δ 2.28 (3 H, s), 6.37 (1 H, br s), 12.06 (1 H, br s); IR (neat): v 1710 cm^{-1}.

Note

The intermediary α,β-unsaturated ester (bp. 116-118 °C/650 mmHg) was also prepared independently, for example, by Poulter and Satterwhite in a similar way.   See Poulter, C. D.; Satterwhite, D. M. *Biochemistry* **1977**, *16*, 5470.

Fukumoto, K., *et al. J. Org. Chem.* **1995**, *60*, 594.

Ethyl (E)-4,4,4-trifluoro-2-methyl-3-phenylbut-2-enoate

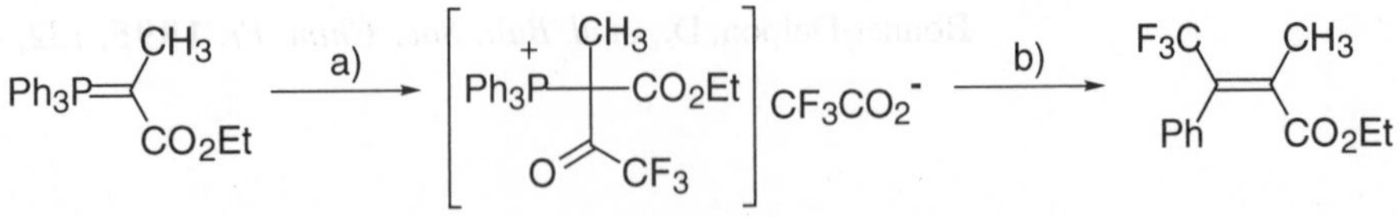

a) (CF$_3$CO)$_2$O / THF,   b) PhLi

Synthetic method

Trifluoroacetic anhydride (0.42 g, 2 mmol) was added dropwise to the solution of (carbethoxy-ethylidene)triphenylphosphorane (0.72 g, 2 mmol) in dry THF (15 mL) at -78 °C under nitrogen.   After stirring for 30 min at this temperature, phenyllithium (4 mmol in Et$_2$O) was slowly added for 30 min.   The mixture was then warmed to room temperature and stirred for a further 3 h.   The product was isolated by column chromatography on silica gel (petroleum ether (60-96 °C):AcOEt=10:1) to give ethyl (E)-4,4,4-trifluoro-2-methyl-3-phenylbut-2-enoate in 66% yield as an 85:15 (E):(Z) mixture.

Spectral data

bp. 80 °C/2 mmHg; ^{1}H NMR (CCl$_4$): δ 1.25 (3 H, t, J=6 Hz), 1.72-1.80 (3 H, m), 4.20 (2 H, q, J=6 Hz), 7.20-7.50 (5 H, m); ^{19}F NMR (CCl$_4$): δ 16.0 (s, (E) isomer), 17.2 (s, (Z) isomer) from CF$_3$CO$_2$H; IR (film): ν 1740, 1670 cm^{-1}.

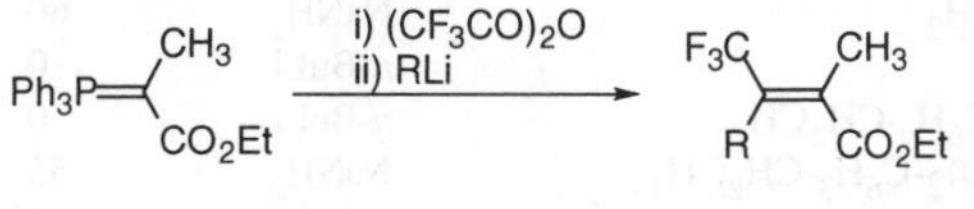

R	Yield (%)	(E):(Z) ratio			bp. (°C/2 mmHg)
Ph-C≡C-	41	77	:	23	88
n-Bu-C≡C-	50	92	:	8	63
n-C$_8$H$_{17}$-C≡C-	46	100	:	0	74
4-Me-C$_6$H$_4$-	66	95	:	5	94
2-Me-C$_6$H$_4$-	59	86	:	14	88
4-MeO-C$_6$H$_4$-	47	100	:	0	92
2-MeO-C$_6$H$_4$-	40	100	:	0	90

(Reproduced with permission from Shen, Y.-C.; Xiang, Y.-J. *J. Fluorine Chem.* **1991**, *52*, 221)

Shen, Y.-C.; Xiang, Y.-J. *J. Fluorine Chem.* **1991**, *52*, 221.

(Z)-2-Ethoxy-1,1,1-trifluoro-5-phenylpent-2-ene

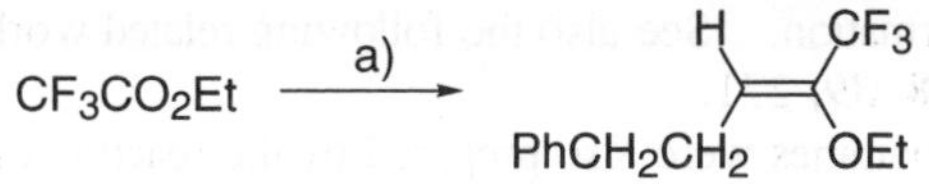

a) Ph(CH$_2$)$_3$PPh$_3$·Br, NaNH$_2$ / THF

Synthetic method

Triphenyl-(3-phenylpropyl)phosphonium bromide (13.83 g, 30 mmol) was added to a suspension of NaNH$_2$ (1.17 g, 30 mmol) in THF (80 mL) under an argon atmosphere.   Then hexamethyldisilazane (0.6 mL) was added *via* a syringe through a septum cap.   The mixture was stirred and heated to reflux until no more NH$_3$ evolved (usually 2-3 h) and then cooled to room temperature.   The red ylide solution was added dropwise into another flask containing ethyl trifluoroacetate (4.26 g, 30 mmol) in THF (10 mL).   After the end of the addition, the reaction medium was stirred again until the red color disappeared (4-6 h) at reflux.   The mixture was concentrated under reduced pressure, and triphenylphosphine oxide was precipitated by the addition of pentane (30 mL).   The solution was filtered through a silica gel column (pentane: Et$_2$O=97:3).   Evaporation of the solvent gave the residue, which was purified by bulb-to-bulb distillation to afford 4.39 g of (Z)-2-ethoxy-1,1,1-trifluoro-5-phenylpent-2-ene in 60% yield.

Spectral data

bp. 80 °C (bath temperature)/10 mmHg; ^{1}H NMR (CDCl$_3$): δ 1.25 (3 H, t, J=7 Hz), 2.51 (2 H, m), 2.68 (2 H, m), 3.80 (2 H, q, J=7 Hz), 6.30 (1 H, t, J=7 Hz), 7.3 (5 H, m); ^{13}C NMR (CDCl$_3$): δ 15.4, 26.7, 35.0, 69.8, 121.7 (q, J=275 Hz), 126.3, 128.3, 128.5, 128.6, 128.7,

141.0, 143.4 (q, J=32 Hz); ^{19}F NMR (CDCl$_3$): δ -68 (s).

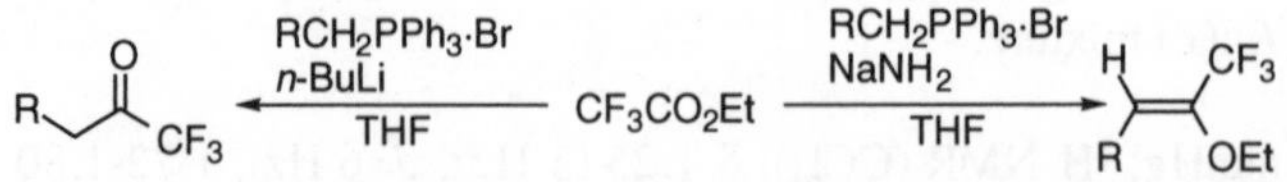

R		Base	Yield (%)	
			Enol ether	Ketone
PhCH$_2$CH$_2$		NaNH$_2$	60	0
		n-BuLi	0	60
4-MeO-C$_6$H$_4$-CH$_2$CH$_2$		n-BuLi	0	40
3,4-(MeO)$_2$-C$_6$H$_3$-CH$_2$CH$_2$		NaNH$_2$	55	0
		n-BuLi	0	40
3,4-(OCH$_2$O)-C$_6$H$_3$-CH$_2$CH$_2$		n-BuLi	0	40
c-C$_6$H$_{11}$		NaNH$_2$	55	0
c-C$_6$H$_{11}$-CH$_2$		NaNH$_2$	50	0
Ph(CH$_2$)$_3$		n-BuLi	0	52
n-C$_6$H$_{13}$		n-BuLi	0	40
Ph		NaNH$_2$	55	0
		n-BuLi	29	0
3-CF$_3$-C$_6$H$_4$		n-BuLi	61	0
EtO$_2$C		NaNH$_2$	30	0

(Reproduced with permission from Bégué, J.-P., *et al. J. Org. Chem.* **1992**, *57*, 3807)

Note

When *n*-BuLi was employed as the base, trifluoromethyl alkyl ketones were obtained except for the case of R=Ar (see the table above).   In this reaction, lithium apparently played a crucial role in the alteration of the reaction path, which was verified by the addition of HMPA to this solution, leading to the enol ether formation.   See also the following related work: Bégué, J.-P.; Mesureur, D. *J. Fluorine Chem.* **1988**, *39*, 271.

The corresponding enamines were also prepared by the reaction of appropriate phosphonium salts and trifluoroacetamides.   See Bégué, J.-P.; Mesureur, D. *Synthesis* **1989**, 309.

Bégué, J.-P., *et al. J. Org. Chem.* **1992**, *57*, 3807.

4.3.6 Rearrangements

Methyl (2S, 3S, 4E)-6-benzyloxy-3-(trifluoromethyl)-2-hydroxyhex-4-enoate

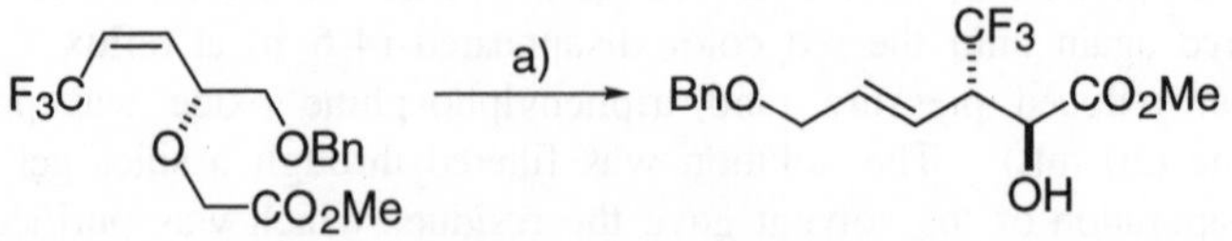

a) LDA / THF, HMPA

Synthetic method

To an LDA (0.62 mmol) solution in THF (5 mL) at -78 ℃, HMPA (0.34 ml, 1.95 mmol) was added, and the resulting clear solution was stirred for 30 min at that temperature.   To this was added a solution of methyl 2-[(S,Z)-1-benzyloxy-5,5,5-trifluoropent-3-en-2-yl]acetate (0.31 mmol) in THF (2 ml), and the mixture was stirred at the same temperature for 10 min.   At this point, the ester disappeared completely (monitored by TLC).   Then the reaction mixture was poured into ice-cooled 3 N HCl aq. and extracted with AcOEt.   The organic layer was dried over anhydrous MgSO₄ and concentrated in vacuo.   After purification by silica gel column chromatography, methyl (2S,3S,4E)-6-benzyloxy-3-(trifluoromethyl)-2-hydroxyhex-4-enoate was obtained in 73% yield.

Spectral data

$[\alpha]_D^{17}$ +10.1 (c 0.3, CHCl₃); ¹H NMR (CDCl₃): δ 2.99 (1 H, d, J=5.37 Hz), 3.27 (1 H, dquint., J=9.04, 1.96 Hz), 3.82 (3 H, s), 3.97-4.07 (2 H, m), 4.48 (2 H, s), 4.66 (1 H, dd, J=5.13, 1.95 Hz), 5.77 (1 H, dd, J=15.6, 9.28 Hz), 5.85 (1 H, dd, J=15.9, 5.13 Hz), 7.20-7.40 (5 H, m); ¹³C NMR (CDCl₃): δ 50.02 (q, J=27.1 Hz), 53.23, 69.15 (q, J= 2.5 Hz), 69.47, 71.91, 120.21 (q, J=2.5 Hz), 125.54 (q, J=281 Hz), 127.71, 127.73, 128.41, 136.00, 137.82, 172.58; ¹⁹F NMR (CDCl₃): δ 93.4 (d, J=7.63 Hz) from C₆F₆; IR (neat) ν 3500, 3150, 3100, 3050, 3000, 2956, 2860, 1746 cm⁻¹.

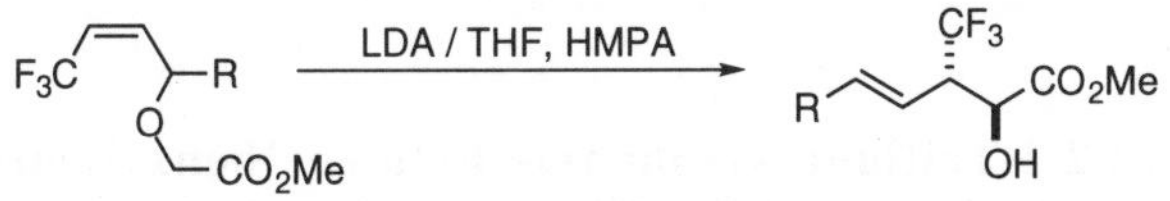

R	Yield (%)	anti:syn		
BnOCH₂CH₂	78	>99	:	1
BnOCH₂	61[a)]	16	:	84[b)]
BnOCH₂CH₂	46[a)]	14	:	86[c)]
n-C₅H₁₁	64	>99	:	1
c-C₆H₁₁	75	>99	:	1

[a)] The corresponding (E) olefin was employed.   [b)] The *syn* isomer consisted of a 1:1 (E),(Z) mixture.   [c)] The *syn* isomer consisted of a 4:1 (E),(Z) mixture.

(Reproduced with permission from Kitazume, T., *et al. J. Org. Chem.* **1997**, *62*, 137)

Note

For the preparation of the starting material, see page 201 of this volume.

Kitazume, T., *et al. J. Org. Chem.* **1997**, *62*, 137.

Ethyl (S,E)-3-(trifluoromethyl)dec-4-enoate

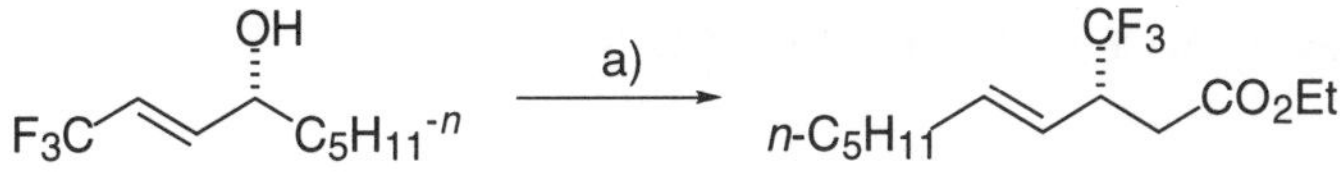

a) CH₃C(OEt)₃, cat. EtCO₂H

Synthetic method

A solution of (R,E)-1,1,1-trifluoronon-2-en-4-ol (0.51 mmol; 87% ee), triethyl orthoacetate (1 mL), and a catalytic amount of propionic acid was stirred for 12 h at 130 °C in a sealed tube.

The reaction mixture was cooled and evaporated *in vacuo*. The resultant materials were purified by column chromatography to afford ethyl (*S*,*E*)-3-(trifluoromethyl)dec-4-enoate in 96% yield.

Spectral data

[α]$_D^{21}$ -14.7 (*c* 0.9, CHCl$_3$; 90% ee); ^{1}H NMR (CDCl$_3$): δ 0.87 (3 H, t, *J*=7.08 Hz), 1.25 (3 H, t, *J*=7.33 Hz), 1.20-1.40 (6 H, m), 2.03 (2 H, q, *J*=6.83 Hz), 2.45 (1 H, dd, *J*=15.38, 9.77 Hz), 2.69 (1 H, dd, *J*=15.62, 4.64 Hz), 3.28 (1 H, dsex, *J*=9.28, 4.64 Hz), 4.14 (2 H, dq, *J*=7.08, 1.47 Hz), 5.28 (1 H, ddt, *J*=15.38, 7.08, 1.46 Hz), 5.74 (1 H, dt, *J*=14.65, 6.84 Hz); ^{13}C NMR (CDCl$_3$): δ 13.96, 14.13, 22.41, 28.47, 31.12, 32.38, 33.95 (q, *J*=2.4 Hz), 43.89 (q, *J*=27.7 Hz), 60.88, 121.44 (q, *J*=2.4 Hz), 126.46 (q, *J*=279.3 Hz), 138.28, 170.26; ^{19}F NMR (CDCl$_3$): δ 90.00 (d, *J*=9.16 Hz) from C$_6$F$_6$; IR (neat): ν 2950, 2932, 2870, 2860, 1743 cm^{-1}.

Note

For the preparation of the starting allylic alcohol, see page 193 of this volume.

Almost complete chilarity transmission and very high (*E*) olefinic stereoselectivity were observed in this process. Usage of a sealed tube was crucial for obtaining high yields, and when such an apparatus was not employed in the above case, the yield dropped to 46%.

Kitazume, T., *et al. J. Fluorine Chem.* **1997**, *86*, 81.

N-(2,2,2-Trifluoro-1-phenylethylidene)benzylamine,
N-Benzylidene-2,2,2-trifluoro-1-phenylethylamine,
2,2,2-Trifluoro-1-phenylethylamine hydrochloride

a) BnNH$_2$ / benzene, b) Et$_3$N, c) 4 *N* HCl / Et$_2$O

Synthetic method

N-(2,2,2-Trifluoro-1-phenylethylidene)benzylamine

2,2,2-Trifluoroacetophenone (typically 30-50 mmol) was first dissolved in 30-50 mL of benzene or toluene in a 100-150-mL round-bottomed flask equipped with a reflux condenser, a Dean-Stark trap, and a magnetic stirring bar. A stoichiometric amount of benzylamine and 1 mol% of *p*-toluenesulfonic acid monohydrate were added to the reaction flask, and the mixture was stirred at reflux. After the reaction was complete (a theoretical amount of H$_2$O was removed; monitored by GLC, TLC, and ^{1}H and ^{19}F NMR), the solvent was removed *in vacuo* and purification by column chromatography afforded *N*-(2,2,2-trifluoro-1-phenylethylidene)benzylamine in 67% yield.

Spectral data

Rf 0.35 (hexane:AcOEt=4:1); ^{1}H NMR (CDCl$_3$): δ 4.61 (2 H, s), 7.26-7.31 (10 H, m); ^{19}F NMR (CDCl$_3$): δ -71.41 (s).

Synthetic method

N-Benzylidene-2,2,2-trifluoro-1-phenylethylamine

N-(2,2,2-Trifluoro-1-phenylethylidene)benzylamine (typically 25-30 mmol) was dissolved in 3-5 mL of triethylamine, and the mixture was stirred at 12-14 °C for 24 h. Progress of the isomerization was monitored by NMR or GLC, and upon completion, any undissolved solid was removed by filtration and triethylamine was evaporated *in vacuo*. The residual material was dried *via* an oil pump to completely remove triethylamine and purified by flash chromatography (hexane:AcOEt=50:3) to give 2,2,2-trifluoro-1-phenylethylamine hydrochloride in 96% yield.

Spectral data

Rf 0.32 (hexane:AcOEt=4:1); ^{1}H NMR (CDCl$_3$): δ 4.80 (1 H, q, J=7.5 Hz), 7.37-7.47 (6 H, m), 7.54-7.58 (2 H, m), 7.82-7.85 (2 H, m), 8.38 (1 H, s); ^{13}C NMR (CDCl$_3$; H, F complete decouple): δ 75.23, 124.64, 128.69, 128.78, 128.92 (2C), 129.02, 131.77, 135.13, 135.49, 165.89; ^{19}F NMR (CDCl$_3$): δ -74.39 (d, J= 7.5 Hz).

Synthetic method

2,2,2-Trifluoro-1-phenylethylamine hydrochloride

N-Benzylidene-2,2,2-trifluoro-1-phenylethylamine (typically 20 mmol) was dissolved in 5 mL of Et$_2$O, then 5 mL of 4 *N* HCl was added under stirring at ambient temperature. Progress of the hydrolysis was monitored by TLC and, upon completion, the aqueous layer was separated, washed with Et$_2$O, and evaporated *in vacuo* to give the crystalline hydrochloride, purified by recrystallization from acetone or MeCN, in 97% yield.

Spectral data

^{1}H NMR (CD$_3$CN:CD$_3$OD=3:1): δ 5.17 (1 H, d, J=7.5 Hz), 7.56 (5 H, s); ^{19}F NMR (CD$_3$CN:CD$_3$OD=3:1): δ -72.69 (d, J=7.5 Hz).

Note

This reaction was reported to be applicable to various fluorinated substrates, but halogeno-methylated (chlorine or bromine) imines led to predominant or exclusive formation of conjugated dienes by further hydrogen halide elimination, respectively.

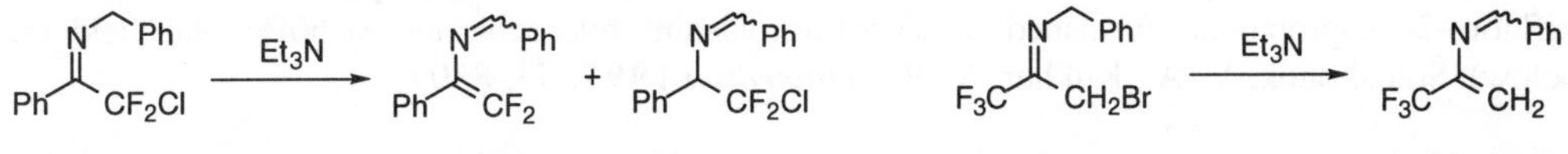

83% yield (4:1 ratio)　　　　　　　　　　　　　　　　　　　　71% yield

This transformation was recently extended to the asymmetric version by use of chiral 2-phenylethylamine, enabling the ready formation of optically active fluorine-containing amines in around 90% ee. See Soloshonok, V. A.; Ono, T. *J. Org. Chem.* **1997**, *62*, 3030.

Soloshonok, V. A., *et al. J. Org. Chem.* **1996**, *61*, 6563.

Methyl 3-(benzylideneamino)-4,4,4-trifluorobutyrate, 3-Amino-4,4,4-trifluorobutyric acid

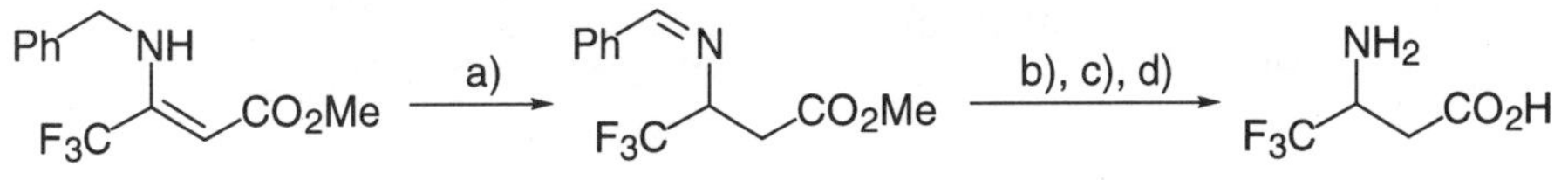

a) Et$_3$N, b) 2 *N* HCl, c) 6 *N* HCl, d) Dowex-50, 0.2 *N* NH$_4$OH

Synthetic method

Methyl 3-(benzylideneamino)-4,4,4-trifluorobutyrate

The isomerization reactions were carried out in triethylamine for 12 h at reflux temperature. After completion of the isomerization (control by GLC or NMR analysis) triethylamine was evaporated in vacuum (40-50 °C/20-30 mmHg) and the residue was distilled to give methyl 3-(benzylidene-amino)-4,4,4-trifluorobutyrate in 84% yield.

Spectral data

bp. 136-139 °C/11 mmHg; ^{1}H NMR (CDCl$_3$): δ 2.78 (1 H, dd, *J*=16.5, 9.6 Hz), 2.92 (1 H, dd, *J*=16.5, 3.3 Hz), 3.52 (3 H, s), 4.29 (2 H, dqd, *J*=9.6, 7.5, 3.3 Hz), 7.35-7.45 (5 H, m), 8.47 (1 H, s); ^{19}F NMR (CDCl$_3$): δ -76.31 (d, *J*=7.5 Hz); IR (CH$_2$Cl$_2$): ν 1730, 1610 cm^{-1}.

Synthetic method

3-Amino-4,4,4-trifluorobutyric acid

A solution of methyl 3-(benzylideneamino)-4,4,4-trifluorobutyrate (0.27 mol) in Et$_2$O (150 mL) was poured under stirring in 2 *N* HCl (200 mL). The mixture was stirred for 1 h more, the Et$_2$O phase was removed and the aqueous phase was washed with Et$_2$O (3x50 mL) and evaporated to dryness. The dry residue was mixed with 6 *N* HCl (70 mL) and heated at 90 °C for 6 h. Evaporation and Dowex-50 (H$^+$-form) column chromatography of the residue yielded 3-amino-4,4,4-trifluorobutyric acid in 77% yield.

Spectral data

mp. 190 °C (dec.; MeOH); ^{1}H NMR (DMSO-d_6): δ 2.77 (1 H, dd, *J*=17.8, 8.8 Hz), 3.00 (1 H, dd, *J*=17.8, 4.2 Hz), 3.52 (3 H, s), 4.39 (1 H, qdd, *J*=10.8, 8.8, 4.2 Hz).

Note

For the preparation of the starting enamine, see page 175 of this volume.

This reaction can be extended to a variety of 3-per (or poly-) fluoroalkylated β-amino acids. See the following closely related work: Soloshonok, V. A., *et al. Tetrahedron:Asym.* **1994**, *5*, 1225. Moreover, employment of a strategy similar to that for the preparation of ethyl 3,3,3-trifluoro-2-oxopropionate furnished 3,3,3-trifluoroalanine hydrochloride in 60% total yield (see below): Soloshonok, V. A.; Kukhar, V. P. *Tetrahedron* **1997**, *53*, 8307.

Soloshonok, V. A.; Kukhar, V. P. *Tetrahedron* **1996**, *52*, 6953.

4.3.7 Nucleophilic Additions

1,1,1-Trifluoroundecan-2-ol

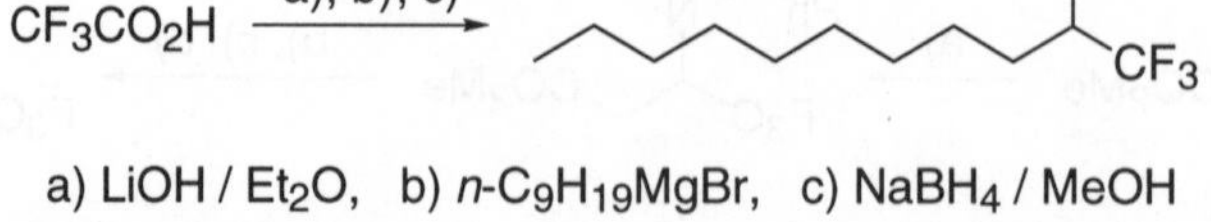

a) LiOH / Et$_2$O,   b) *n*-C$_9$H$_{19}$MgBr,   c) NaBH$_4$ / MeOH

Synthetic method

To an Et$_2$O (83 mL) solution of trifluoroacetic acid (82.8 mmol) was carefully added lithium hydroxide (82.8 mmol) at 0 °C, and the mixture was evaporated to dryness after neutralization was completed. The residue was thoroughly dried under reduced pressure at room temperature overnight and then diluted with 50 mL of Et$_2$O. To this solution was added an Et$_2$O (100 mL) solution of nonylmagnesium bromide (100 mmol) at 0 °C, and the mixture was allowed to warm to room temperature with stirring for 14 h.   The reaction was quenched by addition of 2 N HCl and extracted with Et$_2$O.   The combined organic layers were dried over MgSO$_4$ and evaporated to give 1,1,1-trifluoroundecan-2-one (22.5 g) as a colorless oil.   To a MeOH (100 mL) solution of this ketone was added NaBH$_4$ (3.783, 100 mmol) in several portions at 0 °C.   After being stirred for 2 h at 0 °C, the solvent was evaporated off.   Purification by silica gel flash column chromatography (hexane:AcOEt=10:1) gave 1,1,1-trifluoroundecan-2-ol (12.2 g, 53.8 mmol) in 65% yield.

Spectral data

bp. 65 °C (bath temperature)/4 mmHg; ^{1}H NMR (CDCl$_3$): δ 0.88 (3 H, t, J=6.4 Hz), 1.15-1.43 (14 H, m), 1.51-1.76 (2 H, m), 2.57 (1 H, t, J=4.2 Hz), 3.78-4.00 (1 H, m); ^{13}C NMR (CDCl$_3$): δ 14.04, 20.67, 24.91, 29.01, 29.28, 29.40, 29.45, 29.45, 29.50, 31.87, 70.52 (q, J=30.8 Hz), 125.21 (q, J=280.1 Hz); ^{19}F NMR (CDCl$_3$): δ 81.77 (d, J=6.2 Hz) from C$_6$F$_6$; IR (neat): ν 3350, 2970, 2860, 1460, 1270, 1170, 1140 cm^{-1}.

Note

For related work on the preparation of trifluorinated ketones from trifluoroacetates, see, for example, Creary, X. *J. Org. Chem.* **1987**, *52*, 5026.

Nakamura, K., *et al. J. Org. Chem.* **1996**, *61*, 2332.

6,6,6-Trifluoro-5-oxohexanoic acid

$$CF_3CO_2Et + CH_2(CH_2CO_2Et)_2 \xrightarrow{a), b)}$$

a) EtONa / EtOH,   b) 30% H$_2$SO$_4$

Synthetic method

Sodium (12.6 g, 663 mmol) was dissolved in anhydrous EtOH (140 mL).   At 0 °C, ethyl trifluoroacetate (77.6 g, 546 mmol) was added in one portion to the above EtOH solution.   Diethyl glutarate (25.0 g, 133 mmol) was added to the resulting solution, and the mixture was heated at 70 °C for 3 h.   Then diethyl glutarate (25.0 g, 133 mmol) was added to the mixture again, and the mixture was heated overnight at 70 °C.   After cooling, 20% H$_2$SO$_4$ (300 mL) was added to the mixture, and then the mixture was extracted with Et$_2$O (4x150 mL).   Removal of the solvent under reduced pressure gave a crude product.   Without purification, to the ester obtained was added 30% H$_2$SO$_4$ (300 mL), and the mixture was heated at reflux for 7 h.   After cooling, the mixture was poured into ice H$_2$O and extracted with Et$_2$O (4x100 mL).   The combined extracts were successively washed with brine and H$_2$O.   The Et$_2$O solution was dried over MgSO$_4$, and the solvent was removed under reduced pressure.   Distillation gave 20.5 g of pure 6,6,6-trifluoro-5-oxohexanoic acid as a colorless solid in 42% yield.

Spectral data

bp. 62-65 °C/4 mmHg; mp. 34-37 °C; ^{1}H NMR (CDCl$_3$): δ 2.02 (2 H, quint, J=7.1 Hz), 2.47 (2 H, t, J=7.1 Hz), 2.85 (2 H, t, J=7.1 Hz), 10.20 (1 H, br s); ^{13}C NMR (CDCl$_3$): δ 17.28, 32.38, 35.27, 115.80 (q, J=291 Hz), 179.60, 191.41 (q, J=35 Hz); ^{19}F NMR (CDCl$_3$): δ -79.9 (s); IR (neat): ν 1765, 1713 cm^{-1}.

Eguchi, S., *et al. J. Org. Chem.* **1996**, *61*, 8826.

2-(Trifluoroacetyl)-5-(trimethylsilyl)furan,
2-[1-(2,2,2-Trifluoro-1-hydroxyethyl)]-5-(trimethylsilyl)furan

a) *n*-BuLi / THF,   b) TMSCl,   c) CF$_3$CO$_2$Et,   d) NaBH$_4$ / EtOH

Synthetic method

2-(Trifluoroacetyl)-5-(trimethylsilyl)furan

To a solution of furan (14.5 mL, 199 mmol) in 150 mL of anhydrous THF at -20 °C was added dropwise *n*-BuLi (2.5 *M* in hexane, 84 mL, 210 mmol) under a nitrogen atmosphere.   After the solution was stirred for 30 min, trimethylsilyl chloride (200 mmol) was added, and the whole was stirred for 1 h at room temperature.   After the solution was recooled to -20 °C, *n*-BuLi (84 mL, 210 mmol) was added and the whole was stirred for 30 min.   The reaction mixture was then treated with ethyl trifluoroacetate (26.2 mL, 220 mmol) in THF (50 mL) at -78°C, followed by further stirring for 3 h at room temperature.   The mixture was quenched with 3 *N* HCl aq. (80 mL), and the volatiles were removed under reduced pressure.   After extraction with AcOEt three times, the combined organic layers were washed with sat. NaHCO$_3$ aq. and brine, dried (MgSO$_4$), and concentrated *in vacuo*.   An analytical sample was obtained by distillation.   Usually, the crude product was used in the next step without further purification.

Spectral data

bp. 80-85 °C/15 mmHg; ^{1}H NMR (CDCl$_3$): δ 0.32 (9 H, s), 6.79 (1 H, d, *J*=3.68 Hz), 7.46 (1 H, dq, *J*=3.69, 1.38 Hz); ^{13}C NMR (CDCl$_3$): δ -2.33, 116.51 (q, *J*=290.5 Hz), 121.95, 124.17 (q, *J*=2.9 Hz), 153.41, 172.30, C=O was not observed; ^{19}F NMR (CDCl$_3$): δ 4.5 (s) from CF$_3$CO$_2$H; IR (neat): ν 2950, 1690 cm^{-1}.

Synthetic method

2-[1-(2,2,2-Trifluoro-1-hydroxyethyl)]-5-(trimethylsilyl)furan

To a solution of the above crude 2-(trifluoroacetyl)-5-(trimethylsilyl)furan in EtOH (200 mL) was added slowly sodium borohydride (2.27 g, 60 mmol) at 0 °C, and the mixture was stirred overnight at room temperature.   After the solvent was removed under reduced pressure and the addition of both AcOEt (200 mL) and 3 *N* HCl aq. (200 mL), the organic materials were extracted, and the separated aqueous phase was extracted twice with AcOEt.   The combined organic layers were washed with brine, dried over MgSO$_4$, and concentrated.   39.0 g of 2-[1-(2,2,2-trifluoro-1-hydroxyethyl)]-5-(trimethylsilyl)furan was obtained after purification by distillation in 82% total yield calculated from furan.

Spectral data

bp. 64-67 °C/2.7 mmHg; Rf 0.29 (hexane:AcOEt=7:1); ^{1}H NMR (CDCl$_3$): δ 0.25 (9 H, s), 2.72 (1 H, dd, *J*=7.41, 1.71 Hz), 5.06 (1 H, dq, *J*=6.75, 6.75 Hz), 6.47 (1 H, d, *J*=3.30 Hz), 6.59 (1 H, d, *J*=3.30 Hz); ^{13}C NMR (CDCl$_3$): δ -2.02, 67.36 (q, *J*=34.2 Hz), 109.80 (q, *J*=1.5 Hz), 120.50, 123.62 (q, *J*=282.9 Hz), 151.19 (q, *J*=1.6 Hz), 162.49; ^{19}F NMR (CDCl$_3$): δ 0.5 (d, *J*=6.2 Hz) from CF$_3$CO$_2$H; IR (neat): ν 3400, 2975 cm^{-1}.

Yamazaki, T., *et al. J. Org. Chem.* **1993**, *58*, 4346.

3,3,3-Trifluoro-1-(phenylsulfonyl)propan-2-ol, 3,3,3-Trifluoro-1-(phenylsulfonyl)propene

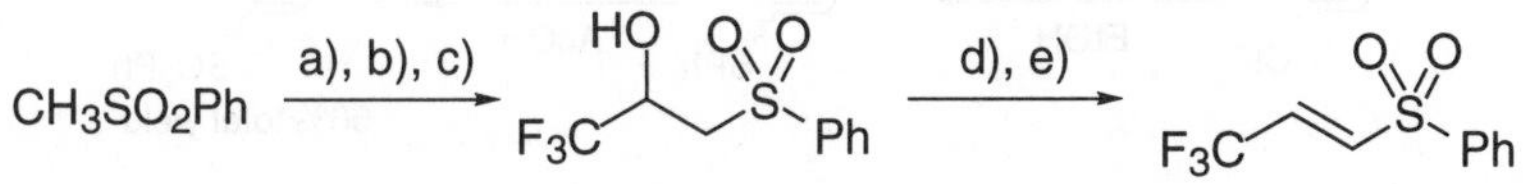

a) NaH / THF,   b) CF$_3$CO$_2$Et,   c) NaBH$_4$ / MeOH,
d) TsCl, Et$_3$N / CH$_2$Cl$_2$,   e) DBU

Synthetic method

3,3,3-Trifluoro-1-(phenylsulfonyl)propan-2-ol

To a stirred THF (20 mL) solution of methyl phenyl sulfone (1.60 g, 10 mmol) under nitrogen was added NaH (60% oil dispersion; 600 mg, 15 mmol) in portions after which the resulting suspension was stirred at 0 ℃ for 10 min, then treated dropwise with ethyl trifluoroacetate (3.60 mL, 30 mmol) at 0 ℃.   After 2 h under reflux, the resulting solution was poured into sat. NaCl aq. (250 mL) and extracted with Et$_2$O (4x100 mL).   The combined extracts were dried (MgSO$_4$) and evaporated under reduced pressure to give 3,3,3-trifluoro-1-(phenylsulfonyl)-propan-2-one (1.80 g).

Without purification, NaBH$_4$ (2.50 g, 10 mmol) was added to a solution of 3,3,3-trifluoro-1-(phenylsulfonyl)propan-2-one (1.80 g) in MeOH (20 mL).   After being stirred overnight at room temperature, the solution was poured into sat. NaCl aq. (200 mL) and extracted with Et$_2$O (4x100 mL).   The combined extracts were dried (MgSO$_4$) and evaporated under reduced pressure.   The residue was recrystallized (hexane:Et$_2$O) to give 3,3,3-trifluoro-1-(phenylsulfonyl)propan-2-ol (1.40 g, 57% from methyl phenyl sulfone).

Spectral data

mp. 73-74 ℃; ^{1}H NMR (CDCl$_3$): δ 3.44 (2 H, m), 3.67-3.82 (1 H, m), 4.61 (1 H, dqd, *J*=10.1, 6.4, 3.8 Hz), 7.59-8.01 (5 H, m); ^{13}C NMR (CDCl$_3$): δ 56.3, 66.2 (q, *J*=33 Hz), 123.9 (q, *J*=280 Hz), 128.4, 130.1, 135.1, 139.1; ^{19}F NMR (CDCl$_3$): δ -79.9 (s); IR (KBr): ν 3447, 1265, 1127 cm^{-1}.

Synthetic method

3,3,3-Trifluoro-1-(phenylsulfonyl)propene

To a stirred solution of 3,3,3-trifluoro-1-(phenylsulfonyl)propan-2-ol (1.40 g, 5.7 mmol) and triethylamine (1.60 mL, 11 mmol) in CH$_2$Cl$_2$ (20 mL) was added solid *p*-toluenesulfonyl chloride (1.08 g, 5.7 mmol) in portions.   After 2 h under reflux, the mixture was treated with DBU (868 mg, 5.7 mmol) and heated under reflux for a further 1 h.   The solution was then poured into sat. NaHCO$_3$ aq. (200 mL) and extracted with CH$_2$Cl$_2$ (4x100 mL).   The combined extracts were washed with 1 *N* HCl aq. (2x100 mL), dried (MgSO$_4$), and evaporated under reduced pressure.   The residue was recrystallized (hexane-Et$_2$O) to give 3,3,3-trifluoro-1-(phenyl-sulfonyl)propene (0.89 g, 67%; 39% from methyl phenyl sulfone).

Spectral data

mp. 66.0-66.5 ℃ (68-69 ℃ obtained from the previous procedure shown below); ^{1}H NMR (CDCl$_3$): δ 6.87 (1 H, dq, *J*=15, 4.7 Hz), 7.15 (1 H, d, *J*=15 Hz), 7.52-8.50 (5 H, m); ^{19}F NMR (CDCl$_3$): δ -0.7 (d, *J*=4.7 Hz) from C$_6$H$_5$CF$_3$; IR (KBr): ν 1325, 1140, 960 cm^{-1}.

Note

3,3,3-Trifluoro-1-(phenylsulfonyl)propene was previously prepared by Kobayashi and his co-workers *via* the procedure depicted below (Kobayashi, Y., *et al. Chem. Pharm. Bull.* **1985**, *33*, 4077; see also *J. Fluorine Chem.* **1988**, *40*, 171) and the above physical data for this material were extracted from the latter article.

$$F_3C \underset{Cl}{\diagup} \xrightarrow[\text{EtOH}]{\text{PhSH, KOH}} F_3C \underset{SPh}{\diagup} \xrightarrow[\text{AcOH}]{\text{30\% H}_2\text{O}_2} F_3C \underset{SO_2Ph}{\diagup}$$

66% total yield

The followings are representative Michael acceptors with a CF_3 group at the β position of various functionalities, basically prepared in a similar manner. See Yamazaki, T., *et al. J. Chem. Soc., Chem. Commun.* **1987**, 1340 (chiral sulfoxide), Tanaka, K., *et al. Bull. Chem. Soc. Jpn.* **1993**, 66, 2432 (α,β-unsaturated nitro compound), Nickson, T. E. *J. Org. Chem.* **1988**, *53*, 3870 (α,β-unsaturated phosphonate), and Gautschi, M.; Seebach, D. *Angew. Chem. Int. Ed. Engl.* **1992**, *31*, 1083 (chiral 6-CF_3-1,3-dioxin-4-one).

$H_3C\overset{O}{\underset{p\text{-Tol}}{\overset{\|}{S}}}$	i) LDA ii) CF_3CO_2Et iii) $NaBH_4$ iv) TsCl, Et_3N → $F_3C\diagup\overset{O}{\underset{p\text{-Tol}}{\overset{\|}{S}}}$

$$F_3C\overset{OH}{\underset{OEt}{\diagdown}} \xrightarrow[\substack{\text{ii) Ac}_2\text{O, H}^+ \\ \text{iii) K}_2\text{CO}_3}]{\text{i) CH}_3\text{NO}_2,\ \text{K}_2\text{CO}_3} F_3C\diagup\diagup NO_2$$

$$H_3C\overset{O}{\underset{OEt}{\overset{\|}{P}}}\text{OEt} \xrightarrow[\substack{\text{iii) NaBH}_4 \\ \text{iv) MsCl, Et}_3\text{N}}]{\substack{\text{i) }n\text{-BuLi} \\ \text{ii) CF}_3\text{CO}_2\text{Et}}} F_3C\diagup P(O)(OEt)_2$$

$$F_3C\overset{OH}{\diagdown}CO_2H \xrightarrow[\substack{\text{iii) Br}_2 \\ \text{iv) DBU}}]{\substack{\text{i) }t\text{-BuCHO, H}^+ \\ \text{ii) }t\text{-BuLi}}} \overset{t\text{-Bu}}{\underset{F_3C\diagdown\overset{}{\diagdown O}}{O\diagdown O}}$$

Eguchi, S., *et al. J. Chem. Soc. Perkin Trans.* 1 **1995**, 2761.

2-(Trifluoromethyl)propiophenone

$$H_3C\overset{}{\underset{F_3C}{\diagdown}}CHO \xrightarrow{a),\ b)} H_3C\overset{O}{\underset{F_3C}{\diagdown}}Ph$$

a) PhLi / E$_2$O, b) PCC / CH$_2$Cl$_2$

Synthetic method

To a solution of 3 mL (28.8 mmol) of 2-(trifluoromethyl)propanal in 30 mL of Et$_2$O was added 14.4 mL of a 2.2 M solution of phenyllithium (31.7 mmol) in Et$_2$O slowly at -78 °C. After being gradually warmed to room temperature, the mixture was quenched with 30 mL of 2 N HCl. The mixture was extracted with Et$_2$O (2x50 mL), and the combined Et$_2$O layers were dried and concentrated to give an oil. This crude product was chromatographed on silica gel (hexane: AcOEt=40:1), to obtain 4.57 g (78%) of a mixture of two diastereomers of 2-(trifluoromethyl)-1-phenylpropan-1-ol.

To a suspension of 5.3 g (24.6 mmol) of PCC and 5.3 g of Celite in 50 mL of CH$_2$Cl$_2$ was added 4.57 g (22.4 mmol) of a mixture of two diastereomers of 2-(trifluoromethyl)-1-phenyl-propan-1-ol without separation. After being stirred for 12 h, the reaction mixture was filtered through Florisil. The Florisil was washed with Et$_2$O and the solvent was removed *in vacuo*. The residue was chromatographed on silica gel (hexane:AcOEt=60:1) to give 2.71 g of 2-(trifluoromethyl)propiophenone in 60% yield.

Spectral data
^{1}H NMR (CDCl$_3$): δ 1.47 (3 H, d, *J*=7.3 Hz), 4.0-4.5 (1 H, m), 7.3-8.1 (5 H, m); IR (neat): ν 3100, 3050, 2950, 1695, 1600, 765, 705, 690 cm^{-1}.

Hanamoto, T.; Fuchikami, T. *J. Org. Chem.* **1990**, *55*, 4969.

Ethyl 2-(trifluoromethyl)-4-oxopentanoate

a) [Rh(OAc)$_2$]$_2$ / Et$_2$O, b) TBAF / THF

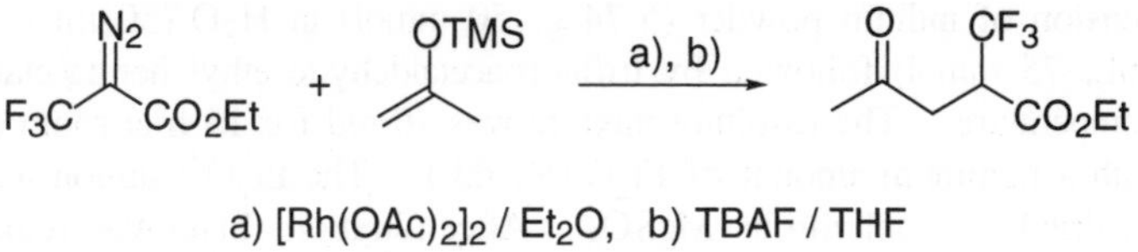

Synthetic method
A solution of ethyl 3,3,3-trifluoro-2-diazopropionate (4 mmol) in dry Et$_2$O (10 mL) was added dropwise within 4 h under nitrogen to a suspension of [Rh(OAc)$_2$]$_2$ (0.005 mmol) in 2-[(trimethylsilyl)oxy]propene (2 mmol) kept at reflux. The resulting reaction mixture was then diluted with dry THF (15 mL) and treated with tetra-*n*-butylammonium fluoride (TBAF) in THF (2 mL) at room temperature for a few minutes. H$_2$O (30 mL) was added, and three extractions with Et$_2$O were carried out. The organic layer was washed with brine, dried over Na$_2$SO$_4$, and concentrated to give a residue which was subjected to silica gel chromatography (petroleum ether:AcOEt=9:1). Ethyl 2-(trifluoromethyl)-4-oxopentanoate was obtained in 83% yield.
Spectral data
^{1}H NMR (CCl$_4$): δ 1.10 (3 H, t, *J*=7 Hz), 1.95 (3 H, s), 2.75-3.90 (3 H, m), 4.23 (2 H, q, *J*=7 Hz); ^{19}F NMR (CCl$_4$): δ 9.0 (d, *J*= 6.6 Hz) from CF$_3$CO$_2$H; IR (neat): ν 1755, 1730 cm^{-1}.

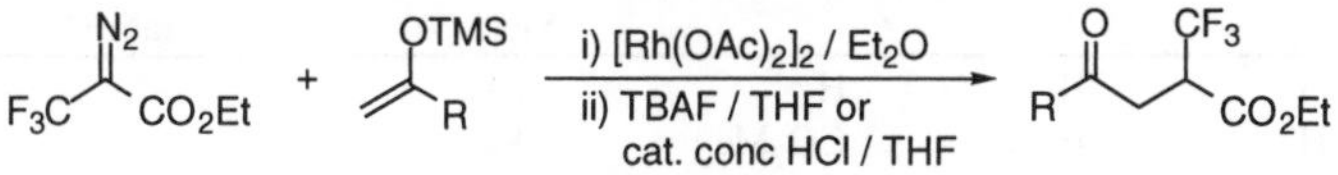

R	Yield (%)	R	Yield (%)
H	73	CH$_2$=CH-	85
Me$_2$C=CH-	87	Ph	90
p-MeO-C$_6$H$_4$	94	(*E*)-PhCH=CH-	97

(Reproduced with permission from Shi, G.-Q.; Xu, Y.-Y. *J. Org. Chem.* **1990**, *55*, 3383)

Note
For the preparation of the starting material, see page 204 of this volume.
For the recent reports using trifluorinated diazopropionates, see i) Xu, Y.-Y., *et al.* *Tetrahedron* **1991**, *47*, 1629, ii) Shi, G.-Q., *et al.* *Tetrahedron* **1995**, *51*, 5011, iii) Burger, K., *et al.* *Tetrahedron Lett.* **1996**, *37*, 615.

Shi, G.-Q.; Xu, Y.-Y. *J. Org. Chem.* **1990**, *55*, 3383.

1,1,1-Trifluoropent-4-en-2-ol

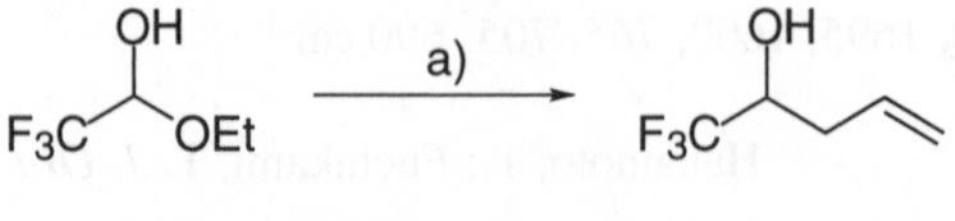

a) In, Allyl-Br / H$_2$O

Synthetic method

To a suspension of indium powder (5.74 g, 50 mmol) in H$_2$O (50 mL) was added allyl bromide (6.50 mL, 75 mmol) followed by trifluoroacetaldehyde ethyl hemiacetal (2.90 mL, 25 mmol) at room temperature. The resulting mixture was stirred for 15 h at room temperature and then extracted with a minimum amount of Et$_2$O (50 mL). The Et$_2$O solution was then washed with H$_2$O, brine, dried over anhydrous MgSO$_4$. After filtration, Et$_2$O was removed by slowly blowing with nitrogen. The crude product was then distilled to afford 3.32 g of pure 1,1,1-trifluoropent-4-en-2-ol in 95% yield.

Spectral data

bp. 35-36/30 mmHg; Rf 0.42 (hexane:AcOEt=4:1); ^{1}H NMR (CDCl$_3$): δ 2.25 (1 H, d, J=5.79 Hz), 2.41 (1 H, m), 2.54 (1 H, m), 4.00 (1 H, m), 5.26 (2 H, m), 5.86 (1 H, m); ^{13}C NMR (CDCl$_3$): δ 34.40, 69.69 (q, J=30.18 Hz), 119.87, 127.15 (q, J=282.98 Hz), 131.69; ^{19}F NMR (CDCl$_3$): δ -3.88 (d, J=7.32 Hz) from CF$_3$CO$_2$H; IR (neat): ν 3386.7, 2927.3, 1643.7, 1279.1, 1174.0, 1100.2 cm^{-1}.

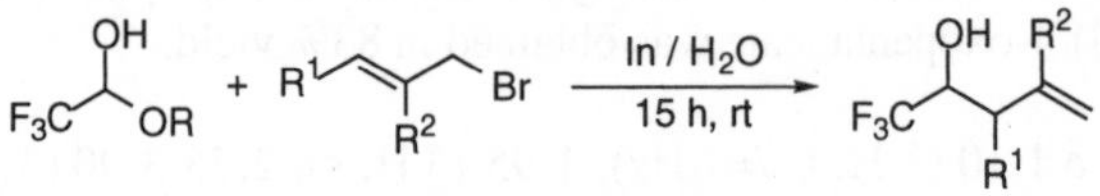

R	R^1	R^2	Isolated Yield (%)	Diastereomeric ratio[a]
H	H	H	81	
H	H	CO$_2$Me	82	
H	Me	H	70	68:32
Et	H	H	0[b]	
Et	H	H	85[c]	
Et	H	CO$_2$Me	87	
Et	H	CO$_2$Me	65[c]	
Et	H	CO$_2$H	65	
H	Me	H	80	65:35
H	Me	H	72[c]	67:33

[a] Diastereomeric ratio. [b] Sn was employed instead of In. [c] Sn/InCl$_3$ (1 equiv) was employed instead of In.

(Reproduced with permission from Loh, T.-P.; Li, X.-R. *Chem. Commun.* **1996**, 1929).

Note

This reaction afforded only the γ-coupled homoallylic alcohols and unreacted allylic bromides were recovered after the usual work-up. Loh and Li have also found that InCl$_3$-promoted tin-mediated allylation proceeded in a similar way, and this alternative method is advantageous in terms of the cost and the amount of reagent employed (2 equiv of In and 3 equiv of allylic bromide can be reduced to 1 equiv of Sn, InCl$_3$, and allylic bromide).

For examples of the *in situ* generation of hemiacetals and the subsequent reaction with an appropriate nucleophile, see i) Ishihara, T., *et al. Tetrahedron Lett.* **1993**, *34*, 5777 (allylic stannanes), ii) Pastor, R., *et al. Tetrahedron Lett.* **1993**, *34*, 2469 (Hörner-Wittig reagent).

Loh, T.-P.; Li, X.-R. *Chem. Commun.* **1996**, 1929.

3,3,3-Trifluoro-2-phenylpropionic acid, 3,3,3-Trifluoro-1,2-diphenylpropan-1-one

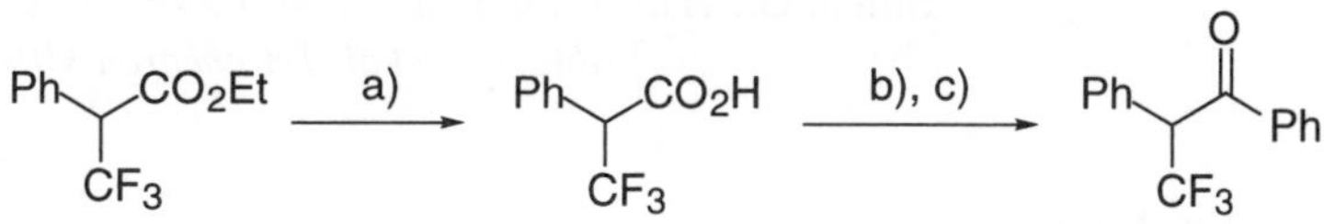

a) conc HCl / dioxane, b) SOCl₂, c) AlCl₃ / benzene

Synthetic method
    3,3,3-Trifluoro-2-phenylpropionic acid
    A mixture of ethyl 3,3,3-trifluoro-2-phenylpropionate (18.8 g, 81 mmol), conc HCl (90 mL) and dioxane (180 mL) was refluxed for 30 h. CH_2Cl_2 (50 mL) was added and the phases separated.   The aqueous phase was extracted with CH_2Cl_2 (2x50 mL).   The combined organic phases were extracted with 1 N NaHCO$_3$ aq. (2x100 mL).   The combined aqueous layers were acidified to pH 1 with a 2 N HCl solution and extracted with CH_2Cl_2 (3x50 mL).   The combined organic phases were evaporated and the residue was crystallized (hexane) to give 12.1 g of 3,3,3-trifluoro-2-phenylpropionic acid in 73% yield.
Spectral data
    mp. 78-79 °C (hexane); ¹H NMR (CDCl₃): δ 4.35 (1 H, q, *J*=8.4 Hz), 7.40-7.50 (5 H, m), 10.76 (1 H, br s); ¹³C NMR (CDCl₃): δ 55.4 (q, *J*=30 Hz), 123.5 (q, *J*=280 Hz), 128.7, 129.2, 129.6, 129.7 (q, *J*=3 Hz), 172.6 (q, *J*=3 Hz); IR (KBr): ν 3045, 1724 cm⁻¹.

Synthetic method
    3,3,3-Trifluoro-1,2-diphenylpropan-1-one
    3,3,3-Trifluoro-2-phenylpropionic acid (10.2 g, 50 mmol) was refluxed with thionyl chloride (50 mL) for 1 h and evaporated.   The solution of the residual oil in benzene (10 mL) was added drop by drop into a mixture of aluminum chloride (8.7 g, 65 mmol) and benzene (12 mL) at 10-15 °C.   After stirring for 1 h at ambient temperature, the mixture was poured on ice (60 g) and extracted with CHCl₃ (3x50 mL), washed with sat. NaHCO₃ aq. (15 mL), dried (MgSO₄) and evaporated.   Recrystallization (petroleum ether (80-100 °C)) gave 9.3 g of 3,3,3-trifluoro-1,2-diphenylpropan-1-one in 70% yield as colorless crystals.
Spectral data
    mp. 81-82 °C; ¹H NMR (CDCl₃): δ 5.30 (1 H, q, *J*=8.2 Hz), 7.35-7.54 (8 H, m), 7.88-7.92 (2 H, m); ¹³C NMR (CDCl₃): δ 56.4 (q, *J*=26.6 Hz), 124.3 (q, *J*=280.2 Hz), 128.7, 128.8, 129.2, 129.3, 129.6, 129.8, 133.8, 135.3, 191.2; IR (KBr): ν 1690 cm⁻¹.
Note
    The starting ester was prepared by the following scheme shown in the next page, whose procedure was already described on page 133 of this volume.
    Alternatively, the same type of esters was prepared from trifluoroacetophenone by way of i) cyanohydrin formation, ii) conversion to ester, iii) reduction of the mesylate.

$$\text{EtO}_2\text{C} \diagdown \text{CO}_2\text{Et} \quad \xrightarrow[\text{ii) CBr}_2\text{F}_2]{\text{i) NaH}} \quad \text{EtO}_2\text{C} \diagdown \text{CO}_2\text{Et} \quad \xrightarrow{\text{iii) KF}} \quad \text{Ph} \diagdown \text{CO}_2\text{Et}$$

Ph   F$_2$BrC   Ph   CF$_3$

$\uparrow$ H$_2$, Pd/C

$$\underset{\text{Ph}}{\overset{O}{\bigvee}}\text{CF}_3 \quad \xrightarrow[\text{ii) EtOH / H}_2\text{SO}_4]{\text{i) NaCN}} \quad \underset{\text{Ph}}{\overset{\text{HO} \quad \text{CO}_2\text{Et}}{\bigvee}}\text{CF}_3 \quad \xrightarrow{\text{MsCl, Et}_3\text{N}} \quad \underset{\text{Ph}}{\overset{\text{MsO} \quad \text{CO}_2\text{Et}}{\bigvee}}\text{CF}_3$$

Simig, G., *et al. J. Fluorine Chem.* **1996**, *76*, 91 (hydrolysis).
Simig, G., *et al. Tetrahedron* **1996**, *52*, 12821.

4.3.8 Electrophilic Additions

anti-2-(Trifluoromethyl)-1-(pyridin-3-yl)but-3-en-1-ol

a) Sn, InCl$_3$, 3-Pyridinecarboxyaldehyde / H$_2$O

Synthetic method

To a suspension of tin powder (47.5 mg, 0.4 mmol), indium(III) chloride (88.4 mg, 0.4 mmol) in H$_2$O (2 mL) was added 4,4,4-trifluorobut-2-en-1-ol (75.6 mg, 0.4 mmol), followed by 3-pyridinecarboxaldehyde (0.019 mL, 0.2 mmol) at room temperature.   The resulting mixture was stirred for 15 h at room temperature and then extracted with AcOEt (3x10 mL).   The solution was then washed with H$_2$O and brine, and dried over anhydrous MgSO$_4$.   After filtration, AcOEt was removed *in vacuo*.   The crude product was purified by silica gel column chromatography (hexane: AcOEt=2:1) to afford 41.2 mg of pure 2-(trifluoromethyl)-1-(pyridin-3-yl)but-3-en-1-ol in 95% yield as a 92:8 *anti:syn* diastereomer mixture.

Spectral data

Rf 0.46 (hexane:AcOEt=1:1); ^{1}H NMR (CDCl$_3$): δ 2.90-2.94 (1 H, m), 5.07 (1 H, d, *J*=17.24 Hz), 5.24 (1 H, d, *J*=3.23 Hz), 5.37 (1 H, d, *J*=10.38 Hz), 5.89-5.96 (1 H, m), 7.28 (1 H, m), 7.70 (1 H, m), 8.45 (2 H, m); ^{13}C NMR (CDCl$_3$): δ 55.84 (q, *J*=24.92 Hz), 68.80, 123.31, 124.26, 125.90 (q, *J*=280.95 Hz), 126.29, 134.48, 137.22, 147.56, 148.79; ^{19}F NMR (CDCl$_3$): δ 8.05 (d, *J*=7.32 Hz) from CF$_3$CO$_2$H; IR (neat): ν 3434.6, 2932.5, 1645.7, 1431.5, 1263.7, 1185.7, 1105.5 cm^{-1}.

Note

For the preparation of the starting allylic bromide, see page 191 of this volume.

One of the advantages of this procedure is that the reaction can occur in an aqueous medium, which enabled Loh and Li to use glyoxylic acid (83% yield) or formaldehyde (90% yield) as the aqueous solution.   Usually this reaction proceeds in a high *anti-* as well as γ-selective manner, while 2-pyridinecarboxaldehyde and glyoxylic acid were the exception and they showed complete *syn* diastereoselectivity, which was explained as the result of the intramolecular interaction of a pyridine ring and a carbonyl group with indium, respectively.   See also Loh, T.-P.; Li, X.-R.

Tetrahedron Lett. **1997**, *38*, 869.
For a similar reaction with 1,1-difluoroallyl bromide, see Momose, T., *et al. Tetrahedron Lett.* **1997**, *38*, 2853.

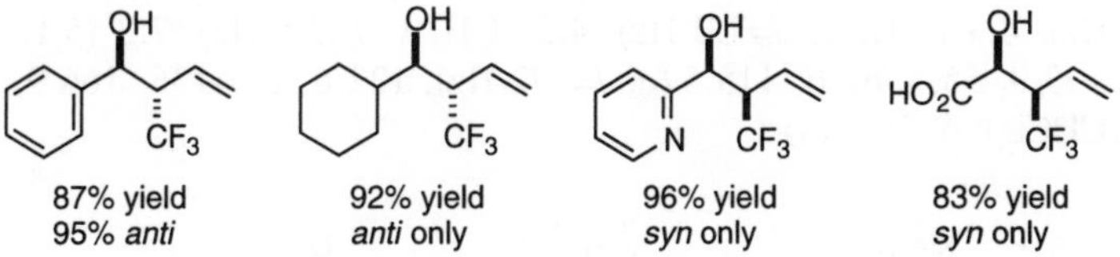

| 87% yield | 92% yield | 96% yield | 83% yield |
| 95% *anti* | *anti* only | *syn* only | *syn* only |

Loh, T.-P.; Li, X.-R. *Angew. Chem. Int. Ed. Engl.* **1997**, *36*, 980.

Ethyl 4,4,4-trifluoro-3-dimethylhydrazonobutyrate, Ethyl 2-benzyl-4,4,4-trifluoro-3-oxobutyrate

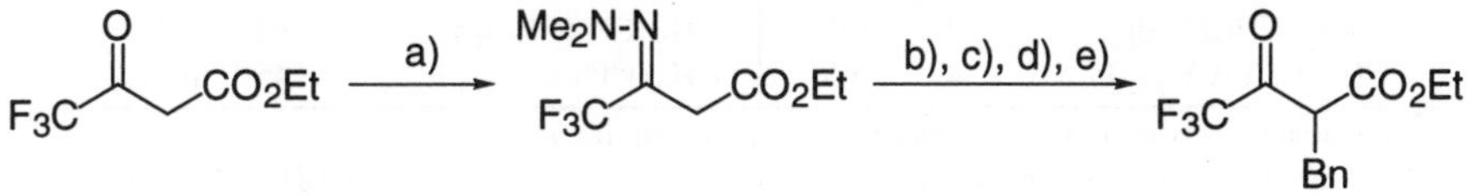

a) H_2N-NMe_2 / EtOH,   b) LDA / THF,   c) HMPA,   d) BnBr,   e) 10% HCl

Synthetic method
Ethyl 4,4,4-trifluoro-3-dimethylhydrazonobutyrate
A solution of ethyl 4,4,4-trifluoro-3-oxobutyrate (50 g, 0.27 mol) and *N*,*N*-dimethylhydrazine (19.8 g, 0.319 mol) in absolute EtOH (100 mL) was heated to reflux for 20 h under an argon atmosphere. The reaction mixture was concentrated *in vacuo*, and distillation afforded 50.5 g of the pure ethyl 4,4,4-trifluoro-3-dimethylhydrazonobutyrate in 82% yield.
Spectral data
bp. 85-86 ℃/17 mmHg; ^{1}H NMR (CDCl$_3$): δ 1.27 (3 H, t, *J*=7 Hz), 2.82 (6 H, s), 3.52 (2 H, s), 4.20 (2 H, q, *J*=7 Hz); ^{13}C NMR (CDCl$_3$): δ 14.1, 33.3, 46.8, 61.7, 121.5 (q, *J*=275 Hz), 136.0 (q, *J*=33 Hz), 168.2; ^{19}F NMR (CDCl$_3$): δ -69.8 (s).

Synthetic method
Ethyl 2-benzyl-4,4,4-trifluoro-3-oxobutyrate
A solution of ethyl 4,4,4-trifluoro-3-dimethylhydrazonobutyrate (10 g, 40 mmol) in 100 mL of THF was added at 0 ℃ to a 1 *M* solution of LDA (44 mmol, 1.1 equiv) in THF under an argon atmosphere. After 2 h, the temperature was decreased to -78 ℃ and hexamethylphosphoramide (HMPA; 19 mL, 3 equiv), then benzyl bromide (1.1 equiv) were added. After 2 h, the temperature was slowly increased to ambient temperature. The reaction mixture was hydrolyzed with 10% HCl aq. for 22 h, then extracted with Et$_2$O. The organic phase was washed with H$_2$O for neutralization, dried over Na$_2$SO$_4$, and concentrated to give, after distillation, the desired ethyl 2-benzyl-4,4,4-trifluoro-3-oxobutyrate in 80% yield (>95% yield by GC analysis of the crude material).
Spectral data
Ethyl 2-benzyl-4,4,4-trifluoro-3-dimethylhydrazonobutyrate
bp. 108-110 ℃/0.2 mmHg; ^{1}H NMR (CDCl$_3$): δ 1.27 (3 H, t, *J*=7 Hz), 2.42 (6 H, s), 2.92 (2 H, m), 4.18 (2 H, q, *J*=7 Hz), 4.25 (1 H, m), 4.78 (1 H, m); ^{13}C NMR (CDCl$_3$): δ 13.9, 34.4,

45.7, 47.1, 61.6, 122.4 (q, J=275 Hz), 126.8, 128.5, 129.1, 138.1, 152.9 (q, J=31 Hz), 169.3; ¹⁹F NMR (CDCl₃): δ -65.7 (s).

Ethyl 2-benzyl-4,4,4-trifluorobutyrate

bp. 105-108 °C/0.5 mmHg; mp. 54-58 °C; ¹H NMR (CDCl₃): δ 1.2 (3 H, t, J=7.2 Hz), 3.2 (2 H, d, J=7.5 Hz), 4.1 (2 H, q, J=7.2 Hz), 4.2 (1 H, t, J=7.5 Hz), 7.3 (5 H, m); ¹³C NMR (CDCl₃): δ 13.8, 33.9, 55.4, 62.6, 115.5 (q, J=292 Hz), 122.6, 127-136, 166.9, 186.7 (q, J=36 Hz); ¹⁹F NMR (CDCl₃): δ -78.7 (s).

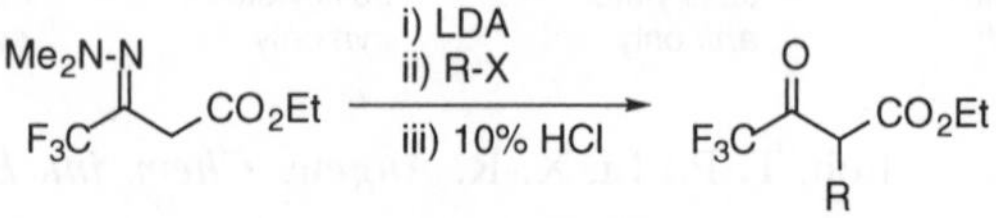

R-X	Yield[a] (%)		R-X	Yield[a] (%)	
n-Pr-I	75	(90)	Bn-Br	80	(>95)
Ph-(CH₂)₂-I	18	(25)	Ph-(CH₂)₃-I	68	(80)
PhCH=CHCH₂-Br	65	(80)	Allyl-Br	85	(>95)
CMe₂=CHCH₂-Br	77	(95)	CH₂=CH-(CH₂)₂-I	35	(45)
CH₂=CPh-(CH₂)₄-I	47	(65)	CH₂=CPh-(CH₂)₃-I	45	(60)

[a] In parentheses was shown the yield determined by capillary GC.
(Reproduced with permission from Langlois, B., et al. J. Fluorine Chem. **1989**, 44, 377)

Note

The following procedure was also described as the alternative route:

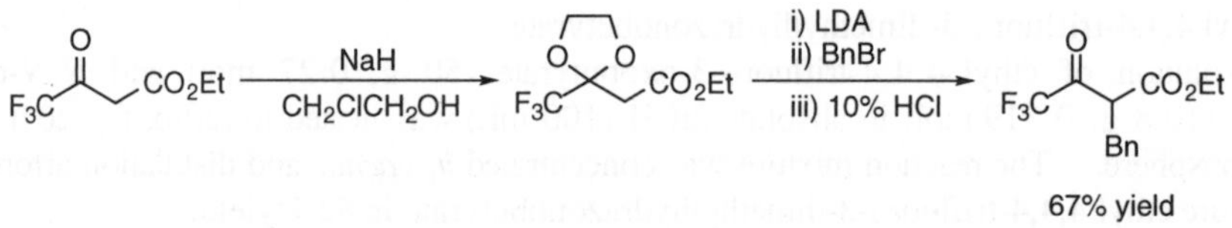

67% yield

The usual alkylation by the action of, for example, MeONa/MeOH, mainly afforded the O-alkylated product.   The corresponding lithium enolate did not react in THF, and potassium enolate yielded the product only in low yield along with the dialkylated product.   See Bégué, J.-P., et al. J. Fluorine Chem. **1989**, 44, 361.

Langlois, B., et al. J. Fluorine Chem. **1989**, 44, 377.

3,3,3-Trifluoro-2-phenylpropene

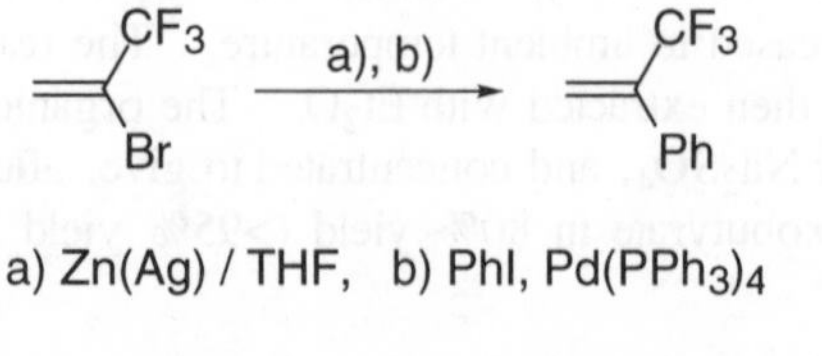

a) Zn(Ag) / THF,   b) PhI, Pd(PPh₃)₄

Synthetic method

To a suspension of Zn(Ag) (2.5 g, 38 mmol) in THF (290 mL) was added trimethylchloro-silane (0.5 mL) during stirring.   After 10 min, tetramethylethylenediamine (6 mL, 25 mmol) was added and then 2-bromo-3,3,3-trifluoropropene (4.4 g, 25 mmol) was added dropwise at room temperature. After the addition, the mixture was heated at 60 °C for 9 h. After being cooled to room temperature, the resulting solution of (3,3,3-trifluoroprop-1-en-2-yl)zinc bromide in THF was transferred to a bottle with a septum under nitrogen and stored. The yield was shown to be 93% as determined by ^{19}F NMR versus $C_6H_5CF_3$. yield. (3,3,3-Trifluoroprop-1-en-2-yl)zinc bromide displayed a single peak at δ 16.0 from CF_3CO_2H in ^{19}F NMR.

To (3,3,3-trifluoroprop-1-en-2-yl)zinc bromide in THF (20 mL, ca. 20 mL) were added tetrakis(triphenylphosphine)palladium(0) (0.2 mmol) and phenyl iodide (10 mmol). The mixture was then allowed to heat at 45 °C for 6 h. After being cooled, the reaction mixture was poured into n-hexane (100 mL), and the solid which precipitated was triturated and filtered. The filtrate was concentrated in vacuo, and the residue thus obtained was purified by short path distillation under reduced pressure to afford 3,3,3-trifluoro-2-phenylpropene in 96% yield as a colorless oil.

Spectral data

^{1}H NMR ($CDCl_3$): δ 5.72 (1 H, s), 5.92 (1 H, s), 7.42 (5 H, m); ^{19}F NMR ($CDCl_3$): δ 11.0 (s) ppm from CF_3CO_2H.

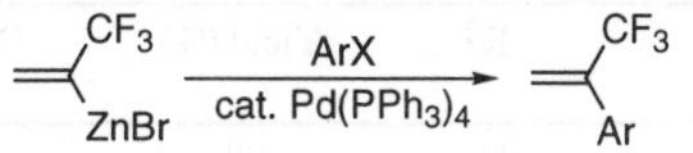

ArX	Temp. (°C)	Time (h)	Yield (%)
4-O_2N-C_6H_4-Br	67	16	93
4-O_2N-C_6H_4-I	45	6	95
3-O_2N-C_6H_4-Br	67	16	88
3-O_2N-C_6H_4-I	50	5	92
2-O_2N-C_6H_4-Br[a]	67	20	80
2,4-$(O_2N)_2$-C_6H_3-Br[a]	67	19	81
4-Br-C_6H_4-I	50	5.5	92
4-Ac-C_6H_4-Br	67	18	90
2-OHC-C_6H_4-Br[a]	67	20	85
2-AcO-C_6H_4-Br	67	20	83
1-Naphthyl-I	67	6	99
2-Naphthyl-Br	67	8	98

[a] Another 2 mol% of Pd(PPh$_3$)$_4$ was added at the end of 10 h.
(Reproduced with permission from Jiang, B.; Xu, Y.-Y. *J. Org. Chem.* **1991**, *56*, 7336).

Jiang, B.; Xu, Y.-Y. *J. Org. Chem.* **1991**, *56*, 7336.

1,1,1-Trifluoro-4-phenylbut-2-yn-4-ol

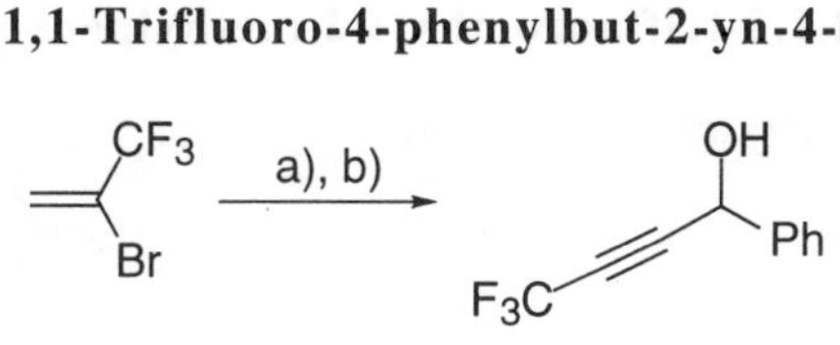

a) LDA / THF, b) PhCHO

Synthetic method

To a solution of LDA (44 mmol) in THF (40 mL) was added dropwise a precooled (-78 °C) solution of 2-bromo-3,3,3-trifluoropropene (3.5 g, 20 mmol) in THF (20 ml) at -78 °C. After the mixture was stirred for 5 min, benzaldehyde (2.5 mL, 24 mmol) was added and the whole was stirred for 30 min.   The reaction mixture was quenched with 1 N HCl aq. (100 mL) and extracted with AcOEt three times.   The organic layer was dried over MgSO$_4$ and concentrated.   The residue was chromatographed on silica gel to afford 3.96 g (19.8 mmol) of 4,4,4-trifluoro-1-phenylbut-2-yn-1-ol in 99% yield.

Spectral data

bp. 71-72 $\degree$C/2 mmHg; Rf 0.46 (hexane:AcOEt=4:1); ^{1}H NMR (CDCl$_3$): δ 2.7-3.2 (1 H, br s), 5.52 (1 H, q, J=2.98 Hz), 7.2-7.5 (5 H, m); ^{13}C NMR (CDCl$_3$): δ 63.94 (q, J=1.4 Hz), 73.45 (q, J=43.0 Hz), 86.45 (q, J=6.41 Hz), 124.14 (q, J=257.4 Hz), 126.71, 129.09, 129.34, 137.87; ^{19}F NMR (CDCl$_3$): δ 27.84 (d, J=2.77 Hz) from CF$_3$CO$_2$H; IR (neat) ν 3350, 2950, 2925, 2850, 2250 cm^{-1}.

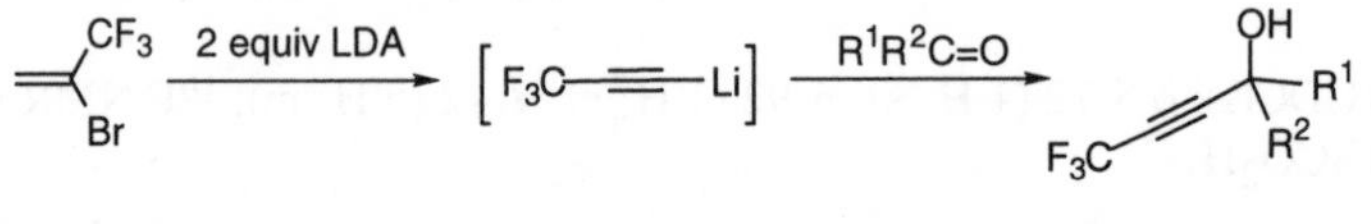

R^1	R^2	Yield (%)	Diastereomeric ratio		
PhCH$_2$CH$_2$	H	99			
BnOCH$_2$CH$_2$	H	92			
BnOCH$_2$	H	90			
n-C$_5$H$_{11}$	H	93			
c-C$_6$H$_{11}$	H	93			
(E)-PhCH=CH	H	99			
n-C$_3$H$_7$CHMe	H	75	50	:	50
CH$_3$CH(OBn)	H	94	64	:	36
PhCHMe	H	97	89	:	11
Ph	CH$_3$	89			
-(CH$_2$)$_5$-		99			

(Reproduced with permission from Yamazaki, T., *et al. J. Org. Chem.* **1995**, *60*, 6046)

Note

The intermediary acetylide was also successfully trapped by dimethylphenylsilyl chloride in 74% yield (determined by ^{19}F NMR with PhCF$_3$).

Yamazaki, T., *et al. J. Org. Chem.* **1995**, *60*, 6046.

Ethyl 2-(2,2,2-trifluoroethyl)-2-methylpent-4-enoate

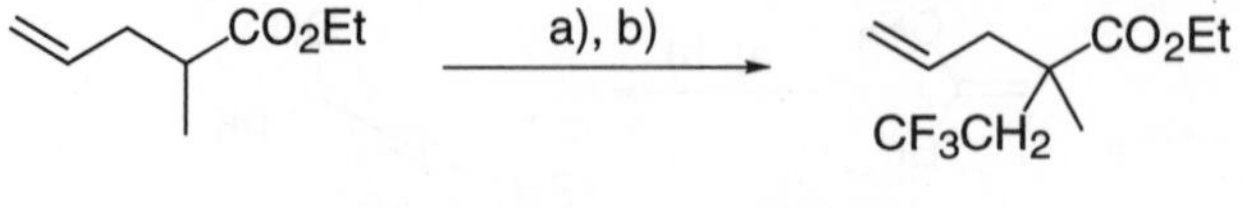

a) LDA / THF,   b) CF$_3$CH$_2$I

Synthetic method

An LDA (50 mmol) solution in dry THF (70 mL) was treated with 50 mmol of ethyl 2-methylpent-4-enoate dissolved in 5 mL of dry THF at -78 $\degree$C.   After 0.5 h, 1.3 equiv of 2,2,2-trifluoroethyl iodide dissolved in 0.5 equiv of HMPA was added from the dropping funnel.   When the addition was complete, stirring was maintained at -78 °C for 4 h.   The reaction was then allowed to warm to room temperature and stirred overnight.   Then, the mixture was quenched by careful addition of 10% HCl aq. (40 mL) followed by dropwise addition of conc HCl until acidic (pH=1-2).   The phases were separated and the aqueous layer was extracted with hexanes (3x40 mL).   The combined organic extract was washed with H_2O, 5% $NaHCO_3$ aq. and brine (20-mL portions of each).   The organic solution was dried over anhydrous $MgSO_4$, and concentrated *in vacuo* to yield the desired crude products as slightly yellow oils.   Purification by chromatography on silica gel (hexane:$CHCl_3$=60:40), followed by vacuum distillation, afforded 6.12 g of ethyl 2-(2,2,2-trifluoroethyl)-2-methylpent-4-enoate in 56% yield as a colorless liquid.

Spectral data
    bp. 91-94 °C/45 mmHg; ^1H NMR (CDCl$_3$) δ 1.24 (3 H, t, *J*=7 Hz), 1.27 (3 H, s), 2.17-2.42 (3 H, m), 2.54-2.71 (1 H, m), 4.15 (2 H, q, *J*=7 Hz), 5.05-5.14 (2 H, m), 5.60-5.74 (1 H, m); ^{19}F NMR (CDCl$_3$) δ -61.1 (t, *J*=11 Hz); IR (neat): ν 3078, 1737, 1642, 1263, 1154, 1111 cm^{-1}.

Note
    Because the above experimental procedure was described as the general method (Covey, D. F., *et al. J. Med. Chem.* **1991**, *34*, 1460), the original procedure was rescaled to the present reaction.

Covey, D. F., *et al. Bioorg. Med. Chem.* **1998**, *6*, 43.

Tributyl(3,3,3-trifluoroprop-1-en-2-yl)tin, 2-(Trifluoromethyl)-1-phenylprop-2-en-1-one

a) (*n*-Bu$_3$Sn)$_2$CuLi / THF,   b) PhC(O)Cl, cat. PdCl(Bn)(PPh$_3$)$_2$ / HMPA

Synthetic method
    Tributyl(3,3,3-trifluoroprop-1-en-2-yl)tin
    To a suspension of copper(I) iodide (1.43 g, 7.5 mmol) in THF (5 mL) cooled at -10 °C was added dropwise a THF solution of *n*-Bu$_3$SnLi (10 mL, 15 mmol), and the dark mixture was stirred at -10 °C for an additional 30 min and was then cooled down to -78 °C.   2-Bromo-3,3,3-trifluoropropene (1.32 g, 7.5 mmol) was introduced *via* a syringe at -78 °C, and the resulting solution was stirred at -78 °C for 0.5 h and then at room temperature for 1 h.   The solvent was removed and the residue dissolved in Et$_2$O (100 mL).   After removal of the solid by filtration, the filtrate was concentrated and distilled at reduced pressure to afford tributyl(3,3,3-trifluoroprop-1-en-2-yl)tin in 90% yield as a colorless oil.

Spectral data
    bp. 70 °C/1 mmHg; ^1H NMR (CDCl$_3$): δ 0.70-1.70 (27 H, m), 5.84 (1 H, s), 6.36 (1 H, s); ^{19}F NMR (CDCl$_3$): δ 16.0.(s) ppm from CF$_3$CO$_2$H.

Synthetic method
    2-(Trifluoromethyl)-1-phenylprop-2-en-1-one
    Tributyl(3,3,3-trifluoroprop-1-en-2-yl)tin (385 mg, 1 mmol) and benzoyl chloride (1.5

mmol) were dissolved in HMPA (3 mL) followed by the introduction of PdCl(Bn)(PPh$_3$)$_2$ (1 mol% based on tributyl(3,3,3-trifluoroprop-1-en-2-yl)tin).   The resulting mixture was stirred at 65 °C for 3 h.   Et$_2$O (30 mL) was added, and the ethereal solution was washed successively with KF aq., dilute NaHCO$_3$ aq. and brine, and dried over anhydrous Na$_2$SO$_4$.   After concentration, the residue was subjected to silica gel column chromatography (petroleum ether:AcOEt or acetone) to afford 2-(trifluoromethyl)-1-phenylprop-2-en-1-one in 90% yield as a colorless oil.

Spectral data
   ^{1}H NMR (CDCl$_3$): δ 6.08 (1 H, s), 6.60 (1 H, s), 7.30-7.85 (5 H, m); ^{19}F NMR (CCl$_4$): δ 12.2 (s) ppm from CF$_3$CO$_2$H; IR (neat): ν 1660, 1580, 1340, 1120, 980, 710 cm^{-1}.

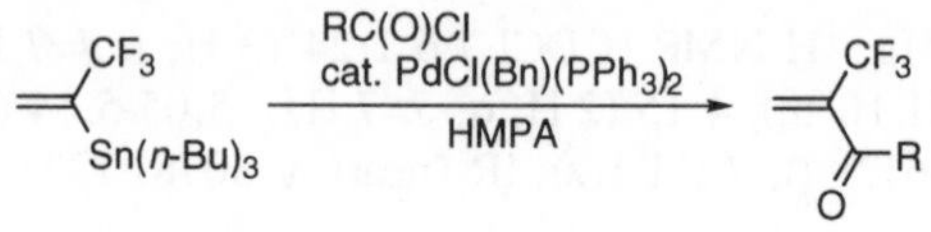

R	Time (h)	Yield (%)	R	Time (h)	Yield (%)
4-Br-C$_6$H$_4$-	5	89	4-Cl-C$_6$H$_4$-	4	93
2-Furyl	10	85	3-MeO-C$_6$H$_4$-	13	81
(E)-PhCH=CH-	4	97	n-C$_7$H$_{15}$-	24	78
Adamantyl	22	83			

(Reproduced with permission from Xu, Y.-Y., *et al. J. Org. Chem.* **1994**, *59*, 2638)

Xu, Y.-Y., *et al. J. Org. Chem.* **1994**, *59*, 2638.

4.3.9 Miscellaneous

1-Ethoxy-1-(trifluoromethyl)-2-*cis*-tetralol

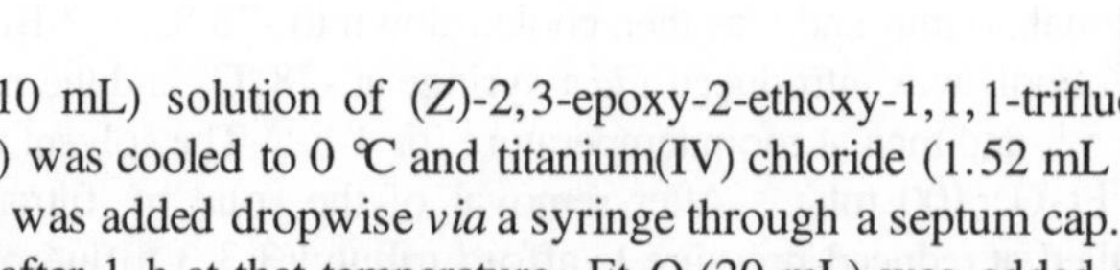

a) TiCl$_4$ / CH$_2$Cl$_2$

Synthetic method
   The CH$_2$Cl$_2$ (10 mL) solution of (Z)-2,3-epoxy-2-ethoxy-1,1,1-trifluoro-5-phenylpentane (200 mg, 0.76 mmol) was cooled to 0 °C and titanium(IV) chloride (1.52 mL of a 1 *M* solution in CH$_2$Cl$_2$, 1.52 mmol) was added dropwise *via* a syringe through a septum cap.   When the starting material disappeared after 1 h at that temperature, Et$_2$O (20 mL) was added and the mixture was hydrolyzed with sat. NH$_4$Cl aq. then allowed to warm to room temperature.   The organic layer was separated, washed with NaHCO$_3$ aq. until neutral, then washed twice with brine, dried over MgSO$_4$, and concentrated under reduced pressure.   The crude product was further purified by column chromatography (pentane:Et$_2$O) to give 150 mg of 1-ethoxy-1-(trifluoromethyl)-2-*cis*-tetralol in 75% yield.

Spectral data
   ^{1}H NMR (CDCl$_3$): δ 1.24 (3 H, t, *J*=7 Hz), 1.87 (1 H, m, *J*=13.5, 9.2 Hz), 2.2 (1 H, m, 13.5, 6.6, 3.6 Hz), 2.59 (1 H, d, *J*=1.2 Hz), 2.72 (1 H, m), 2.94 (1 H, m), 3.53 (1 H, m), 3.7 (1

H, m), 4.40 (1 H, ddq, $J=9.2$, 3.6 Hz), 7.2 (3 H, m), 7.68 (1 H, m); ^{13}C NMR (CDCl$_3$): δ 15.4, 25.9, 27.5, 61.0, 67.0, 78.6 (q, $J=25.4$ Hz), 125.0 (q, $J=286$ Hz), 126.2, 128.5, 128.9, 129.1, 129.5, 139.2; ^{19}F NMR (CDCl$_3$): δ -74.6 (s); IR (neat): ν 3600 cm^{-1}.

Note

For the preparation of the starting epoxide, see page 184 of this volume.

The reaction products in the present procedure were highly dependent on the substrates used or reaction conditions (*e.g.* temperature, Lewis acid) employed. Representative examples below describe such a trend.

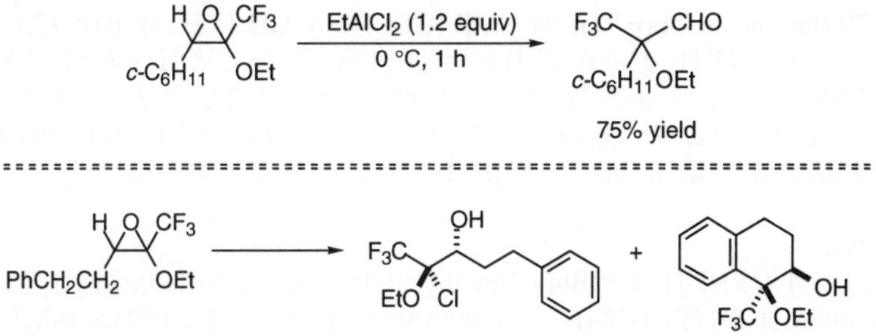

75% yield

Lewis acid (equiv)	Temp. (°C)	Time (h)	Yield of product (%)[a]	
			Acyclic	Cyclic
EtAlCl$_2$ (1.2)	0	1	56 (34)[b]	44 (28)
EtAlCl$_2$ (1.2)	-78	1	89	11
EtAlCl$_2$ (5)	20	72	52[b]	48
TiCl$_4$ (2)	-78	1[c]	69	31
TiCl$_4$ (2)	0	1	0	90 (75)
(CH$_3$)$_3$Al (2)	0	1	96 (70)[d]	0

[a] GC ratio. The isolated yield is shown in parentheses. [b] A 1:1 diastereomer mixture was obtained. [c] Reaction was performed at -78 °C and then 2 h at 20 °C, selectively providing the cyclic product in 95% (by GC analysis). [d] The product contained a methyl group instead of chlorine.

(Reproduced with permission from Bégué, J.-P., *et al. J. Org. Chem.* **1995**, *60*, 5029)

Bégué, J.-P., *et al. J. Org. Chem.* **1995**, *60*, 5029.

2-{(S)-1-[1-(*tert*-Butyldimethylsilyloxy)-2,2,2-trifluoroethyl]}-5-(trimethylsilyl)furan, (S)- and (R)-4-{(S)-1-[1-(*tert*-Butyldimethylsilyloxy)-2,2,2-trifluoroethyl]}but-2-en-4-olide

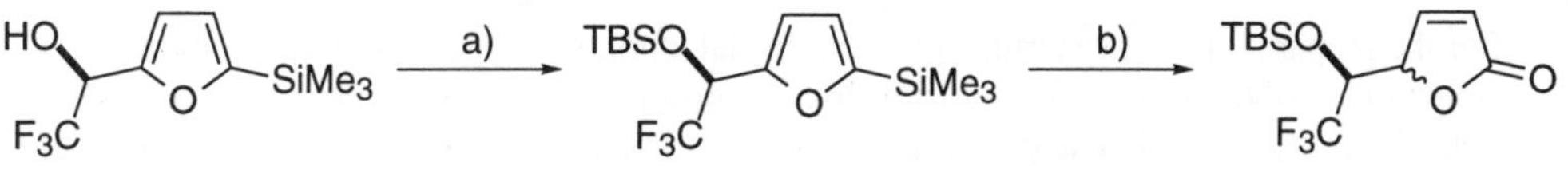

a) TBSCl, imidazole / CH$_2$Cl$_2$, b) MMPP / AcOH

Synthetic method

2-{(S)-1-[1-(*tert*-Butyldimethylsilyloxy)-2,2,2-trifluoroethyl]}-5-(trimethylsilyl)furan

A 0.5 *M* CH$_2$Cl$_2$ solution of 2-[(S)-1-(2,2,2-trifluoro-1-hydroxyethyl)-5-(trimethylsilyl)-

furan was treated with imidazole and *tert*-butyldimethylsilyl chloride (both 1.2 equiv) under a nitrogen atmosphere at 0 °C. The reaction mixture was stirred overnight at room temperature. The reaction was quenched with 1 N HCl aq., and the aqueous phase, after separation of the resulting CH_2Cl_2 phase, was extracted with CH_2Cl_2 twice. Then the combined organic layers were washed with sat. $NaHCO_3$ aq. and brine, dried ($MgSO_4$), and evaporated. Purification by distillation afforded 2-{(S)-1-[1-(*tert*-butyldimethylsilyloxy)-2,2,2-trifluoroethyl]}-5-(trimethyl-silyl)furan in 93% yield.

Spectral data

[α]$_D^{27}$ +36.72 (c 1.15, MeOH) from the starting alcohol with 94.3% ee; bp. 67-69 °C/0.6 mmHg; Rf 0.79 (hexane:AcOEt=7:1); ^1H NMR ($CDCl_3$): δ 0.03 (3 H, s), 0.08 (3 H, s), 0.24 (9 H, s), 0.86 (9 H, s), 5.03 (1 H, q, J=6.37 Hz), 6.41 (1 H, d, J=3.25 Hz), 6.57 (1 H, dq, J=3.26, 0.45 Hz); ^{13}C NMR ($CDCl_3$): δ -5.72, -5.60, -1.99, 17.95, 25.27, 68.22 (q, J=34.4 Hz), 109.29 (q, J=1.7 Hz), 120.4, 123.75 (q, J=283.2 Hz), 152.96 (q, J=1.6 Hz), 161.63; ^{19}F NMR ($CDCl_3$): δ 0.5 (d, J=6.3 Hz) from CF_3CO_2H; IR (neat): ν 2950, 2875 cm^{-1}.

Synthetic method

(S)- and (R)-4-{(S)-1-[1-(*tert*-Butyldimethylsilyloxy)-2,2,2-trifluoroethyl]}but-2-en-4-olide

To a solution of 2-{(S)-1-[1-(*tert*-butyldimethylsilyloxy)-2,2,2-trifluoroethyl]}-5-(trimethyl-silyl)furan was added magnesium monoperoxyphthalate (MMPP; 3 equiv) in AcOH (0.3 M) at room temperature, and the mixture was heated to 85 °C for 12 h. After cooling to room temperature, the reaction was quenched with dimethyl sulfide and the resulting solid was removed by filtration. After the usual extractive work-up with AcOEt, a mixture of butenolides (66% yield, a 1:1 diastereomeric ratio) and the starting material (29% recovery) were obtained, which were purified by silica gel column chromatography.

Spectral data

(S,S)-isomer (*anti* form): [α]$_D^{27}$ +98.24 (c 1.00, MeOH) starting from the alcohol with 94.3% ee; bp. 95-100 °C (bath temperature)/0.5 mmHg; Rf 0.27 (hexane:AcOEt=7:1); ^1H NMR ($CDCl_3$): δ 0.05 (3 H, s), 0.07 (3 H, s), 0.80 (9 H, s), 4.41 (1 H, qd, J=7.09, 2.40 Hz), 5.21 (1 H, m), 6.20 (1 H, ddd, J=5.86, 2.03, 0.67 Hz), 7.45 (1 H, m); ^{13}C NMR ($CDCl_3$): δ -5.80, -5.47, 17.75, 25.15, 70.97 (q, J=31.0 Hz), 80.82 (q, J=1.8 Hz), 123.65 (q, J=284.5 Hz), 123.95, 151.95 (q, J=2.1 Hz), 172.12; ^{19}F NMR ($CDCl_3$): δ 2.1 (d, J=6.3 Hz) from CF_3CO_2H; IR (neat): ν 2975, 2950, 2900, 2875, 1780 cm^{-1}.

(S,R)-isomer (*syn* form): [α]$_D^{27}$ +87.86 (c 0.66, MeOH) starting from the alcohol with 97.6% ee; mp. 64.5-65.5 °C; Rf 0.11 (hexane:AcOEt=7:1); ^1H NMR ($CDCl_3$): 0.10 (3 H, s), 0.11 (3 H, s), 0.87 (9 H, s), 4.09 (1 H, dq, J=6.45, 5.01 Hz), 5.11 (1 H, dt, J=6.45, 5.02 Hz), 6.22 (1 H, dd, J=5.78, 2.12 Hz), 7.44 (1 H, m); ^{13}C NMR ($CDCl_3$): δ -5.21, -4.89, 18.22, 25.60, 72.04 (q, J=31.2 Hz), 81.78 (q, J=1.7 Hz), 123.90 (q, J=284.7 Hz), 124.05, 152.04 (q, J=1.7 Hz), 172.37; ^{19}F NMR ($CDCl_3$): δ 2.4 (d, J=6.0 Hz) from CF_3CO_2H; IR (neat): ν 2975, 2950, 2900, 2875, 1770 cm^{-1}.

Note

For the preparation of the starting silylated furylalcohol, see page 160 of this volume.

Among two silyl groups, use of a trimethylsilyl group as the hydroxy protection furnished the deprotected hydroxy butenolide and a TBS moiety at the furan ring migrated to the α position of the resulting carbonyl group.

Treatment of the crude reaction mixture with 1.1 equiv of LDA in THF (about a 1 M solution) at -78 °C for 30 min, followed by the solution quenched with 3 mL of AcOH affected the isomerization of butenolide, giving a 75:25 *anti:syn* mixture.

Yamazaki, T., *et al. J. Org. Chem.* **1993**, *58*, 4346.

4.4 Reactions with Hetero Nucleophiles

4.4.1 Reactions with Nitrogen Nucleophiles

Methyl (Z)-3-benzylamino-4,4,4-trifluorobut-2-enoate

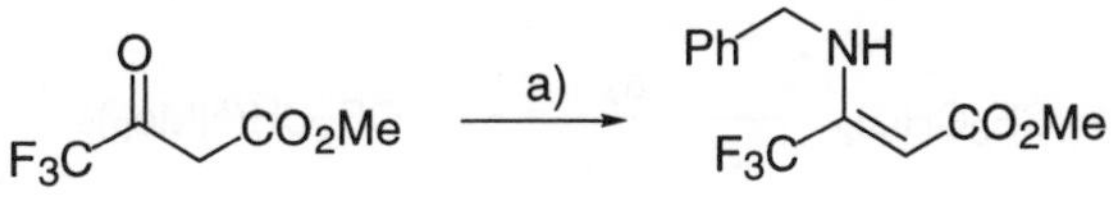

a) BnNH₂, Dowex-50 / benzene

Synthetic method

Benzylamine (0.55 mol) was added dropwise to a solution of methyl 4,4,4-trifluoro-3-oxobutyrate (0.5 mol) in benzene (400 mL) at room temperature. When the addition was over, 1 g of cation exchange resin Dowex-50 (H⁺-form) was added to the solution and the mixture was boiled using the Dean-Stark device until the theoretical amount of H₂O was removed. Filtration, evaporation in vacuum (40-50 °C/20-30 mmHg), and distillation of the reaction mixture yielded methyl (Z)-3-benzylamino-4,4,4-trifluorobut-2-enoate in 80% yield as a colorless oil.

Spectral data

bp. 139-141 °C/11 mmHg; ^{1}H NMR (CDCl$_3$): δ 3.62 (3 H, s), 4.50 (2 H, d, J=6.6 Hz), 5.14 (1 H, s), 7.30-7.42 (5 H, m), 8.63 (1 H, br t, J=6.6 Hz); ^{13}C NMR (CDCl$_3$): δ 47.9, 51.3, 84.8 (q, J=5.8 Hz), 120.0 (q, J=292.0 Hz), 127.4, 127.9, 128.9, 137.6, 148.9 (q, J=30.8 Hz), 169.9; ^{19}F NMR (CDCl$_3$): δ -65.9 (s); IR (CH$_2$Cl$_2$): ν 3200, 1670, 1630 cm^{-1}.

Soloshonok, V. A.; Kukhar, V. P. *Tetrahedron* **1996**, *52*, 6953.

N-(2,2,2-Trifluoroethylidene)anisidine

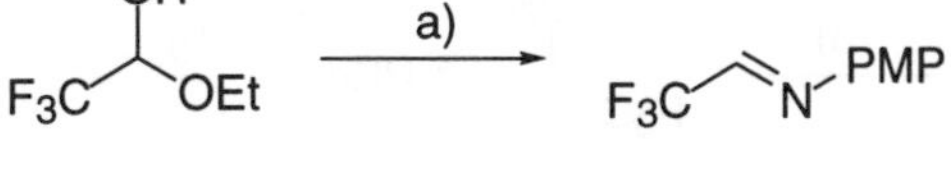

a) *p*-Anisidine, cat. TsOH / toluene

Synthetic method

A solution of trifluoroacetaldehyde ethyl hemiacetal (15 g, 100 mmol), *p*-methoxyaniline (10.26 g, 83 mmol), *p*-toluenesulfonic acid (50 mg) in toluene (120 mL) was refluxed under argon for 1.5 h in a Dean-Stark apparatus. The solution was washed with NaHCO$_3$ aq. and then with brine and dried over MgSO$_4$. The solution was evaporated under reduced pressure, and the residue was distilled, leading to 10.4 g of N-(2,2,2-trifluoroethylidene)anisidine in 64% yield.

Spectral data

bp. 60 °C/0.5 mmHg; ^{1}H NMR (CDCl$_3$): δ 3.84 (3 H, s), 7.1 (4 H, q), 7.8 (1 H, q, J=3.7 Hz); ^{13}C NMR (CDCl$_3$): δ 55.0, 119.5 (q, J=279 Hz), 114.5, 123.1, 139.8, 44.2 (q, J=33 Hz), 160.0; ^{19}F NMR (CDCl$_3$): δ -70.7 (d, J=3.7 Hz).

Note

   Similar reactions towards difluoro- or chlorodifluorinated hemiacetals realized the formation of the corresponding imines in 60 or 64% yields, respectively.

Bégué, J.-P., *et al. J. Org. Chem.* **1997**, *62*, 8826.

2,2,2-Trifluoro-1,1-bis(dimethylamino)ethane

$$CF_3CHCl_2 \xrightarrow{\text{a)}} CF_3CH(NMe_2)_2$$

a) Cu, Me$_2$NH

Synthetic method

   A mixture of 1,1-dichloro-2,2,2-trifluoroethane (50 g, 0.33 mol) and copper powder (43 g, 0.68 mol) in an autoclave was cooled to -78 °C.   The air in the autoclave was removed under vacuum.   Then 100 g of dimethylamine was transferred to the autoclave.   The reaction mixture was allowed to warm to room temperature, then heated at 60 °C for 12 h, after which the mixture was allowed to cool to room temperature.   The liquid was separated, and the solid was washed with 30 mL of CH$_2$Cl$_2$.   The combined liquid phases were distilled to give 33.8 g of 2,2,2-trifluoro-1,1-bis-(dimethylamino)ethane in 60% yield.

Spectral data

   bp. 92-93 °C; ^{1}H NMR (CDCl$_3$): δ 2.39 (12 H, q, *J*=1.2 Hz), 3.18 (1 H, q, *J*=7.3 Hz); ^{13}C NMR (CDCl$_3$): δ 41.3, 83.7 (q, *J*=26.2 Hz), 126.9 (q, *J*=293.6 Hz); ^{19}F NMR (CDCl$_3$): δ -67.0 (d, *J*=7.3 Hz); IR (neat): ν 1688, 1259 cm^{-1}.

Xu, Y.-L.; Dolbier, Jr., W. R. *J. Org. Chem.* **1997**, *62*, 6503.

(*S*)-3-(Benzylamino)-1,1,1-trifluoropropan-2-ol, (*R*)-1-Benzyl-2-(trifluoromethyl)aziridine

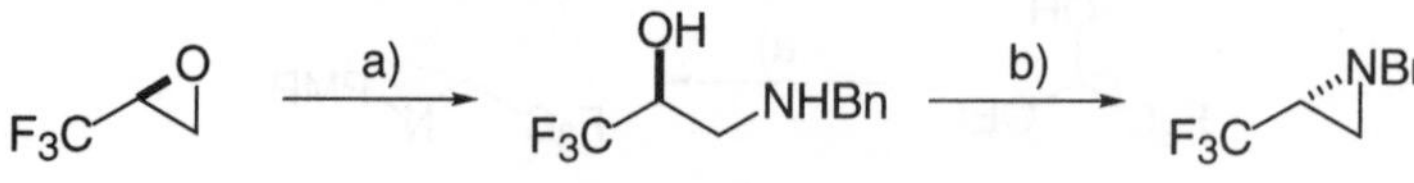

a) BnNH$_2$ / MeCN,   b) Ph$_3$PCl$_2$, Et$_3$N / MeCN

Synthetic method

   (*S*)-3-(Benzylamino)-1,1,1-trifluoropropan-2-ol

   To an ice-cooled solution of benzylamine (48 mmol) in 12 mL of MeCN, (*S*)-1,1,1-trifluoro-2,3-epoxypropane (49 mmol) was added dropwise.   The reaction mixture was stirred at room temperature for 2 d.   After removal of the solvent under reduced pressure, the residue was purified by column chromatography on silica gel.   (*S*)-3-(Benzylamino)-1,1,1-trifluoropropan-2-ol was obtained in 85% yield.

Spectral data

   [α]$_D^{28}$ -9.24 (*c* 0.50, CHCl$_3$; for the sample of >99.5% ee); mp. 103 °C (hexane:Et$_2$O); ^{1}H NMR (CDCl$_3$): δ 2.88 (1 H, dd, *J*=12.6, 5.0 Hz), 2.98 (1 H, dd, *J*=13.0, 6.0 Hz), 3.82 (2 H, s), 3.95 (1 H, m), 7.31 (5 H, m); ^{13}C NMR (CDCl$_3$): δ 46.9, 53.6, 67.8 (q, *J*=31 Hz), 116.5, 122.1,

127.5, 127.7, 128.1, 128.6, 136.2 (q, J=282 Hz); ^{19}F NMR (CDCl$_3$): δ 83.2 (d, J=7.1 Hz) from C$_6$F$_6$; IR (KBr): ν 3316, 3092, 1434, 1376, 1280, 1268, 1144, 1084, 1068, 1030, 922, 844, 754, 706, 692 cm^{-1}.

Synthetic method

(*R*)-1-Benzyl-2-(trifluoromethyl)aziridine

To a solution of dichlorotriphenylphosphorane (14 mmol) in 10 mL of MeCN, (*S*)-3-(benzyl-amino)-1,1,1-trifluoropropan-2-ol (5 mmol) was added. Then, triethylamine (52 mmol) was dropped into the above mixture with ice cooling. The reaction mixture was stirred for 3 h at reflux. (*R*)-1-Benzyl-2-(trifluoromethyl)aziridine was isolated as a colorless liquid in 85% yield by distillation of filtrate from the precipitated triethylamine hydrochloride and triphenylphosphine oxide.

Spectral data

[α]$_D^{24}$ +20.1 (*c* 0.139, CHCl$_3$; for the sample of >99.5% ee); bp. 110 ℃/20 mmHg; ^{1}H NMR (CDCl$_3$): δ 1.68 (1 H, d, J=5.9 Hz), 2.17 (2 H, m), 3.56 (2 H, s), 7.34 (5 H, m); ^{19}F NMR (CDCl$_3$): δ 90.7 (d, J=5.3 Hz) from C$_6$F$_6$; IR (KBr): ν 1498, 1432, 1364, 1280, 1156, 1070, 1028, 848, 736, 698, 658 cm^{-1}.

Note

For the preparation of the starting epoxide, see page 178 of this volume.

The optical purity of the first product, aminoalcohol, was improved by recrystallization from a mixture of hexane:Et$_2$O. In the present case, Uneyama and his co-workers used (*S*)-1,1,1-trifluoro-2,3-epoxypropane with the optical activity of 75% ee, and twice recrystallization was allowed to obtain the optically pure aminoalcohol.

The trifluoromethylated aziridine was also transformed into the regioisomeric aminoalcohol by ring opening with trifluoroacetic acid with retention of stereochemistry.

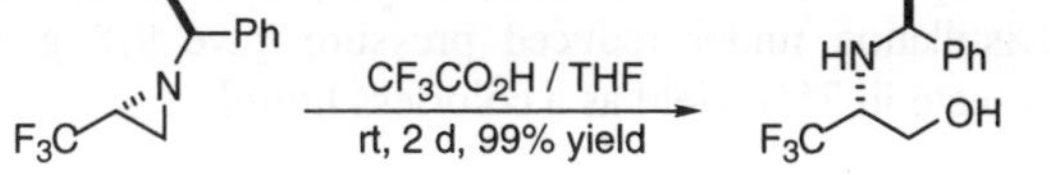

Uneyama, K., *et al. Tetrahedron: Asym.* **1997**, *8*, 2933.

4.4.2 Reactions with Oxygen Nucleophiles

2-(Trifluoromethyl)-3-hydroxypropionic acid

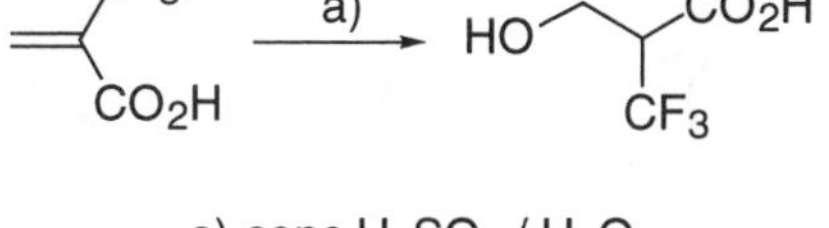

a) conc H$_2$SO$_4$ / H$_2$O

Synthetic method

A solution of 2-(trifluoromethyl)acrylic acid (50.0 g, 0.357 mol) and conc H$_2$SO$_4$ (4.0 mL) in H$_2$O (1 L) was heated at reflux for 2 d. The aqueous solution was saturated with NaCl and extracted with Et$_2$O (5x300 mL). The combined organic layers were dried over MgSO$_4$. Evaporation of the solvent by rotary evaporator and drying of the resulting waxy solid overnight under high vacuum (0.1-0.01 mmHg) gave 48.2 g of the hygroscopic 2-(trifluoromethyl)-3-hydroxypropionic acid in 85% yield.

Spectral data

^{1}H NMR (Acetone-d_6): δ 3.55 (1 H, ddq, J=9.0, 6.7, 5.2 Hz), 4.04 (1 H, dd, J=11.1, 5.2 Hz), 4.07 (1 H, dd, J=11.1, 6.7 Hz); ^{13}C NMR (Acetone-d_6): δ 53.27 (q, J=25.5 Hz), 58.96, 125.5 (q, J=279.2 Hz), 167.58; ^{19}F NMR (Acetone-d_6): δ -65.9 (d, J=9.2 Hz); IR (neat): ν 3300, 1734, 1467, 1323, 1176, 1123, 1040 cm^{-1}.

Götzö, S. P.; Seebach, D. *Chimia* **1996**, *50*, 20.

Ethyl 2-chloro-3,3,3-trifluoro-2-methoxypropionate

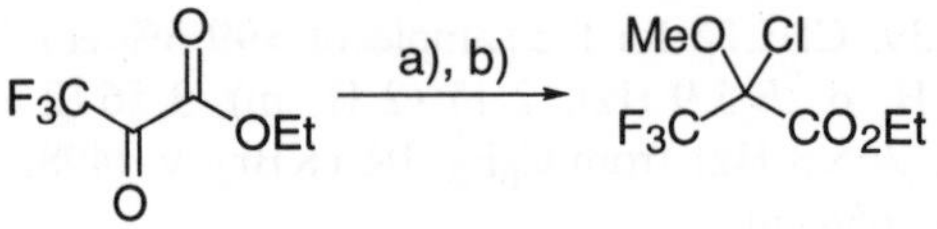

a) MeOH / benzene,   b) SOCl$_2$, pyridine

Synthetic method

Freshly distilled ethyl 3,3,3-trifluoropyruvate (10.2 g, 60 mmol) and methanol (2.5 mL, 60 mmol) were mixed in dry benzene (100 mL).   To the resulting solution cooled with an ice bath was added pyridine (14.4 mL, 180 mmol), followed by thionyl chloride (6.5 mL, 90 mmol) added dropwise over 10 min.   After being stirred at 0 ℃ for 30 min, the reaction mixture was poured into ice H$_2$O (100 mL) and extracted with Et$_2$O (3x50 mL).   The combined organic phase was washed successively with H$_2$O (100 mL), sat. NaHCO$_3$ aq. (2x100 mL) and brine (100 mL), then dried over Na$_2$SO$_4$.   Distillation under reduced pressure gave 9.8 g of ethyl 2-chloro-3,3,3-trifluoro-2-methoxypropionate in 75% yield as a colorless liquid.

Spectral data

bp. 60 ℃/23 mmHg; ^{1}H NMR (CCl$_4$): δ 1.31 (3 H, t, J=7.1 Hz), 3.62 (3 H, s), 4.30 (2 H, q, J=7.1 Hz); ^{19}F NMR (CCl$_4$): δ 0.3 (s) from CF$_3$CO$_2$H.

Note

For the preparation of trifluorinated pyruvate, see page 179 of this volume.

Shi, G.-Q., *et al. J. Org. Chem.* **1995**, *60*, 6608.

(S)-1,1,1-Trifluoro-2,3-epoxypropane

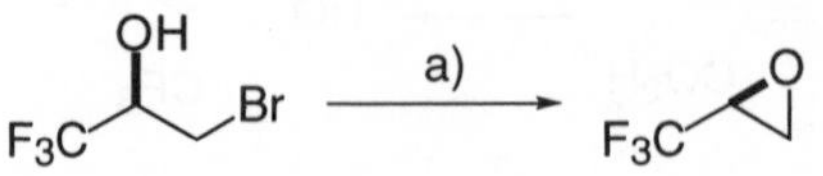

a) NaOH aq.

Synthetic method

(R)-3-Bromo-1,1,1-trifluoropropan-2-ol (96% ee; 2.9 g, 15 mmol) was added dropwise to an ice-cold solution of NaOH aq. (15 N, 60 mmol) contained in a 15-mL round-bottomed flask fitted with a distillation set-up.   The mixture was stirred at 95-100 ℃; then (S)-1,1,1-trifluoro-2,3-

epoxypropane was distilled and collected in a flask cooled in a dry ice-acetone bath.   Yield: 1.38 g (82%).   Optical purity was determined to be 91% in favor of the (S)-isomer on the basis of the maximum rotation reported in the literature.   However, the material was 96% ee as confirmed by ring-cleavage reactions.

Spectral data

$[\alpha]_D^{22}$ -10.92 (c 5.0, CHCl$_3$); bp. 37-38 °C; ^{1}H NMR (CDCl$_3$): δ 2.90-3.02 (2 H, m), 3.40-3.50 (1 H, m); ^{13}C NMR (CDCl$_3$): δ 43.11, 48.16 (q, J=41.7 Hz), 122.84 (q, J=274.8 Hz); ^{19}F NMR (CDCl$_3$): δ -75.04 (d, J=4.6 Hz).

Note

For the preparation of the starting alcohol, see page 189 of this volume.

The opposite mode of reaction was reported by use of trimethylsilyl bromide, followed by hydrolysis under acidic conditions.   See Sznaidman, M. L.; Beauchamp, L. M. *Nucleosides Nucleotides* **1996**, *15*, 1315.

Brown, H. C., *et al. J. Org. Chem.* **1995**, *60*, 41.

Methyl 2,3,3,3-tetrafluoro-2-methoxypropionate, Methyl 3,3,3-trifluoropyruvate

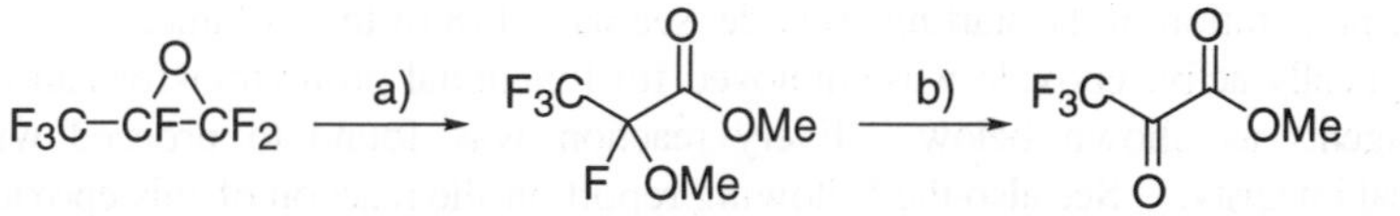

a) MeOH,   b) conc H$_2$SO$_4$, silica gel

Synthetic method

Methyl 2,3,3,3-tetrafluoro-2-methoxypropionate

Hexafluoropropene oxide (56.6 g) was bubbled with cooling during 4-5 h through 300 mL of anhydrous MeOH at such a rate that the gas was completely absorbed by the reaction mixture. The resulting mixture was poured into 1 L of H$_2$O, the lower layer was separated, washed twice with H$_2$O, dried over Na$_2$SO$_4$ and distilled.   The yield of methyl 2,3,3,3-tetrafluoro-2-methoxypropionate was 62 g (96%).   bp. 40-41 °C/21 mmHg.

Synthetic method

Methyl 3,3,3-trifluoropyruvate

In a flask fitted with a reflux condenser and a CaCl$_2$ tube, are placed 12 mL of conc H$_2$SO$_4$, 1 g of finely divided silica gel,   and 9.5 g (50 mmol) of methyl 2,3,3,3-tetrafluoro-2-methoxypropionate.   The mixture was heated to 140 °C in an oil bath, then allowed to stand at 140-150 °C for 20-30 min before vigorous boiling started.   The flask was cooled, connected to a downward condenser and the product was distilled in an oil bath.   A fraction boiling at 105-110 °C was redistilled.   The yield of methyl 3,3,3-trifluoropyruvate was 6.5 g (83%).   bp. 84-86 °C.

Sianesi, D., *et al. J. Org. Chem.* **1966**, *31*, 2312
(for the preparation of methyl 2,3,3,3-tetrafluoro-2-methoxypropionate).

Knunyants, I. L., *et al. Dokl. Acad. Nauk SSSR* **1966**, *169*, 594
(for the preparation of methyl 3,3,3-trifluoropyruvate).

(S)-3,3,3-Trifluoropropane-1,2-diol

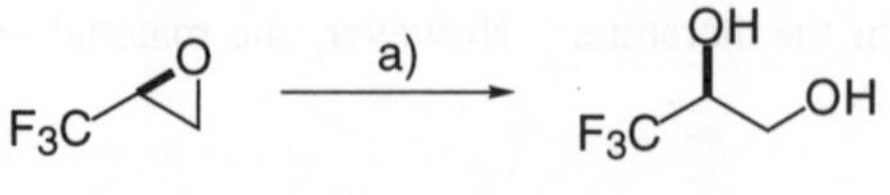

a) 1% H_2SO_4 aq.

Synthetic method

A mixture of (S)-1,1,1-trifluoro-2,3-epoxypropane (96% ee; 0.79 g, 7.0 mmol) and H_2O (3.5 mL) containing 1% H_2SO_4 was heated in a sealed tube at 100 °C for 2 h. The product mixture was extracted with AcOEt, dried over $MgSO_4$, concentrated, and distilled using a Kügelrohr apparatus to provide 0.73 g of (S)-3,3,3-trifluoropropane-1,2-diol in 80% yield. Analysis of the TFA derivative on a Chiraldex-GTA capillary column showed it to be 96% ee.

Spectral data

$[\alpha]_D^{24}$ -10.95 (c 1.4, MeOH); bp. 100 °C (bath temperature)/25 mmHg; ^1H NMR ($CDCl_3$): δ 3.5 (1 H, br s), 3.75-3.90 (2 H, m), 4.0-4.15 (1 H, m), 4.58 (1 H, br s); ^{13}C NMR ($CDCl_3$+acetone-d_6): δ 60.27, 70.40 (q, J=30.1 Hz), 124.26 (q, J=282.0 Hz); ^{19}F NMR ($CDCl_3$): δ -77.82 (d, J=6.8 Hz); IR (neat) ν 3373 cm^{-1}.

Note

For the preparation of the starting epoxide, see page 178 of this volume.

This optically active epoxide was employed for functional group transformation with various types of reagents as shown below. Every reaction was found to proceed without loss of stereochemical integrity. See also the following report on the reaction of this epoxide with carbon nucleophiles: Takahashi, O., et al. Tetrahedron Lett. **1990**, 31, 7031.

A similar reaction was also possible using cyclic sulfate. Vanhessche, K. P. M.; Sharpless, K. B. Chem. Eur. J. **1997**, 3, 517.

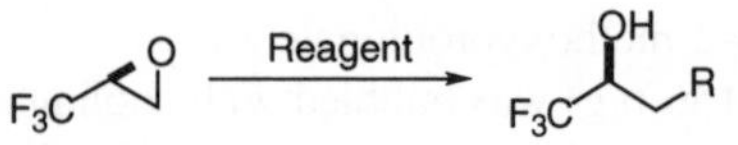

Reagent	R	Temp. (°C)	Time (h)	Yield (%)
$KN(SiMe_3)_2$[a)]	NH_2	25	12	61
NaN_3	N_3	25	5	65
Et_2NH	Et_2N	55	1	88
NaCN	CN	25	12	65
$LiAlH_4$	H	25	1.5	70
n-$C_5H_{11}MgBr$	n-C_5H_{11}	-20→25	3	75
PhH[b)]	Ph	0→25	3	72
EtOH[c)]	EtO	80	12	81

a) The product was hydrolyzed by HCl/Et_2O.   b) In the presence of $AlCl_3$ as a catalyst.
c) In the presence of dilute H_2SO_4.
(Reproduced with permission from Brown, H. C., et al. J. Org. Chem. **1995**, 60, 41)

Brown, H. C., et al. J. Org. Chem. **1995**, 60, 41.

4.4.3 Reaction with a Sulfur Nucleophile

syn-1,1,1-Trifluoro-3-(phenylsulfenyl)nonan-2-ol,
(E)-1,1,1-Trifluoronon-3-en-2-ol

a) PhSH, NaH / THF,   b) NaBH$_4$ / EtOH,   c) NaIO$_4$ / MeOH-H$_2$O,   d) Heating

Synthetic method
syn-1,1,1-Trifluoro-3-(phenylsulfenyl)nonan-2-ol
Sodium hydride (300 mg, 80% dispersion in oil, 10 mmol) was placed in an argon-flushed three-necked flask and washed twice with dry pentane, after which THF (50 mL) was added *via* a syringe through a septum cap.   The suspension was cooled to 0 °C and thiophenol (1.03 mL, 10 mmol) was added dropwise.   The mixture was stirred for *ca.* 0.5 h at room temperature and (Z)-2,3-epoxy-2-ethoxy-1,1,1-trifluorononane (2.54 g, 9 mmol) was then added in one portion to the suspension.   After 10 min the precipitate disappeared.   The reaction mixture was quenched with NH$_4$Cl aq. and extracted twice with Et$_2$O.   The organic phase was washed with brine, dried over MgSO$_4$, and concentrated.   EtOH (20 mL) was added and the solution obtained was cooled to 0 °C.   NaBH$_4$ (0.76 g, 20 mmol) was added in portions over 15 min and the reaction mixture was then stirred at room temperature for 2 h.   Saturated NH$_4$Cl aq. (20 mL) was added at 0 °C and the mixture was extracted with CH$_2$Cl$_2$ (3x20 mL).   The combined organic layers were dried over MgSO$_4$, concentrated *in vacuo* and purified by column chromatography on silica gel (pentane: Et$_2$O=20:1, 5:1) to afford 2.49 g of 1,1,1-trifluoro-3-(phenylsulfenyl)nonan-2-ol in 78% yield as a mixture of *syn* and *anti* isomers in a ratio of 91:9.
Spectral data
^{1}H NMR (CDCl$_3$): δ 0.91 (3 H, t, J=7.0 Hz), 1.33 (6 H, br s), 1.65 (4 H, m), 2.82 (1 H, d, J=8.0 Hz; *anti* isomer), 3.32 (1 H, dt, J=7.0, 6.7 Hz), 3.41 (1 H, d, J=7.5 Hz; *syn* isomer), 3.85 (1 H, ddq, J=7.5, 7.1, 7.0 Hz), 7.35 (3 H, m), 7.5 (2 H, m); ^{13}C NMR (CDCl$_3$): δ 14.1, 22.7, 27.1, 28.9, 31.7, 32.4, 46.1, 52.2, 71.9 (q, J=30 Hz), 124.8 (q, J=292 Hz), 128.3, 129.3, 133.2; ^{19}F NMR (CDCl$_3$): δ -75.4 (d, J=7.1 Hz; *syn* isomer), -74.5 (d, J=7.5 Hz; *anti* isomer).

Synthetic method
(E)-1,1,1-Trifluoronon-3-en-2-ol
To a solution of syn-1,1,1-trifluoro-3-(phenylsulfenyl)nonan-2-ol (3.06 g, 10 mmol) in MeOH (100 mL) was added in one portion NaIO$_4$ (2.27 g, 10.5 mmol) in H$_2$O (35 mL) with stirring.   The reaction mixture was heated at 70 °C for 4 h, cooled and extracted with CH$_2$Cl$_2$ (3x100 mL).   The combined organic layers were dried over MgSO$_4$, concentrated *in vacuo* and purified by column chromatography on silica gel (pentane: Et$_2$O=5:1, 1:1) to afford 2.90 g of 1,1,1-trifluoro-3-(phenylsulfinyl)nonan-2-ol in 90% yield as a mixture of diastereomers in an approximate ratio of 45:45:5:5.
1,1,1-Trifluoro-3-(phenylsulfinyl)nonan-2-ol (1.68 g, 5 mmol) thus obtained was heated at 135-140 °C for 3 h, cooled and purified by column chromatography on silica gel (pentane:Et$_2$O= 10:1, 2:1) to afford 0.85 g of (E)-1,1,1-trifluoronon-3-en-2-ol in 81% yield with traces of the (Z)-isomer confirmed by GC.

Spectral data
    ^{1}H NMR (CDCl$_3$): δ 0.91 (3 H, t, *J*=7.5 Hz), 1.3 (6 H, m), 2.05 (2 H, dt, *J*=7.0, 6.9 Hz), 2.2 (1 H, s), 4.32 (1 H, dq, *J*=7.0, 6.8 Hz), 5.45 (1 H, dd, *J*=14.2, 7.0 Hz), 5.92 (1 H, dt, *J*=14.2, 7.1 Hz); ^{13}C NMR (CDCl$_3$): δ 13.9, 22.3, 28.2, 31.1, 32.1, 71.5 (q, *J*=32 Hz), 121.8, 124.3 (q, *J*=280 Hz), 139.1; ^{19}F NMR (CDCl$_3$): δ -79.6 (d, *J*=6.8 Hz).

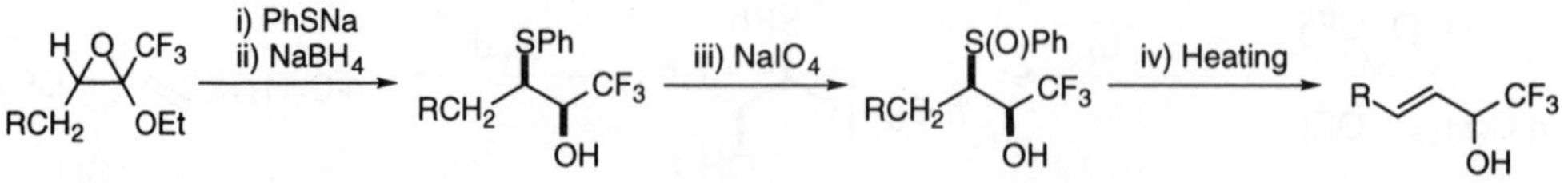

| | Yield (%) | | |
R	PhS-OH	PhS(O)-OH	Allyl-OH
c-C$_6$H$_{11}$	77	90	91
p-MeO-C$_6$H$_4$-CH$_2$-	71	93	86
C$_6$H$_5$-	71	55[a]	85
C$_6$H$_5$-CH$_2$-	76	87	87

[a] Formation of this sulfoxide was accompanied by the allylic alcohol (the final product) which was isolated in 43% yield.
(Reproduced with permission from Bégué, J.-P., *et al. J. Fluorine Chem.* **1996**, *80*, 13)

Note
    For the preparation of the starting epoxide, see page 184 of this volume.
    Bégué and his co-workers also reported the isolation of the above intermediate, α-phenyl-sulfenylated trifluoroketones.   However, this reaction was applicable only for the trifluoro-methylated materials, and, as shown below, elimination of fluorine-containing groups occurred to give nonfluorinated esters.   See Bégué, J.-P., *et al. Synthesis* **1996**, 529.
    For a similar ring opening by MgBr$_2$ giving the corresponding α-bromoketones, see Bégué, J.-P., *et al. Synthesis* **1993**, 1083.

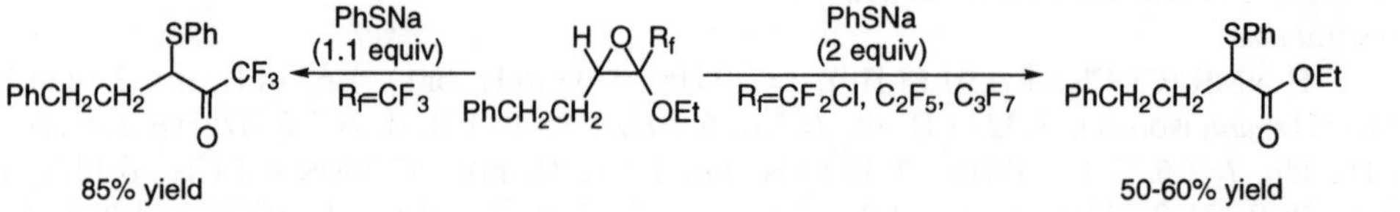

Bégué, J.-P., *et al. J. Fluorine Chem.* **1996**, *80*, 13.

4.5 Dehydrations

Ethyl (*E*)-4,4,4-trifluorobut-2-enoate

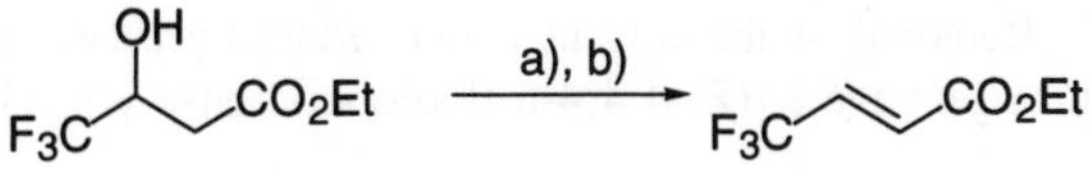

a) PPh₃ / Et₂O, b) EtO₂CN=NCO₂Et

Synthetic method
    A mixture of ethyl 4,4,4-trifluoro-3-hydroxybutanoate (186 g, 1 mol) and triphenyl-phosphine (262 g, 1 mol) in anhydrous Et₂O (1 L) was stirred at ambient temperature until it was completely dissolved.   The mixture was cooled to 0 °C and diethyl azodicarboxylate (174 g, 1 mol) was slowly added, forming a thick precipitate of Ph₃PO.   The cooling bath was removed and the mixture was stirred at room temperature overnight.   The phosphine oxide was removed by filtration and washed with Et₂O (2x250 mL).   The ether was removed from the combined filtrate, which was washed by distillation through a short column, and ethyl (*E*)-4,4,4-trifluorobut-2-enoate was isolated as a colorless liquid by distillation (161 g) in 96% yield.
Spectral data
    bp. 110-112 °C; ^{1}H NMR (CDCl₃): δ 1.33 (3 H, t, *J*=7 Hz), 4.25 (2 H, q, *J*=7 Hz), 6.49 (1 H, dq, *J*=16, 1.5 Hz), 6.76 (1 H, dq, *J*=16, 6.1 Hz); ^{19}F NMR (CDCl₃): δ -66.25 (d, *J*=6.1, 1.5 Hz); IR (neat): ν 1727, 1667, 1380, 1310, 1280, 1137, 1020, 980 cm^{-1}.

Roberts, J. L., *et al. J. Org. Chem.* **1984**, *49*, 1430.

1-[(*E*)-4,4,4-Trifluorobut-2-en-2-yl]cyclohexene,
1-[(*Z*)-4,4,4-Trifluorobut-2-en-2-yl]cyclohexene

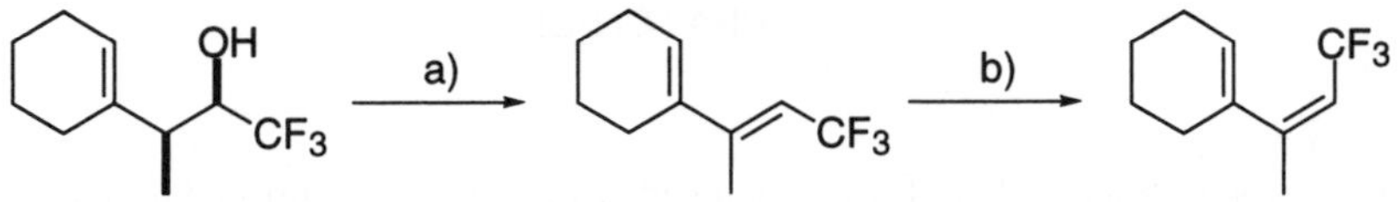

a) POCl₃ / pyridine, b) hv, benzophenone / Et₂O

Synthetic method
    1-[(*E*)-4,4,4-Trifluorobut-2-en-2-yl]cyclohexene
    To a solution of *syn*-3-(cyclohex-1-en-1-yl)-1,1,1-trifluorobutan-2-ol (208 mg, 1 mmol) in pyridine (0.5 mL) phosphorus oxychloride (0.09 mL, 1 mmol) was added at 0 °C, and the mixture was stirred at 110 °C for 48 h.   After cooling, the mixture was treated with ice H₂O and extracted with Et₂O, and dried over MgSO₄.   Removal of the solvent under reduced pressure followed by chromatographic separation by silica gel gave 1-[(*E*)-4,4,4-trifluorobut-2-en-2-yl]cyclohexene in 72% yield.
Spectral data
    ^{1}H NMR (CDCl₃): δ 1.59 (2 H, m), 1.68 (2 H, m), 2.00 (3 H, s), 2.12-2.19 (4 H, m), 5.57

(1 H, q, J=8.80 Hz), 6.17 (1 H, br s); ^{13}C NMR (CDCl$_3$): δ 14.6, 21.8, 22.6, 25.6, 26.0, 112.4 (q, J=33 Hz), 124.8 (q, J=227 Hz), 129.7, 136.2, 148.8.

1-[(Z)-4,4,4-Trifluorobut-2-en-2-yl]cyclohexene

To a solution of 1-[(E)-4,4,4-trifluorobut-2-en-2-yl]cyclohexene (95 mg, 0.5 mmol) in Et$_2$O (5 mL) was added benzophenone (5 mg). The mixture was irradiated with a high-pressure mercury lamp for 1 h. Removal of the solvent under reduced pressure followed by chromatographic separation by silica gel gave 1-[(Z)-4,4,4-trifluorobut-2-en-2-yl]cyclohexene in quantitative yield.

Spectral data

^{1}H NMR (CDCl$_3$): δ 1.56-1.75 (4 H, m), 1.89 (3 H, br s), 2.00-2.15 (4 H, m), 5.34 (1 H, m), 5.53 (1 H, m); ^{13}C NMR (CDCl$_3$): δ 21.8, 22.5, 23.9, 24.9, 27.1, 113.8 (q, J=33 Hz), 123.3 (q, J=269 Hz), 124.6, 136.5, 153.6.

Note

For the preparation of the starting alcohol, see page 147 of this volume.

Mikami, K., et al. Tetrahedron 1996, 52, 85.

4.6 Oxidations and Reductions

4.6.1 Oxidations

(Z)-2,3-Epoxy-2-ethoxy-1,1,1-trifluorononane

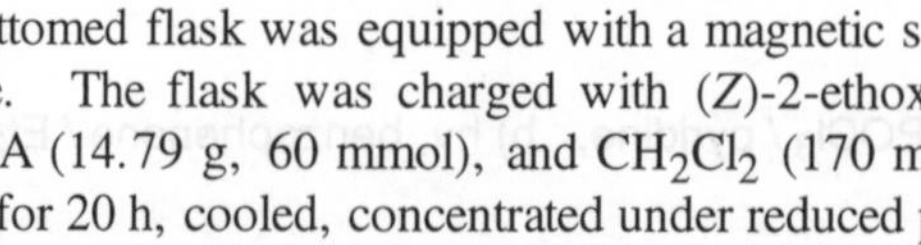

a) mCPBA / CH$_2$Cl$_2$

Synthetic method

A 250-mL round-bottomed flask was equipped with a magnetic stirring bar and a condenser with a CaCl$_2$ drying tube. The flask was charged with (Z)-2-ethoxy-1,1,1-trifluoronon-2-ene (9.2 g, 40 mmol), mCPBA (14.79 g, 60 mmol), and CH$_2$Cl$_2$ (170 mL). The reaction mixture was refluxed with stirring for 20 h, cooled, concentrated under reduced pressure to a volume of 50-60 mL, diluted with pentane (200 mL) and passed through a short silica gel column. Then the column was washed with a mixture of pentane:Et$_2$O=5:1 (100 mL). Evaporation of solvents under reduced pressure and bulb-to-bulb distillation provided 9.15 g of (Z)-2,3-epoxy-2-ethoxy-1,1,1-trifluorononane in 90% yield as a clear colorless liquid.

Spectral data

bp. 95 ℃ (bath temperature)/10 mmHg; ^{1}H NMR (CDCl$_3$): δ 0.85 (3 H, t, J=7.5 Hz), 1.25 (3 H, t, J=7.0 Hz), 1.32 (8 H, m), 1.61 (2 H, m), 3.12 (1 H, t, J=6.2 Hz), 3.82 (2 H, q, J=7.0 Hz); ^{13}C NMR (CDCl$_3$): δ 14.1, 15.4, 22.6, 25.7, 26.5, 26.6, 29.0, 31.7, 62.2, 63.8, 81.7 (q, J=40 Hz), 122.0 (q, J=282 Hz); ^{19}F NMR (CDCl$_3$): δ -77.1 (s).

Note

For the preparation of the starting olefin, see page 150 of this volume.

Bégué, J.-P., *et al. Synthesis* **1996**, 529.

(*R*)-3,3,3-Trifluoro-2-phenylpropane-1,2-diol

a) 0.2 mol% OsO$_4$, 1 mol% (DHPD)$_2$-DPP,
K$_3$Fe(CN)$_6$, K$_2$CO$_3$ / *tert*-BuOH:H$_2$O (1:1)

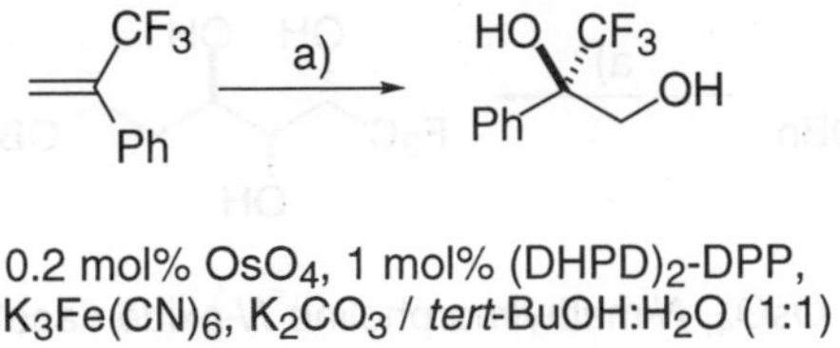

Synthetic method

A solution of potassium ferricyanide (K$_3$Fe(CN)$_6$, 28.8 g, 87 mmol), K$_2$CO$_3$ (11.9 g, 87 mmol), (DHQD)$_2$-DPP (0.275 g, 1.0 mol%), OsO$_4$ (0.58 mL of a 0.1 *M* solution in toluene, 0.2 mol%), *tert*-BuOH (145 mL) and H$_2$O (145 mL) was cooled to 0 °C and under vigorous stirring, α-(trifluoromethyl)styrene (5.0 g, 29 mmol) was added.   The reaction mixture was stirred at 0 °C for 72 h then solid Na$_2$S$_2$O$_5$ (43 g) was added slowly.   After stirring for 45 min at 25 °C, AcOEt (200 mL) was added and the organic layer was separated.   The aqueous layer was extracted with AcOEt (3x25 mL).   The organic fractions were combined, washed with H$_2$O (2x50 mL), brine (2x50 mL) then dried over MgSO$_4$.   The solvent was evaporated giving a yellow residue which was taken up in AcOEt (100 mL; organic layer A) and washed with 1 *N* H$_2$SO$_4$ (2x25 mL; aqueous layer A).   The aqueous layer A was neutralized using 0.1 *N* NaOH, then extracted with AcOEt (3x 30 mL).   This organic layer was dried over MgSO$_4$ and evaporated to give 0.26 g (95% recovery) of ligand as a yellow solid.   The organic layer A was washed with H$_2$O (2x25 mL) and brine (2x25 mL), dried over MgSO$_4$ and evaporated to give a thick clear oil which was virtually pure by TLC.   Bulb-to-bulb distillation gave 5.61 g (94% yield) of (*R*)-3,3,3-trifluoro-2-phenylpropane-1,2-diol in 91% ee, which solidified upon cooling.

Spectral data

[α]$_D$ -14.8 (*c* 2.05, MeOH); bp. 145 °C (bath temperature)/0.02 mmHg; mp. 33-37 °C; Rf 0.5 (hexane:AcOEt=1:1); ^{1}H NMR (CDCl$_3$): δ 2.29 (1 H, br s), 3.75 (1 H, br s), 3.89 (1 H, d, *J*=12.0 Hz), 4.25 (1 H, d, *J*=12.0 Hz), 7.23 (3 H, m), 7.56 (2 H, m); ^{13}C NMR (CDCl$_3$): δ 64.8, 76.2, 123.6, 126.0, 126.4, 128.4, 129.0, 135.0; ^{19}F NMR (CDCl$_3$): δ -77.7 (s); IR (neat) ν 3434, 2957, 1165 cm^{-1}.

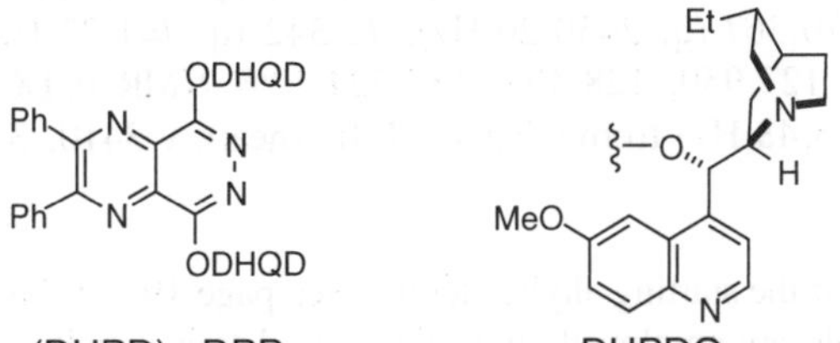

(DHPD)$_2$-DPP   DHPDO-

Note

For the preparation of β-(trifluoromethyl)styrene, see page 168 of this volume.
See also the following article for a similar reaction with 3,3,3-trifluoropropene: Vanhessche,

K. P. M.; Sharpless, K. B. *Chem. Eur. J.* **1997**, *3*, 517.

Sharpless, K. B., *et al. Tetrahedron: Asym.* **1994**, *5*, 1473.

(2R, 3S, 4S)-6-(Benzyloxy)-1,1,1-trifluorohexane-2,3,4-triol, (2S, 3R, 4S)-6-(Benzyloxy)-1,1,1-trifluorohexane-2,3,4-triol

a) cat. OsO_4, *N*-methylmorpholine *N*-oxide / acetone-H_2O

Synthetic method

To a solution of *N*-methylmorpholine *N*-oxide hydrate (NMO; 2.7 g, *ca.* 20 mmol) in 66% acetone-H_2O (12 mL) was added a 2.5 wt% solution of osmium tetroxide (OsO_4) in *tert*-BuOH (0.48 mL, 0.3 mol%) under a nitrogen atmosphere at 0 ℃, followed by the addition of (*R*)-1-(benzyloxy)-6,6,6-trifluorohex-4-en-3-ol (12.87 mmol). The reaction was quenched after 2 days at ambient temperature by addition of Na_2SO_3 aq. (10 mL) and the residue was then removed through a pad of Celite, the filtrate was extracted with AcOEt, dried over by $MgSO_4$, and evaporated. Purification by silica gel column chromatography gave two separable stereoisomers of triols, (2R,3S,4S)- and (2S,3R,4S)-6-(benzyl-oxy)-1,1,1-trifluorohexane-2,3,4-triol in 86% combined yield in a ratio of 87:13.

Spectral data

(2R,3S,4S)-6-(Benzyloxy)-1,1,1-trifluorohexane-2,3,4-triol

$[\alpha]_D^{16}$ -9.03 (*c* 1.01, CHCl$_3$), 99.6% ee; mp. 99.0-99.5 ℃; Rf 0.39 (hexane:AcOEt=1:1); ^{1}H NMR (CDCl$_3$ with a few drops of DMSO-d_6): δ 1.7-1.9 (1 H, m), 2.106 (1 H, dddd, *J*=14.75, 6.60, 4.27, 2.48 Hz), 3.6-3.8 (4 H, m), 3.8-4.1 (3 H, m), 4.303 (1 H, dq, *J*=7.78, 1.06 Hz), 4.523 (2 H, s), 7.2-7.4 (5 H, m); ^{13}C NMR (CDCl$_3$ with a few drops of DMSO-d_6): δ 32.626, 67.800 (q, *J*=29.59 Hz), 68.660, 70.619, 70.183 (q, *J*= 1.83 Hz), 73.317, 125.333 (q, *J*=282.96 Hz), 127.714, 127.795, 128.456, 137.743; ^{19}F NMR (CDCl$_3$ with a few drops of DMSO-d_6): δ 2.41 (d, *J*=7.56 Hz) from CF$_3$CO$_2$H; IR (neat): ν 3335, 2960, 2880, 2865 cm^{-1}.

(2S,3R,4S)-6-(Benzyloxy)-1,1,1-trifluorohexane-2,3,4-triol

$[\alpha]_D^{16}$ -21.20 (*c* 0.77, MeOH), 99.6% ee; mp. 79.5-80.0 ℃; Rf 0.30 (hexane:AcOEt=1:1); ^{1}H NMR (CDCl$_3$ with a few drops of DMSO-d_6): δ 1.6-2.1 (2 H, m), 3.5-4.3 (8 H, m), 4.512 (2 H, s), 7.2-7.4 (5 H, m); ^{13}C NMR (CDCl$_3$ with a few drops of DMSO-d_6): δ 32.498, 68.007, 69.780 (q, *J*=1.68 Hz), 70.367 (q, *J*=30.20 Hz), 72.342 (q, *J*=1.27 Hz), 73.381, 124.358 (q, *J*=282.67 Hz), 127.766, 127.950, 128.498, 137.324; ^{19}F NMR (CDCl$_3$ with a few drops of DMSO-d_6): δ 1.94 (d, *J*=5.48 Hz) from CF$_3$CO$_2$H; IR (neat): ν 3410, 3355, 2955, 2930, 2900, 2875 cm^{-1}.

Note

For the preparation of the starting allylic alcohol, see page 193 of this volume.

When allylic alcohols are employed, this dihydroxylation usually furnishes triols with *anti* selectivity between the original hydroxy group and the newly created neighboring OH moiety (for example, the relative stereochemistry between C^1-C^2 centers as shown in the next page). However, starting from the above Z isomers, low to moderate *syn* selectivity was obtained. See the following review for a few such exceptional examples. Cha, J.-K.; Kim, N.-S. *Chem. Rev.*

1995, *95*, 1761.

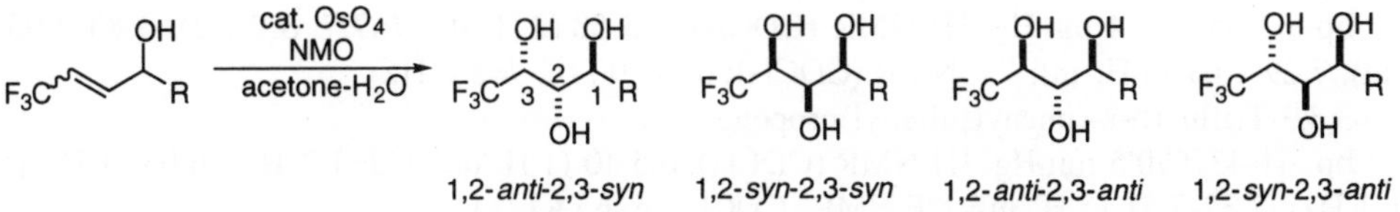

cat. OsO_4
NMO
acetone-H_2O
1,2-*anti*-2,3-*syn*
1,2-*syn*-2,3-*syn*
1,2-*anti*-2,3-*anti*
1,2-*syn*-2,3-*anti*

R	(*E*) or (*Z*)	Yield (%)	diastereomeric ratio			
$PhCH_2CH_2$	(*E*)	83	85	15	0	0
$PhCH_2CH_2$	(*Z*)	83	0	0	9	91
$BnOCH_2CH_2$[a]	(*Z*)	74	0	0	26	74
$BnOCH_2$	(*E*)	76	80	20	0	0
$BnOCH_2$[a]	(*Z*)	66	0	0	42	58

[a] 19 and 34% of starting materials were recovered, respectively.
(Reproduced with permission from Yamazaki, T., *et al. J. Org. Chem.* **1995**, *60*, 6046)

Yamazaki, T., *et al. J. Org. Chem.* **1995**, *60*, 6046.

3,3,3-Trifluoro-2-(phenylsulfenyl)propene, 3,3,3-Trifluoro-2-(phenylsulfinyl)propene, 3,3,3-Trifluoro-2-(phenylsulfonyl)propene

c)
CF_3
a), b)
CF_3
SPh
c)
CF_3
S(O)Ph
CF_3
SO_2Ph

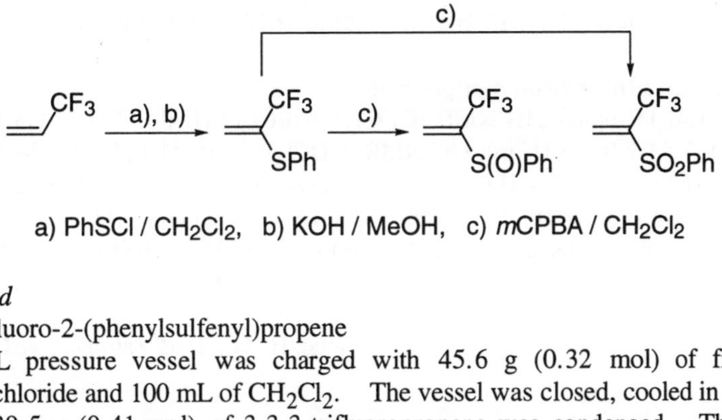

a) PhSCl / CH_2Cl_2, b) KOH / MeOH, c) *m*CPBA / CH_2Cl_2

Synthetic method

3,3,3-Trifluoro-2-(phenylsulfenyl)propene

A 500-mL pressure vessel was charged with 45.6 g (0.32 mol) of freshly distilled phenylsulfenyl chloride and 100 mL of CH_2Cl_2. The vessel was closed, cooled in dry ice-acetone, evacuated and 39.5 g (0.41 mol) of 3,3,3-trifluoropropene was condensed. The mixture was heated overnight at 70 ℃. The vessel was cooled to room temperature and vented to atmospheric pressure. The contents were concentrated on a rotary evaporator to afford a dark oil. Distillation through a short path still gave 66.0 g of 3-chloro-1,1,1-trifluoro-2-(phenylsulfenyl)propane in 86% yield.

A mixture of 13.2 g (0.24 mol) of KOH pellets and 50 mL of dry MeOH was stirred under an argon atmosphere until KOH dissolved. The mixture was cooled in a dry ice-acetone bath to -40 ℃. To this vigorously stirred solution was added 37.8 g (0.16 mol) of 3-chloro-1,1,1-trifluoro-2-(phenylsulfenyl)propane in one portion. The resulting mixture was stirred for 30 min at -10 ℃ and then poured into 100 mL of ice H_2O containing 2 mL of conc HCl. The aqueous mixture was extracted with Et_2O (3x20 mL). The combined extracts were washed with H_2O, dried ($MgSO_4$) and concentrated on a rotary evaporator to give 30.7 g of 3,3,3-trifluoro-2-(phenylsulfenyl)-propene in 96% yield as a colorless oil after Kügelrohr distillation.

Spectral data

    3-Chloro-1,1,1-trifluoro-2-(phenylsulfenyl)propane
    bp. 63-65 °C/0.4 mmHg; ^{1}H NMR (CDCl$_3$): δ 3.58 (2 H, dq, J=7.7, 5.0 Hz), 3.83-3.95 (1
H, m), 7.29-7.61 (5 H, m); ^{19}F NMR (CDCl$_3$): δ -69.00 (d, J=8.0 Hz).
    3,3,3-Trifluoro-2-(phenylsulfenyl)propene
    bp. 41-43 °C/0.5 mmHg; ^{1}H NMR (CDCl$_3$): δ 5.40 (1 H, quint, J=1.3 Hz), 6.03 (1 H, quint,
J=1.3 Hz), 7.32-7.51 (5 H, m); ^{19}F NMR (CDCl$_3$): δ -65.97 (s).

Synthetic method

    3,3,3-Trifluoro-2-(phenylsulfinyl)propene, 3,3,3-Trifluoro-2-(phenylsulfonyl)propene
    A mixture of 10 g (50 mmol) of 3,3,3-trifluoro-2-(phenylsulfenyl)propene and 100 mL of
dry CH$_2$Cl$_2$ was cooled in a dry ice-acetone bath to -40 °C.   To this vigorously stirred solution
was added *m*CPBA in several portions (1 equiv for the synthesis of 3,3,3-trifluoro-2-(phenylsulfinyl)propene or 2 equiv for the synthesis of 3,3,3-trifluoro-2-(phenylsulfonyl)propene).   The
resulting mixture was stirred for 5 h at room temperature and then filtered through alkaline Al$_2$O$_3$.
The combined extracts were dried over MgSO$_4$ and concentrated on a rotary evaporator to give
3,3,3-trifluoro-2-(phenylsulfinyl)propene in 78% yield or 3,3,3-trifluoro-2-(phenylsulfonyl)-
propene in 88% yield after Kügelrohr distillation.

Spectral data

    3,3,3-Trifluoro-2-(phenylsulfinyl)propene
    bp. 60-65 °C/0.15 mmHg; ^{1}H NMR (CDCl$_3$): δ 6.47 (1 H, m), 6.67 (1 H, quint, J=0.9 Hz),
7.43-7.70 (5 H, m); ^{13}C NMR (CDCl$_3$): δ 121.15 (qdd, J=277.3, 13.0, 6.3 Hz), 123.74 (tq,
J=166.4, 4.5 Hz), 125.60 (dt, J=165.2, 6.9 Hz), 141.51, 146.12 (q, J=31.6 Hz); ^{19}F NMR
(CDCl$_3$): δ -60.99 (d, J=6.3 Hz); IR (neat): ν 2950, 1650, 1280-1140, 1170-1050, 1080, 990,
720, 675 cm^{-1}.
    3,3,3-Trifluoro-2-(phenylsulfonyl)propene
    bp. 70-75 °C/0.15 mmHg; ^{1}H NMR (CDCl$_3$): δ 6.67 (1 H, dq, J=3.1, 1.3 Hz), 6.99 (1 H,
quint, J=1.3 Hz), 7.53-7.95 (5 H, m); ^{13}C NMR (CDCl$_3$): δ 119.41 (qdd, J=274.7, 12.6, 6.4 Hz),
128.36 (ddd, J=162.9, 7.7, 5.4 Hz), 129.29 (dd, J=165.9, 7.2 Hz), 133.22 (ddq, J=168.8,
166.4, 4.3 Hz), 134.38 (dt, J=162.9, 7.2 Hz), 138.43 (t, J=8.6 Hz), 141.99 (q, J=33.5 Hz); ^{19}F
NMR (CDCl$_3$): δ -61.99 (s); IR (neat): ν 2950, 1650, 1390-1185, 1160-1025, 990-850, 730, 670
cm^{-1}.

                                        Viehe, H. G., *et al*. *Tetrahedron* **1996**, *52*, 4383.

4.6.2 Reductions

Trifluoroacetaldehyde

$$CF_3CO_2H \xrightarrow{\text{a), b)}} CF_3CHO$$

a) LiAlH$_4$ / Et$_2$O,   b) conc H$_2$SO$_4$

Synthetic method

    In a three-necked flask provided with a stirrer, a dropping funnel, and a reflux condenser,
was placed a solution of 350 g (3 mol) of trifluoroacetic acid in 2 L of anhydrous Et$_2$O.   The flask

was flushed with dry nitrogen, cooled to -12 °C, and a suspension of 67 g (1.77 mol) of lithium aluminum hydride in 800 mL of anhydrous Et_2O was added at that temperature in portions, with vigorous stirring over a period of 1.5 h.   The reaction mixture was stirred at -10 to -5 °C for 2.5 h, whereupon 530 mL of H_2O was dropped into the flask with stirring and cooling, and a solution of 244 mL of conc H_2SO_4 in 1 L of H_2O was added.   The Et_2O layer was separated and the aqueous layer was extracted with Et_2O.   The combined extracts were dried over $MgSO_4$, Et_2O was partially distilled off and the residue was distilled on a column, and a boiling fraction was collected at 99-106 °C.   The product was heated with stirring, with an equal volume of conc H_2SO_4 in an oil bath at 120 °C.   The liberated gas was condensed in a trap cooled to -78 °C and distilled (bp. -18 °C). The yield of trifluoroacetaldehyde was 156 g (50%).

Note

Because of the high electrophilicity of trifluoroacetaldehyde, this compound is readily trans-formed to the corresponding hydrate form and the dehydration process by hot conc H_2SO_4 is required for the second step.   So commercially available trifluoroacetaldehyde ethyl hemiacetal or the corresponding hydrate are directly employed for the second step.   From our experience, it is better to add hydrate or hemiacetal at an approximate rate of 0.5 drops/sec *via* a dropping funnel.

See also Pierce, O. R.; Kane, T. G. *J. Am. Chem. Soc.* **1954**, *76*, 300.

Braid, M., *et al. J. Am. Chem. Soc.* **1954**, *76*, 4027.

(*R*)-3-Bromo-1,1,1-trifluoropropan-2-ol

a) (-)-DIP-Cl / Et_2O, b) $HN(CH_2CH_2OH)_2$

Synthetic method

An oven-dried, 100-mL round-bottom flask equipped with a side arm, a magnetic stirring bar, and a connecting tube was cooled to room temperature in a stream of nitrogen.   (-)-DIP-Chloride (8.8 g, 27.5 mmol) was transferred to the flask in a glove bag and dissolved in Et_2O (25 mL). The solution was cooled to -25 °C, and 3-bromo-1,1,1-trifluoropropan-2-one (4.7 g, 25 mmol) was added using a syringe.   The reaction was followed by ^{11}B NMR after aliquots were methanolyzed at the reaction temperature at periodic intervals.   When the reaction was complete ([11]B δ: 32 ppm, 96 h), the mixture was warmed to 0 °C and diethanolamine (5.3 mL, 55 mmol) was added dropwise.   The mixture was then warmed to room temperature and stirred for 2 h, whereupon the boranes precipitated as a complex which was filtered and washed with pentane. The solvents were removed by distillation, and the residue was passed through a silica gel column to separate α-pinene and the product (pentane: CH_2Cl_2).   The fraction containing the product was concentrated and distilled to yield 2.9 g (60%) of 3-bromo-1,1,1-trifluoropropan-2-ol.   The alcohol was further purified by preparative GC using an SE-30 column.   The optical purity was determined to be 96% ee by conversion to the corresponding MTPA ester, followed by capillary GC analysis.

Spectral data

$[\alpha]_D^{22}$ -51.06 (neat); bp. 122-123 °C; ^{1}H NMR ($CDCl_3$): δ 2.72-2.82 (1 H, m), 3.48 (1 H, dd, *J*=11.2, 8.5 Hz), 3.64 (1 H, dd, *J*=11.2, 3.0 Hz), 4.20-4.30 (1 H, m); ^{13}C NMR ($CDCl_3$): δ 29.69, 70.53 (q, *J*=31.8 Hz), 123.39 (q, *J*= 282.8 Hz); ^{19}F NMR ($CDCl_3$): δ -78.07 (d, *J*=5.0

Hz); IR (neat): ν 3380 cm⁻¹.

Note

Brown *et al.* also reported the enantioselective reduction of trifluorinated ketones as shown in the table below.   See Brown, H. C., *et al. Tetrahedron* **1993**, *49*, 1725 and *Tetrahedron: Asym.* **1994**, *5*, 1061, 1075.

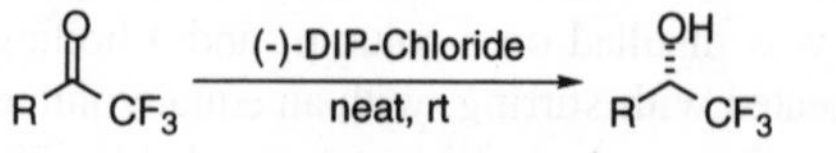

R	Time (h)	Yield (%)	Optical purity (% ee)
Ph	24	90	90
1-Naph	72	93	78
2-Naph	72	92	91
9-Anthryl[a)	30d	48	82
Me	3	70	89
Me[b)	96	72	96
n-C$_7$H$_{15}$	8	80	92
n-C$_8$H$_{17}$	8	87	91
c-C$_6$H$_{11}$	12	75	87
PhC≡C[c)	0.25	81	98
n-BuC≡C[c)	1	76	≥99

[a) 60% conversion was completed in 30 days.   [b) Reaction was carried out at -25 °C.   [c) Reaction was carried out in Et$_2$O (1 *M*) at -25 °C.
(Reproduced with permission from Brown, H. C., *et al. Tetrahedron* **1993**, *49*, 1725)

Brown, H. C., *et al. J. Org. Chem.* **1995**, *60*, 41.

(1*S**,2*R**)-2-(Trifluoromethyl)-1-phenylpropan-1-ol

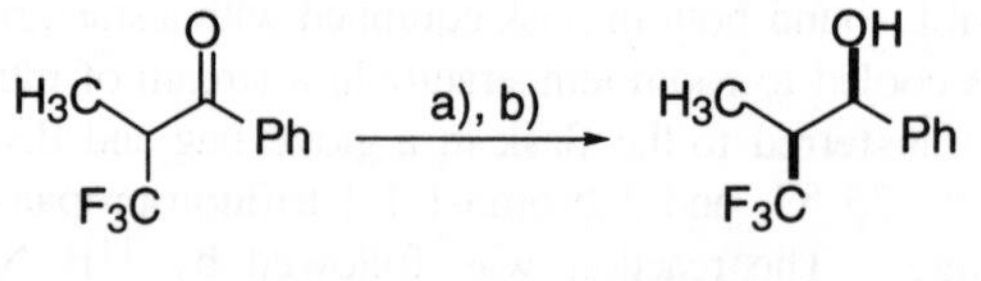

a) MeAlCl₂ / Et₂O,   b) LiBH₄

Synthetic method

To a solution of 26.9 mg (0.13 mmol) of 2-(trifluoromethyl)propiophenone in 2 mL of Et$_2$O was added 0.11 mL of a 1.77 *M* solution of MeAlCl$_2$ (0.19 mmol) in hexane slowly at -78 °C under an argon atmosphere.   After 15 min at -78 °C, 14.5 mg (0.67 mmol) of LiBH$_4$ was added.   After being gradually warmed to room temperature, the mixture was quenched by the slow addition of 2 mL of 2 *N* HCl.   The mixture was extracted with Et$_2$O (2x2 mL) and the combined organic layers were dried and concentrated.   The residue was chromatographed on preparative TLC (hexane:AcOEt= 4:1) to give 20.4 mg of (1*S**,2*R**)-2-(trifluoromethyl)-1-phenylpropan-1-ol in 75% yield as a 99:1 *syn:anti* mixture.

Spectral data

¹H NMR (CDCl₃): δ 0.87 (3 H, d, *J*=7.2 Hz), 2.17 (1 H, d, *J*=3.0 Hz), 2.58-2.70 (1 H, m), 4.81 (1 H, dd, *J*=8.1, 3.0 Hz), 7.32-7.40 (5 H, m); IR (neat): ν 3450, 1465, 1260, 1170, 765,

710 cm^{-1}.

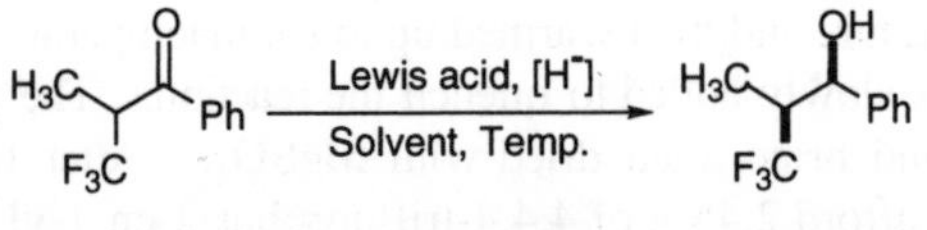

Lewis acid	[H$^-$]	Solvent	Temp (°C)	syn:anti	Yield (%)
none	NaBH$_4$	EtOH	0→rt	86 : 14	56
none	LiAlH$_4$	Et$_2$O	0→rt	84 : 16	63
none	KBHBu$_3$$^{-sec}$	Et$_2$O	-78→rt	99 : 1	41
none	LiBH$_4$	Et$_2$O	-78→rt	77 : 23	88
ZnCl$_2$	LiBH$_4$	Et$_2$O	-78→rt	87 : 13	85
Et$_3$Al	LiBH$_4$	Et$_2$O	-78→rt	96 : 4	86
TiCl$_4$	LiBH$_4$	Et$_2$O	-78→rt	98 : 2	49
MeAlCl$_2$	LiBH$_4$	CH$_2$Cl$_2$	-78→rt	79 : 21	62
MeAlCl$_2$	LiBH$_4$	Et$_2$O	-78→rt	99 : 1	75
MeAlCl$_2$	LiBH$_4$	THF	-78→rt	86 : 14	68

(Reproduced with permission from Hanamoto, T.; Fuchikami, T. *J. Org. Chem.* **1990**, *55*, 4969)

Note

For the preparation of the starting ketone, see page 162 of this volume.

Hanamoto, T.; Fuchikami, T. *J. Org. Chem.* **1990**, *55*, 4969.

4,4,4-Trifluorobut-2-en-1-ol, 4,4,4-Trifluorobut-2-en-1-yl bromide

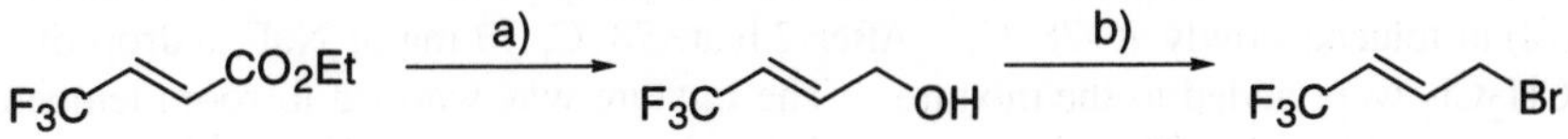

a) LiAlH$_4$, AlCl$_3$ (3:1) / Et$_2$O, b) PBr$_3$ / Et$_2$O

Synthetic method

4,4,4-Trifluorobut-2-en-1-ol

To a suspension of anhydrous AlCl$_3$ (17 mmol, 2.27 g) in Et$_2$O (30 mL) was added an Et$_2$O (40 mL) solution of LiAlH$_4$ (1.90 g, 50 mmol) at 0 °C.   The resulting mixture was stirred for 15 min at 0 °C.   A solution of ethyl (*E*)-4,4,4-trifluorobut-2-enoate (2.99 mL, 20 mmol) in Et$_2$O (10 mL) was then added at 0 °C and stirred for another 2 h.   The resulting mixture was quenched with a sat. Na$_2$SO$_4$ solution at 0 °C.   The resulting suspension was dried with anhydrous MgSO$_4$, filtered and washed with Et$_2$O.   Et$_2$O was then removed *in vacuo* to afford the crude 4,4,4-trifluorobut-2-en-1-ol, which was purified by distillation (2.27 g, 90% yield).

Spectral data

bp. 128-129 °C; Rf 0.40 (hexane:AcOEt=4:1); ^{1}H NMR (CDCl$_3$): δ 1.64 (1 H, br s), 4.10 (2 H, m), 5.95 (1 H, m), 6.53 (1 H, m); ^{13}C NMR (CDCl$_3$): δ 61.00, 117.86 (q, *J*=33.96 Hz), 123.20 (q, *J*=269.15 Hz), 138.94 (q, *J*=6.29 Hz); ^{19}F NMR (CDCl$_3$): δ 11.43 (m) from CF$_3$CO$_2$H; IR (neat): ν 3426.5, 1691.9, 1637.9, 1300.3, 1267.5, 1131.9, 1086.2, 1018.7 cm^{-1}.

Synthetic method

4,4,4-Trifluorobut-2-en-1-yl bromide

4,4,4-Trifluorobut-2-en-1-ol (2.52 g, 20 mmol) was added into Et$_2$O (40 mL), then PBr$_3$

(2.85 mL, 30 mmol) was added dropwise into the solution at -10 °C.   The reaction mixture was stirred for 1 h at that temperature and then warmed up to room temperature.   After being stirred for an additional 2 h, H_2O was slowly added to quench the reaction.   The product was extracted with Et_2O, washed with H_2O and brine, then dried with $MgSO_4$.   After filtration, the Et_2O solution was fractionally distilled to afford 2.45 g of 4,4,4-trifluorobut-2-en-1-yl bromide in 65% yield.

Spectral data
　　bp. 95-97 °C; Rf 0.60 (hexane:AcOEt=4:1); ^1H NMR ($CDCl_3$): δ 3.97 (2 H, m), 5.91 (1 H, m), 6.52 (1 H, m); ^{13}C NMR ($CDCl_3$): δ 61.01, 122.02 (q, J=33.96 Hz), 122.27 (q, J=270.41 Hz), 135.06 (q, J=6.29 Hz); ^{19}F NMR ($CDCl_3$): δ 10.91 (m) from CF_3CO_2H; IR (neat): ν 2976.9, 1691.9, 1429.6, 1321.6, 1259.8, 1180.7, 1118.9, 1045.7, 964.6, 621.2 cm^{-1}.

Loh, T.-P.; Li, X.-R. Angew. Chem. Int. Ed. Engl. 1997, 36, 980.

(1R*,2R*)-2-(Trifluoromethyl)-1-phenylpropan-1-ol

a) *n*-Bu₃SnH, Et₃Al / toluene

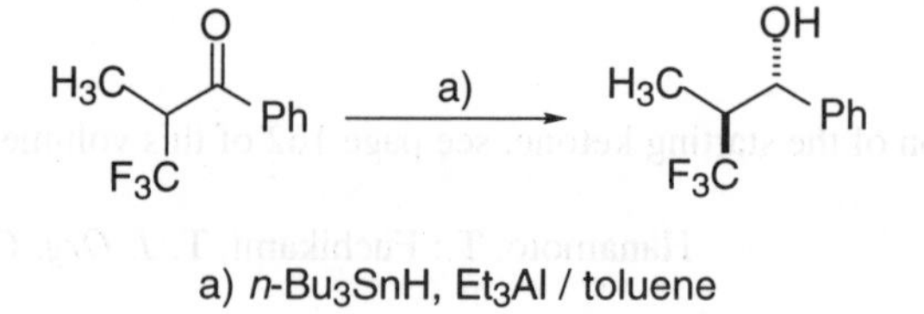

Synthetic method
　　To a solution of 15.6 mg (0.077 mmol) of 2-(trifluoromethyl)propiophenone and 0.083 mL (0.31 mmol) of *n*-Bu₃SnH in 2 mL of toluene was added 0.14 mL of a 1.1 *M* solution of Et₃Al (0.15 mmol) in toluene slowly at -78 °C.   After 2 h at -78 °C, 30 mg of NaF, a drop of H_2O, and 1 mL of CH_2Cl_2 were added to the mixture.   The mixture was warmed to room temperature and the precipitate was filtered.   The filtrate was dried and concentrated.   The residue was chromato-graphed on preparative TLC (hexane:AcOEt=4:1) to give 10.5 mg of (1R*,2R*)-2-(trifluoro-methyl)-1-phenylpropan-1-ol in 67% yield as a 15:85 *syn:anti* mixture.

Spectral data
　　^1H NMR ($CDCl_3$): δ 1.09 (3 H, d, J=7.1 Hz), 1.94 (1 H, d, J=3.5 Hz), 2.40-2.52 (1 H, m), 5.24 (1 H, t, J=3.0 Hz), 7.25-7.40 (5 H, m); IR (neat): ν 3500, 1275, 1140, 990, 760, 710 cm^{-1}.

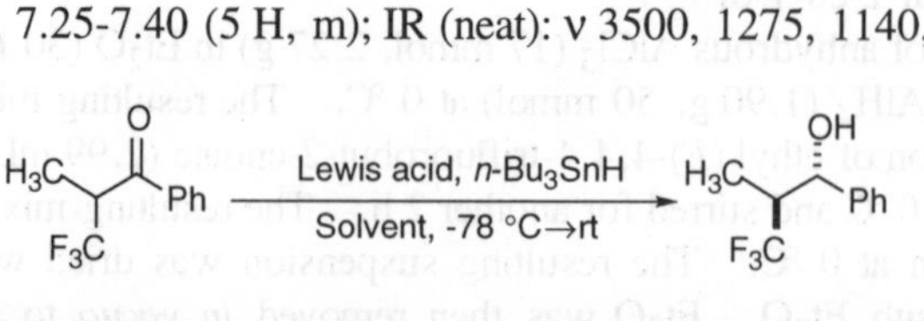

Lewis acid	Solvent	syn:anti		Yield (%)
none	CH_2Cl_2			0
$TiCl_2(OPr^{-i})_2$	CH_2Cl_2	72 : 28		87
$MeAlCl_2$	CH_2Cl_2	64 : 36		70
Et_2AlCl	CH_2Cl_2	48 : 52		72
Et_3Al	CH_2Cl_2	32 : 68		71
Et_3Al	hexane	25 : 75		67
Et_3Al	toluene	15 : 85		64

(Reproduced with permission from Hanamoto, T.; Fuchikami, T. *J. Org. Chem.* **1990**, *55*, 4969)

Note

    For the preparation of the starting ketone, see page 160 of this volume.

Hanamoto, T.; Fuchikami, T. J. Org. Chem. 1990, 55, 4969.

(E)-1-(Benzyloxy)-6,6,6-trifluorohex-4-en-3-ol

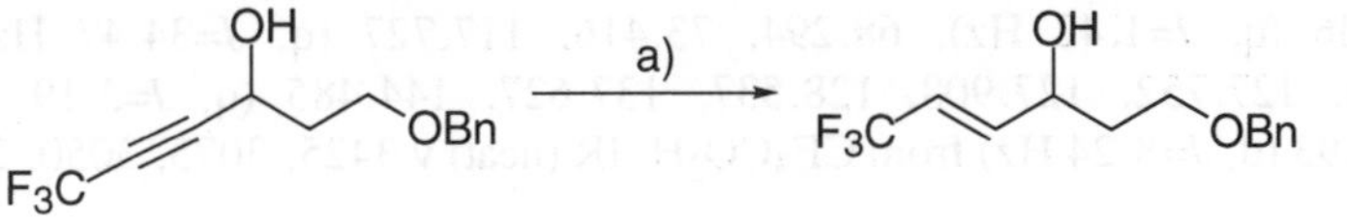

a) Red-Al / toluene

Synthetic method

    To a stirring solution of Red-Al (2.5 mmol) in toluene (3 mL) at -78 °C was added 1-(benzyl-oxy)-6,6,6-trifluorohex-4-yn-3-ol (2.12 mmol). After the reaction mixture was stirred for 3 h at that temperature, it was quenched with 1 *N* HCl aq. (10 mL) and extracted with AcOEt three times. The organic layer was dried over MgSO$_4$ and concentrated. The residue was chromatographed on silica gel to afford (*E*)-1-(benzyloxy)-6,6,6-trifluorohex-4-en-3-ol in 92% yield.

Spectral data

    Rf 0.38 (hexane:AcOEt=4:1); ^{1}H NMR (CDCl$_3$): δ 1.7-2.1 (2 H, m), 3.414 (1 H, d, *J*=3.81 Hz), 3.654 (1 H, dt, *J*=9.32, 4.27 Hz), 3.734 (1 H, dt, *J*=9.38, 4.04 Hz), 4.4-4.6 (1 H, m), 4.521 (2 H, s), 5.950 (1 H, dqd, *J*=15.62, 6.57, 2.01 Hz), 6.376 (1 H, ddq, *J*=15.60, 4.01, 2.01 Hz), 7.2-7.4 (5 H, m); ^{13}C NMR (CDCl$_3$): δ 35.385 (q, *J*=1.32 Hz), 68.330, 69.788, 73.535, 117.907 (q, *J*=33.65 Hz), 123.375 (q, *J*=268.84 Hz), 127.779, 128.011, 128.581, 137.445, 141.789 (q, *J*=6.30 Hz); ^{19}F NMR (CDCl$_3$): δ 15.01 (d, *J*=5.53 Hz) from CF$_3$CO$_2$H; IR (neat) ν 3450, 3075, 3050, 2950, 2875 cm^{-1}.

Note

    For the preparation of the starting propargylic alcohol, see page 169 of this volume.

    This compound was obtained as the chiral form after enzymatic optical resolution. $[\alpha]_D^{18}$ +8.33 (*c* 1.55, CHCl$_3$) from the (*S*)-substrate with 99.6% ee. See also the following articles: i) Kobayashi, Y., *et al. Tetrahedron Lett.* **1984**, *25*, 4749, ii) Kitazume, T., *et al. J. Org. Chem.* **1997**, *62*, 137.

Yamazaki, T., et al. J. Org. Chem. 1995, 60, 6046.

(Z)-1-(Benzyloxy)-6,6,6-trifluorohex-4-en-3-ol

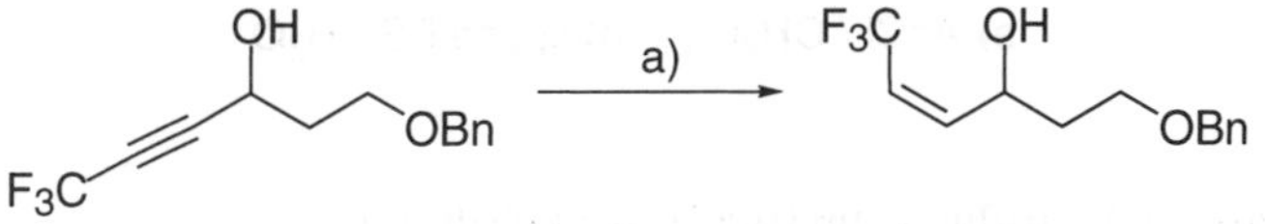

a) Lindlar cat. / hexane

Synthetic method

    A solution of 1-(benzyloxy)-6,6,6-trifluorohex-4-yn-3-ol (2.0 mmol) and a catalytic amount of Lindlar catalyst in hexane (20 mL) was stirred under hydrogen.   After removal of the catalyst by filtration, concentration under reduced pressure and chromatography of the residue on silica gel gave (Z)-1-(benzyloxy)-6,6,6-trifluorohex-4-en-3-ol in 96% yield.

Spectral data

Rf 0.23 (hexane:AcOEt=4:1); ^{1}H NMR (CDCl$_3$): δ 1.6-2.1 (2 H, m), 3.316 (1 H, d, *J*=3.25 Hz), 3.661 (1 H, ddd, *J*=8.91, 4.90, 4.06 Hz), 3.739 (1 H, ddd, *J*=9.38, 4.88, 2.56 Hz), 4.485 (1 H, d, *J*=11.90 Hz), 4.558 (1 H, d, *J*=11.72 Hz), 4.7-5.0 (1 H, m), 5.589 (1 H, ddq, *J*=11.94, 8.76, 1.23 Hz), 6.033 (1 H, dd, *J*=11.94, 8.81 Hz), 7.2-7.5 (5 H, m); ^{13}C NMR (CDCl$_3$): δ 36.184, 67.536 (q, *J*=1.42 Hz), 68.294, 73.416, 117.727 (q, *J*=34.47 Hz), 122.879 (q, *J*=272.00 Hz), 127.752, 127.908, 128.537, 137.627, 144.485 (q, *J*=5.19 Hz); ^{19}F NMR (CDCl$_3$): δ 20.93 (d, *J*=8.24 Hz) from CF$_3$CO$_2$H; IR (neat) ν 3425, 3075, 3050, 2950, 2875, 675 cm^{-1}.

Note

For the preparation of the starting propargylic alcohol, see page 169 of this volume.

This compound was obtained as the chiral form after enzymatic optical resolution. $[\alpha]_D^{16}$ +5.40 (*c* 1.00, CHCl$_3$) from the (*S*)-substrate with 98.5% ee.   See also the following articles: i) Kobayashi, Y., *et al. Tetrahedron Lett.* **1984**, *25*, 4749, ii) Kitazume, T., *et al. J. Org. Chem.* **1997**, *62*, 137.

Yamazaki, T., *et al. J. Org. Chem.* **1995**, *60*, 6046.

4.7 Optical Resolutions

4.7.1 Bioorganic Methods

2-[(*S*)-1-(2,2,2-Trifluoro-1-hydroxyethyl)]-5-(trimethylsilyl)furan,
2-[(*R*)-1-(1-Acetoxy-2,2,2-trifluoroethyl)]-5-(trimethylsilyl)furan

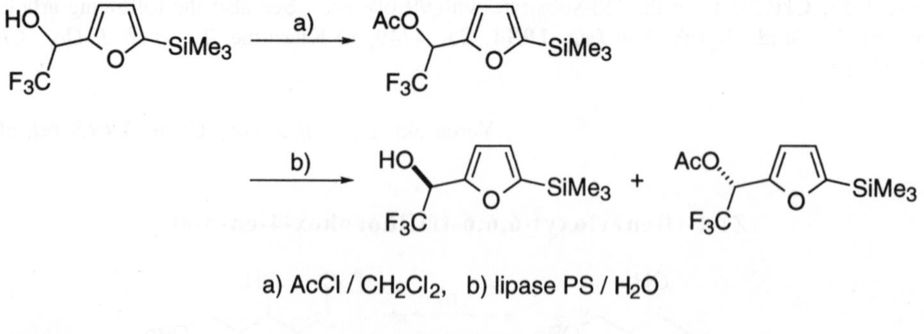

a) AcCl / CH$_2$Cl$_2$, b) lipase PS / H$_2$O

Synthetic method

2-[1-(1-Acetoxy-2,2,2-trifluoroethyl)]-5-(trimethylsilyl)furan

To a 0.5 *M* solution of 2-[1-(2,2,2-trifluoro-1-hydroxyethyl)]-5-(trimethylsilyl)furan in CH$_2$Cl$_2$ under a nitrogen atmosphere were added pyridine and acetyl chloride (both 1.2 equiv) at 0 °C, and the reaction mixture was stirred overnight.   The reaction was quenched with 1 *N* HCl aq., and the aqueous phase, after separation of the resulting CH$_2$Cl$_2$ phase, was extracted with CH$_2$Cl$_2$

twice.   Then the combined organic layers were washed with sat. NaHCO₃ aq. and brine, dried
(MgSO₄), and evaporated.   Purification by distillation afforded 2-[1-(1-acetoxy-2,2,2-trifluoro-
ethyl)]-5-(trimethylsilyl)furan in 95% yield.

Spectral data

bp. 66-68 °C/2.3 mmHg; Rf 0.53 (hexane:AcOEt=7:1); ¹H NMR (CDCl₃): δ 0.24 (9 H, s),
2.15 (3 H, s), 6.33 (1 H, q, *J*=6.72 Hz), 6.48 (1 H, d, *J*=3.29 Hz), 6.53 (1 H, d, *J*=3.32 Hz);
¹³C NMR (CDCl₃): δ -2.05, 20.31, 65.62 (q, *J*=35.2 Hz), 111.65 (q, *J*=1.3 Hz), 120.35, 121.60
(q, *J*=281.4 Hz), 148.47, 163.14, 168.97; ¹⁹F NMR (CDCl₃): δ 3.2 (d, *J*=6.6 Hz) from
CF₃CO₂H; IR (neat): ν 2975, 1770 cm⁻¹.

Synthetic method

2-[(*S*)-1-(2,2,2-Trifluoro-1-hydroxyethyl)]-5-(trimethylsilyl)furan

To a 0.1 *M* solution of 2-[1-(1-acetoxy-2,2,2-trifluoroethyl)]-5-(trimethylsilyl)furan in H₂O
was added lipase PS (*Pseudomonas cepacia*, Amano Pharmaceutical Co. Ltd.; 6000 units for 1
mmol of the substrate), and the reaction mixture, maintained at about pH 7 by titration with 1 *N*
NaOH aq., was stirred at 40 °C.   When the conversion was found to reach to about 50%,
flocculant (P-713, Daiichi Kogyo Seiyaku, Japan) was added.   The whole was filtered through
Celite 545 to remove the enzyme, and the aqueous phase was extracted with AcOEt three times,
dried (MgSO₄), and the solvent was evaporated.   Isolation by silica gel column chromatography
afforded the optically active alcohol (97.6% ee) and the ester (94.3% ee).   Their enantiomeric
excesses were evaluated by capillary GC analysis after conversion to the corresponding MTPA
ester.

Spectral data

[α]$_D^{27}$ +7.45 (*c* 1.10, MeOH).

2-[(*R*)-1-(1-Acetoxy-2,2,2-trifluoroethyl)]-5-trimethylsilylfuran

[α]$_D^{29}$ -102.68 (*c* 1.28, MeOH).

Note

For the preparation of the starting alcohol and its physical properties, see page 160 of this
volume.

Yamazaki, T., *et al. J. Org. Chem.* **1993**, *58*, 4346.

Ethyl (2*S*, 3*S*)-3-(trifluoromethyl)pyroglutamate

(racemic)   (optically active)

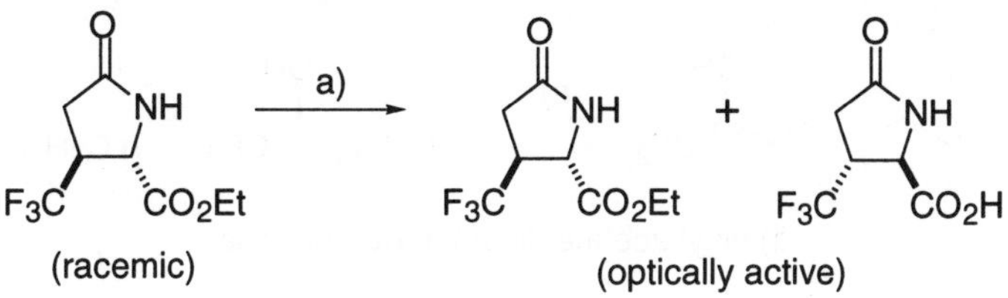

a) α-Chymotrypsin / H₂O

Synthetic method

Racemic ethyl *trans*-3-(trifluoromethyl)pyroglutamate (600 mg, 2.66 mmol) was added to 0.1
M potassium phosphate buffer (100 mL containing NaCl (0.1 *M*)), pH 7.5 at 37 °C and treated with
α-chymotripsin (240 mg, from bovine pancreas, 74.6 U/mg, purchased from Fluka) with vigorous
mechanical stirring.   Hydrolysis was stopped after 3 h, and the aqueous phase was extracted with

CH_2Cl_2 (5x80 mL); the combined organic layers were dried over $MgSO_4$ and concentrated under reduced pressure; the crude residue was purified by silica gel column chromatography (hexane:AcOEt=1:1) affording 196 mg of ethyl $(2S,3S)$-3-(trifluoromethyl)pyroglutamate in 33% yield as a solid.

Spectral data

$[\alpha]_D$ +20.3 (c 1.1, $CHCl_3$), >96% ee; mp. 50-51 °C.

The aqueous phase was centrifuged (10 min, 3000 rpm) and the enzyme was filtered off. The pH of the solution was adjusted to pH 1 (HCl 10%), and the solvent removed under reduced pressure. The crude residue was desiccated overnight (anhydrous calcium chloride) then refluxed with anhydrous EtOH (10 mL) and thionyl chloride (2 mL) for 2 h. The suspension was concentrated *in vacuo*, diluted with CH_2Cl_2 (30 mL), washed with sat. $NaHCO_3$ aq. and H_2O, then dried over $MgSO_4$. The solvent was removed and the residue purified by silica gel column chromatography (hexane:AcOEt=1:1) affording 274 mg of ethyl $(2R,3R)$-3-(trifluoromethyl)pyroglutamate in 46% yield as a solid. This sample was crystallized (hexane:EtOH=95:5) affording 151 mg of racemic ethyl *trans*-3-(trifluoromethyl)pyroglutamate (mp. 96-97 °C) and from the mother liquor, 103 mg of ethyl $(2R,3R)$-3-(trifluoromethyl)pyroglutamate were recovered.

Spectral data

$[\alpha]_D$ -19.94 (c 1.0, $CHCl_3$), >96% ee; mp. 50-51 °C. Spectroscopic data were identical to those reported for the racemic form.

Note

For the preparation of the starting ester and its physical properties, see page 140 of this volume.

Prati *et al.* also tried to construct the optically active adducts using the chiral enolate, while Michael addition afforded four possible diastereomers in 63% yield in a ratio of 52:31:13:4. See Laurent, A. J., *et al. J. Fluorine Chem.* **1996**, *80*, 27.

See also the following article for the construction of $(2S,3S)$-3-(trifluoromethyl)-pyroglutamic acid by stereoselective Michael addition reaction: Soloshonok, V. A., *et al. Tetrahedron Lett.* **1997**, *38*, 4903.

Prati, F., *et al. Tetrahedron:Asym.* **1996**, *7*, 3309.

(*R*)-1,1,1-Trifluoroundecan-2-ol,
(*S*)-1,1,1-Trifluoroundecan-2-yl acetate

a) vinyl acetate, lipase CAL / hexane

Synthetic method

(±)-1,1,1-Trifluoroundecan-2-ol (3.432 g, 15 mmol), vinyl acetate (12.9 g, 150 mmol), lipase CAL (787 mg, Novo Nordisk Co., Ltd.) in hexane (150 mL), and 4 Å molecular sieves (3.75 g) were stirred at room temperature for 11 days. The mixture was filtered through a glass sintered filter with a Celite pad, and the filtrate was chromatographed on silica gel flash column (hexane:AcOEt=100:1 to 10:1) and gave (*S*)-1,1,1-trifluoroundecan-2-yl acetate (1.29 g, 4.80 mmol, 32%) and unreacted 1,1,1-trifluoroundecan-2-ol (2.05 g, 9.06 mmol, 60%), respectively.

The optical purities were found to be >99% ee (S) and 46.7% ee (R), respectively, by GPC analysis using a capillary column on the chiral phase.

Spectral data

(R)-1,1,1-Trifluoroundecan-2-ol

$[\alpha]_D^{19}$ +8.16 (c 0.80, Et$_2$O).

(S)-1,1,1-Trifluoroundecan-2-yl acetate

$[\alpha]_D^{17}$ +5.6 (c 1.09, Et$_2$O); bp. 65 °C (bath temperature)/4 mmHg; ^{1}H NMR (CDCl$_3$): δ 0.87 (3 H, t, J=6.3 Hz), 1.17-1.41 (14 H, m), 1.63-1.80 (2 H, m), 2.13 (3 H, s), 5.28 (1 H, dt, J=20.1, 6.8 Hz); ^{13}C NMR (CDCl$_3$): δ 14.01, 20.37, 22.64, 24.52, 27.77, 29.05, 29.27, 29.41, 31.83, 69.54 (q, J=32.0 Hz), 123.84 (q, J=280.6 Hz); ^{19}F NMR (CDCl$_3$): δ 84.57 (d, J=6.8 Hz) from C$_6$F$_6$; IR (neat): ν 2950, 2850, 1760, 1460, 1220, 1020 cm^{-1}.

Note

For the preparation of the starting alcohol, see page 158 of this volume.

For other examples of lipase-catalyzed optical resolution in an organic medium, see the following recent articles: i) Kato, K., *et al. J. Ferment. Bioeng.* **1993**, *76*, 178, ii) Guerrero, A.; Gaspar, J. *Tetrahedron: Asym.* **1995**, *6*, 231, iii) Yamazaki, T., *et al. J. Org. Chem.* **1995**, *60*, 6046, iv) Fujisawa, T., *et al. Bull. Chem. Soc. Jpn.* **1996**, *69*, 2655, v) Itoh, T., *et al. Tetrahedron Lett.* **1996**, *37*, 5001, vi) Guerrero, A., *et al. Tetrahedron: Asym.* **1996**, *7*, 2135, vii) Kitazume, T., *et al. J. Org. Chem.* **1997**, *62*, 1370.

Nakamura, K., *et al. J. Org. Chem.* **1996**, *61*, 2332.

(S)-5,5,5-Trifluoro-4-hydroxypentan-2-one

a) Baker's yeast, glucose, allyl alcohol / H$_2$O

Synthetic method

43 g of fresh baker's yeast (FALA, Strasbourg) was suspended in distilled H$_2$O (380 mL) and 1.14 g of allylic alcohol was added. The mixture was kept at 27-30 °C and shaken with a magnetic stirrer for 1 h. Subsequently, 1,1,1-trifluoropentane-2,4-dione (2.3 g, 15 mmol) and 23 g of glucose were added and the suspension was allowed to stand in the same condition for 20-24 h. The reaction mixture was then extracted four times with AcOEt. The organic extracts were dried over Na$_2$SO$_4$, filtered and concentrated *in vacuo*. The crude product, showing an absence of the starting diketone, was chromatographed on a silica gel column (petroleum ether:Et$_2$O=1:4) and 1.8 g of (S)-5,5,5-trifluoro-4-hydroxypentan-2-one was obtained in 77% yield.

Spectral data

$[\alpha]_D^{20}$ -29.0 (c 1.0, CHCl$_3$, 85% ee); bp. 84-85 °C/20 mmHg; ^{1}H NMR (CDCl$_3$): δ 2.25 (3 H, s), 2.76 (1 H, dd), 2.89 (1 H, dd), 3.35 (1 H, br s), 4.50 (1 H, m).

Note

The addition of allyl bromide completely changed the enantioselection to give, for example, 70% ee of the (R)-hydroxyketone with 86% conversion. See the following reports on the selective inhibition of specific enzymes contained in baker's yeast: i) Nakamura, K., *et al. Bull. Chem. Soc. Jpn.* **1989**, *62*, 875, ii) Nakamura, K., *et al. Tetrahedron Lett.* **1990**, *31*, 267.

For other examples of the baker's yeast reduction, see i) Seebach, D. *et al. Helv. Chim. Acta*

1984, *67*, 1843, ii) Guanti, G., *et al. J. Chem. Soc., Chem. Commun.* **1986**, 138, iii) Kitazume, T., *et al. J. Fluorine Chem.* **1987**, *35*, 537, iv) Fujisawa, T., *et al. Tetrahedron: Asym.* **1994**, *5*, 1095, v) Resnati, G., *et al. Tetrahedron* **1998**, *54*, 2809.

Forni, A., *et al. Tetrahedron* **1994**, *50*, 11995.

4.7.2 Chemical Method

(1*S*, 2*S*)-(1,3-Dihydroxy-1-phenylpropan-2-yl)ammonium (*S*)-2-(trifluoromethyl)-3-hydroxypropionate, (*S*)-2-(Trifluoromethyl)-3-hydroxypropionic acid

a) (1*S*,2*S*)-2-amino-1-phenylpropane-1,3-diol,   b) recrystallization,   c) 2 *N* HCl / H₂O

Synthetic method

[(1*S*,2*S*)-1,3-Dihydroxy-1-phenylpropan-2-yl]ammonium (*S*)-2-(trifluoromethyl)-3-hydroxypropionate

A warm suspension (*ca.* 60 °C) of (1*S*,2*S*)-2-amino-1-phenylpropane-1,3-diol (26.4 g, 0.158 mol) in AcOEt (420 mL) was added with stirring to a warm solution (*ca.* 65-70 °C) of racemic 2-(trifluoromethyl)-3-hydroxypropionic acid (25.0 g, 0.158 mol) in AcOEt (320 mL). The product precipitated after a few seconds from the clear solution.   After heating under reflux for 15 min, the mixture was allowed to cool to room temperature and stirred overnight.   Filtration and drying of the resulting solid under high vacuum (0.1-0.01 mmHg) gave [(1*S*,2*S*)-1,3-dihydroxy-1-phenylpropan-2-yl]ammonium (*S*)-2-(trifluoromethyl)-3-hydroxypropionate (24.9 g) in 48% yield with a diastereomeric ratio of 92:8.

Dissolving this salt in boiling MeOH (36 mL) and subsequent dropwise addition of AcOEt (215 mL) led to spontaneous crystallization.   Cooling to room temperature and stirring overnight, filtering and drying of the resulting solid under high vacuum gave the salt (18.0 g) in 35% yield from the racemic hydroxy acid and the diastereomeric ratio was increased to 99:1.   A further recrystallization with MeOH (25 mL) and AcOEt (165 mL) gave [(1*S*,2*S*)-1,3-dihydroxy-1-phenyl-propan-2-yl]ammonium (*S*)-2-(trifluoromethyl)-3-hydroxypropionate (15.0 g) in 29% yield from the racemic hydroxy acid, with a diastereomeric ratio of 99.5:0.5.

Spectral data

[α]$_D^{rt}$ +27.7 (*c* 1.31, EtOH); mp. 144.2-144.4 °C; ¹H NMR (CD₃OD): δ 3.11 (1 H, qdd, *J*=9.5, 6.7, 5.7 Hz), 3.27-3.32 (1 H, m), 3.41 (1 H, dd, *J*=11.7, 6.1 Hz), 3.54 (1 H, dd, *J*=11.7, 3.7 Hz), 3.88 (1 H, dd, *J*=11.2, 5.7 Hz), 3.95 (1 H, dd, *J*=11.2, 6.7 Hz), 4.74 (1 H, d, *J*=8.8 Hz), 7.31-7.44 (5 H, m); ¹³C NMR (CD₃OD): δ 56.3 (q, *J*=23.8 Hz), 59.9, 60.2, 60.4, 72.3, 127.0 (q, *J*=279.2 Hz), 128.0 (2C), 129.6, 129.8 (2C), 142.1, 172.8; ¹⁹F NMR (CD₃OD): δ -65.9 (d, *J*=10.0 Hz); IR (neat): ν 3280, 2965, 2769, 2687, 2615, 1638, 1570, 1541, 1406, 1372, 1324, 1240, 1214, 1170, 1122, 1038 cm⁻¹.

Synthetic method

(*S*)-2-(Trifluoromethyl)-3-hydroxypropionic acid

A solution of 2 *N* HCl (30 mL) containing [(1*S*,2*S*)-1,3-dihydroxy-1-phenylpropan-2-yl]-ammonium (*S*)-2-(trifluoromethyl)-3-hydroxypropionate (5.0 g, 15.4 mmol) was saturated with NaCl and extracted with Et$_2$O (5x20 mL).   The combined organic layers were dried over MgSO$_4$, evaporated and dried under high vacuum to give 2.43 g of hygroscopic (*S*)-2-(trifluoromethyl)-3-hydroxypropionic acid in 99% yield, whose enantiomeric excess was found to be 99% determined by chiral GC after derivatization into the corresponding methyl ester.

Spectral data

[α]$_D^{rt}$ +18.9 (*c* 1.29, MeOH); mp. 43.8-44.2 °C; IR (neat): ν 3030, 1734, 1467, 1323, 1125, 1040 cm^{-1}.   Other physical properties were identical to those of the racemic material.

Note

For the preparation of the starting racemic hydroxy carboxylic acid, see page 177 of this volume.

For other examples, see i) Feigl, D. M.; Mosher, H. S. *J. Org. Chem.* **1968**, *33*, 4242, ii) Ugi, I., *et al. Angew. Chem. Int. Ed. Engl.* **1973**, *12*, 25, iii) Pirkle, W. H.; Hauske, J. R. *J. Org. Chem.* **1977**, *42*, 2436, iv) Weinges, K.; Kromm, E. *Liebig Ann. Chem.* **1985**, 90, v) Kobayashi, Y., *et al. Tetrahedron Lett.* **1986**, *27*, 5117, vi) Seebach, D., *et al. Chem. Ber.* **1992**, *125*, 2795.

Götzö, S. P.; Seebach, D. *Chimia* **1996**, *50*, 20.

4.8 Miscellaneous Reactions

N-(4-Anisyl)-2,2,2-trifluoroacetimidoyl   chloride,
N-(4-Anisyl)-2,2,2-trifluoroacetimidoyl   iodide

a) Et$_3$N, PPh$_3$ / CCl$_4$, b) *p*-Anisidine, c) NaI / acetone

Synthetic method

N-(4-Anisyl)-2,2,2-trifluoroacetimidoyl chloride

A 200-mL two-necked flask equipped with a septum cap, a condenser, and a Teflon-coated magnetic stir bar was charged with triphenylphosphine (34.5 g, 132 mmol), triethylamine (7.3 mL, 53 mmol), CCl$_4$ (21.1 mL, 220 mmol), and 2,2,2-trifluoroacetic acid (3.4 mL, 44 mmol).   After the solution was stirred for about 10 min (ice bath), 4-anisidine (6.48 g, 53 mmol) dissolved in CCl$_4$ (21.1 mL, 220 mmol) was added.   The mixture was then refluxed under stirring for 3 h. Solvents were removed under reduced pressure, and the residue was diluted with hexane and filtered.   The residual solid triphenylphosphine, the corresponding oxide, and triethylamine hydrochloride were washed with hexane   several times.   The filtrate was concentrated under reduced pressure, and the residue was distilled to afford N-(4-anisyl)-2,2,2-trifluoroacetimidoyl chloride (9.5 g) in 91% yield as a yellow oil.

Spectral data

bp. 97-98 °C/14 mmHg; ^{1}H NMR (CDCl$_3$): δ 3.85 (3 H, s), 6.93-7.02 (2 H, m), 7.30-7.37 (2 H, m); ^{13}C NMR (CDCl$_3$): δ 55.2, 114.2 (2C), 117.0 (q, J=276.6 Hz), 124.4 (2C), 127.9 (q, J=43.1 Hz), 135.3, 159.6; ^{19}F NMR (CDCl$_3$): δ -71.8 (s); IR (neat): ν 1682 cm^{-1}.

Synthetic method

N-(4-Anisyl)-2,2,2-trifluoroacetimidoyl iodide

A mixture of sodium iodide (1.9 g, 12.7 mmol) and N-(4-anisyl)-2,2,2-trifluoroacetimidoyl chloride (1.0 g, 4.2 mmol) in acetone (10 mL) was stirred under a nitrogen atmosphere at room temperature in the dark overnight. The mixture was washed with Na$_2$S$_2$O$_3$ aq. and extracted with AcOEt. The extracts were washed with brine, dried over MgSO$_4$, and concentrated under reduced pressure. The residue was purified by silica gel column chromatography (hexane) to give N-(4-anisyl)-2,2,2-trifluoroacetimidoyl iodide quantitatively.

Spectral data

^{1}H NMR (CDCl$_3$): δ 3.85 (3 H, s), 6.94-7.08 (4 H, m); ^{13}C NMR (CDCl$_3$): δ 55.3, 111.8 (q, J=41.7 Hz), 114.4 (2C), 115.2 (q, J=278.2 Hz), 120.6 (2C), 141.0, 159.1; ^{19}F NMR (CDCl$_3$): δ -70.3 (s); IR (neat): ν 1682 cm^{-1}.

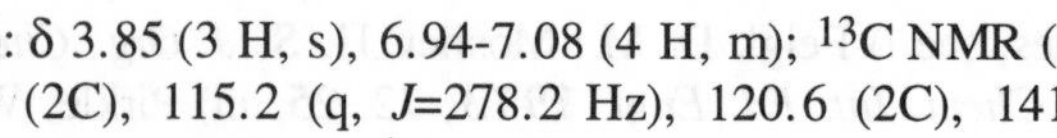

$$CF_3CO_2H \ + \ RNH_2 \ \xrightarrow{CCl_4, \ PPh_3, \ Et_3N} \ F_3C-C(=N-R)Cl$$

R	Yield (%)	R	Yield (%)
3-MeO-C$_6$H$_4$-	90	2-MeO-C$_6$H$_4$-	87
4-Me-C$_6$H$_4$-	86	2,6-Me$_2$-C$_6$H$_3$-	84
Ph-	73	4-Cl-C$_6$H$_4$-	91
3-Cl-C$_6$H$_4$-	87	3,4-Cl$_2$-C$_6$H$_3$-	95
4-F-C$_6$H$_4$-	77	4-O$_2$N-C$_6$H$_4$-	77
1-Naph-	88	C$_6$H$_5$-CH$_2$CH$_2$-	94
n-C$_6$H$_{13}$-	77	4-MeO-C$_6$H$_4$-[a]	84

[a] CBr$_4$ in CH$_2$Cl$_2$ was employed instead of CCl$_4$, giving the corresponding imidoyl bromide.

(Reproduced with permission from Uneyama, K., *et al. J. Org. Chem.* **1993**, *58*, 32)

Note

The corresponding bromide was prepared in a similar manner to the case of the corresponding chloride except for the usage of CBr$_4$ in CH$_2$Cl$_2$ instead of CCl$_4$.

Uneyama, K., *et al. J. Org. Chem.* **1993**, *58*, 32.

1,1,1-Trifluoro-2-[(2-methoxyethoxy)methoxy]ethane

$$CF_3CH_2OH \ \xrightarrow{a), \ b)} \ CF_3CH_2OMEM$$

a) NaH / THF, b) MEM-Cl

Synthetic method

2,2,2-Trifluoroethanol (34.5 mL, 47.4 g, 0.474 mol) in dry THF (945 mL) was added dropwise over 1 h to NaH (18.7 g at 60% dispersion from which the oil had been removed with toluene) in dry THF (45 mL) at 0 °C.   The olive green suspension was stirred for 1 h, then MEMCl (71.2 mL, 0.568 mol assuming 90% purity) in dry THF (75 mL) was added dropwise over 90 min at 0 °C.   The white suspension was allowed to warm to room temperature, and stirred overnight.   Cautious addition of H_2O (450 mL) and extractive work-up with Et_2O (3x200 mL) followed.   The combined Et_2O extracts were washed with H_2O (2x250 mL), dried (MgSO$_4$), and concentrated *in vacuo*.   1,1,1-Trifluoro-2-[(2-methoxyethoxy)methoxy]ethane (71.2 g, 80%) was isolated as a colorless liquid.

Spectral data

    bp. 110 °C/100 mmHg; ^1H NMR (CDCl$_3$): δ 3.30 (3 H, s), 3.44-3.50 (2 H, m), 3.62-3.67 (2 H, m), 3.80 (2 H, q, *J*=9.0 Hz), 4.70 (2 H, s); ^{13}C NMR (CDCl$_3$): δ 58.6, 64.6 (q, 34.6 Hz), 67.3, 71.4, 95.4, 123.9 (q, *J*=277.8 Hz); ^{19}F NMR (CDCl$_3$): δ -77.4 (t, *J*=9.0 Hz); IR (CHCl$_3$): ν 3422 cm^{-1}.

Percy, J. M., *et al. Tetrahedron* **1995**, *51*, 9201.

Methyl 2-[(*S,Z*)-1-benzyloxy-5,5,5-trifluoropent-3-en-2-yloxy]acetate

a) NaH / THF,   b) CH$_2$BrCO$_2$H,
c) (CO$_2$Cl)$_2$, cat. DMF,   d) MeOH, pyridine

Synthetic method

    To a suspension of NaH (*ca.* 0.16 g, 4 mmol) in THF (10 mL) was added dropwise a solution of (*S,Z*)-1-benzyloxy-5,5,5-trifluoropent-3-en-2-ol (2 mmol) in THF (5 mL) at 0 °C, and the mixture was stirred for 10 min.   To this was added a solution of bromoacetic acid (0.278 g, 2 mmol) in THF (5 mL) at 0 °C.   The reaction mixture was stirred overnight at room temperature and poured into ice-cooled H_2O.   The organic layer was washed with 1 *N* NaOH aq. three times and the combined aqueous layers were acidified with 3 *N* HCl, and the whole was then extracted with Et_2O three times.   The organic solution was dried over anhydrous MgSO$_4$, then evaporated. To a solution of crude carboxylic acid in CH$_2$Cl$_2$ (10 mL) was added (COCl)$_2$ (0.349 mL, 4 mmol) and a catalytic amount of DMF (two drops) at 0 °C, and the mixture was stirred overnight at ambient temperature.   The solvent and an excess amount of (COCl)$_2$ were removed under reduced pressure, CH$_2$Cl$_2$ (10 mL), MeOH (1 mL) and pyridine (0.34 mL, 4.19 mmol) were added at 0 °C, the reaction mixture was stirred overnight at room temperature, poured into 3 *N* HCl aq., and extracted with CH$_2$Cl$_2$ three times.   The organic extracts were dried over anhydrous MgSO$_4$ and evaporated.   The residue was purified by silica gel chromatography (hexane:CH$_2$Cl$_2$=1:3) to afford methyl 2-[(*S,Z*)-1-benzyloxy-5,5,5-trifluoropent-3-en-2-yloxy]acetate in 85% total yield.

Spectral data

    [α]$_D^{16}$ -36.7 (*c* 0.8, CHCl$_3$); ^1H NMR (CDCl$_3$) δ 3.62 (1 H, dd, *J*=10.74, 4.15 Hz), 3.65 (1 H, dd, *J*=10.50, 5.38 Hz), 3.74 (3 H, s), 4.12 (1 H, d, *J*=16.60 Hz), 4.18 (1 H, d, *J*=16.61 Hz), 4.57 (1 H, d, *J*=12.21 Hz), 4.60 (1 H, d, *J*=12.20 Hz), 4.6-4.7 (1 H, m), 5.84 (1 H, ddq, *J*=11.96, 7.57, 0.98 Hz), 6.05 (1 H, dq, *J*=11.96, 9.52 Hz), 7.2-7.4 (5 H, m); ^{13}C NMR (CDCl$_3$) δ 51.64, 66.35, 71.92, 73.25, 75.03, 121.57 (q, *J*=34.2 Hz), 122.44 (q, *J*=272.2 Hz),

127.46, 127.54, 128.22, 137.69, 139.07 (q, J=5.1 Hz), 170.26; ^{19}F NMR (CDCl$_3$) δ 103.98 (d, J=9.16 Hz) from C$_6$F$_6$; IR (neat) ν 3065, 3033, 3006, 2925, 2916, 2863, 1758, 1676 cm^{-1}.
Note
    For the preparation of the starting (*Z*)-allylic alcohol, see page 193 of this volume.

Kitazume, T., et al. J. Org. Chem. 1997, 62, 137.

(Z)-1-Ethoxy-3,3,3-trifluoroprop-1-ene,
(Z)-2-Bromo-1-ethoxy-3,3,3-trifluoroprop-1-ene

a) EtOH, KOH,   b) Br$_2$ / CH$_2$Cl$_2$,   c) Et$_3$N

Synthetic method
    (*Z*)-1-Ethoxy-3,3,3-trifluoroprop-1-ene
    A reaction flask equipped with a dry ice condenser and an addition funnel was charged with a solution of potassium hydroxide (33.3g, 0.59 mol) in EtOH.   2-Bromo-3,3,3-trifluoropropene (34.9 g, 0.20 mol) was added over 5 min, during which the reaction mixture started to reflux. After being stirred for 1 h, the reaction mixture was poured into H$_2$O (150 mL).   The organic layer was separated, dried over Na$_2$SO$_4$, and distilled to afford 27.4 g of (*Z*)-1-ethoxy-3,3,3-trifluoro-prop-1-ene in 96% yield.
Spectral data
    bp. 102-104 °C; ^{1}H NMR (CDCl$_3$): δ 1.32 (3 H, t, J=7.0 Hz), 4.0 (2 H, q, J=7.0 Hz), 4.56-4.74 (1 H, m), 6.32 (1 H, d, J=6.8 Hz); ^{19}F NMR (CDCl$_3$): δ 20.00 (d, J=6.7 Hz) from CF$_3$CO$_2$H.

Synthetic method
    (*Z*)-2-Bromo-1-ethoxy-3,3,3-trifluoroprop-1-ene
    To a solution of (*Z*)-1-ethoxy-3,3,3-trifluoroprop-1-ene (5.6 g, 40 mmol) in CH$_2$Cl$_2$ (40 mL) cooled to -20 °C was added dropwise a solution of bromine (12.8 g, 80 mmol) in CH$_2$Cl$_2$ (10 mL).   After the reaction mixture was kept at 0 °C for 30 min and then recooled to -20 °C, triethyl-amine (8.1 g, 80 mmol) was added over 10 min.   The reaction mixture was stirred at room temperature for 1 h and then poured into H$_2$O (100 mL).   The organic layer was separated and the aqueous phase was extracted with CH$_2$Cl$_2$ (3x20 mL).   The combined organic phase was washed with 2 N HCl to neutral and dried over MgSO$_4$.   Distillation under reduced pressure gave 2-bromo-1-ethoxy-3,3,3-trifluoroprop-1-ene in 77% yield with a (*Z*):(*E*) ratio of ≥98:2.
Spectral data
    bp. 60-62 °C/37 mmHg (for the same compound with a (*Z*):(*E*) ratio of 60:40)
    ^{1}H NMR (CDCl$_3$): δ 1.39 (3 H, t, J=7.1 Hz), 4.13 (2 H, q, J=7.1 Hz), 7.23 (1 H, q, J=1.6 Hz); ^{19}F NMR (CDCl$_3$): δ 13.5 (d, J=1.6 Hz) from CF$_3$CO$_2$H.
    (*E*)-isomer
    ^{1}H NMR (CDCl$_3$): δ 1.32 (3 H, t, J=7.1 Hz), 4.05 (2 H, q, J=7.1 Hz), 6.72 (1 H, s); ^{19}F NMR (CDCl$_3$): δ 17.0 (s) from CF$_3$CO$_2$H.
Note
    When the second step was carried out with one equivalent of bromine, the product obtained

was a (*Z*):(*E*) mixture in a ratio of 60:40 in 80% yield.   Moreover, when this reaction (1 equiv of Br$_2$) was worked up before the addition of triethylamine, a 3:2 *syn*:*anti* mixture of 1,2-dibromo-1-ethoxy-3,3,3-trifluoroprop-1-ene could be isolated quantitatively.

See the following recent application of the above materials: i) Hong, F.; Hu, C.-M. *J. Chem. Soc., Chem. Commun.* **1996**, 57, ii) Hong, F.; Hu, C.-M. *J. Chem. Soc., Perkin Trans. 1* **1997**, 1909, iii) Hu, C.-M., *et al. J. Chem. Soc., Perkin Trans. 1* **1998**, 279.

Shi, G.-Q., *et al. J. Org. Chem.* **1996**, *61*, 3200.

N-(4-Anisyl)-2,2,2-(trifluoroacetimidoyl)trimethylsilane

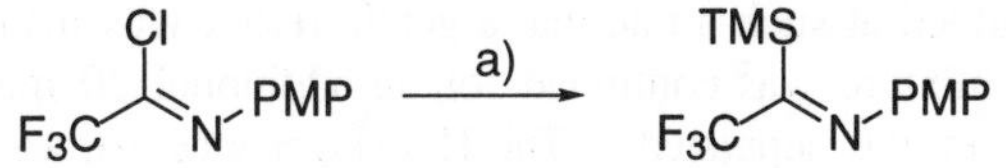

a) TMSLi, CuCN / THF, HMPA

Synthetic method

MeLi (0.6 mL, 0.8 mmol, 1.4 *M* in hexane) was added dropwise into a solution of hexamethyldisilane (0.2 mL, 1.2 mmol) in 0.5 mL of anhydrous HMPA under an argon atmosphere. After 15 min, 1.5 mL of THF was added to the mixture which was cooled to 0 ℃.   Then the solution of TMSLi was added to 38 mg (0.4 mmol) of copper(I) cyanide in one portion.   After being stirred for 30 min, the mixture was added to an equimolar amount of trifluoroacetoimidoyl chloride (0.4 mmol) in 1.5 mL of anhydrous THF cooled at -55 ℃.   The mixture was stirred for 10 min, then quenched with NH$_4$Cl aq. and extracted with Et$_2$O.   The organic layer was washed with brine, dried over Na$_2$SO$_4$, filtered, and condensed.   Column chromatography of the residue gave *N*-(4-anisyl)-2,2,2-(trifluoroacetimidoyl)trimethylsilane in 55% yield.

Spectral data

^{1}H NMR (CDCl$_3$): δ 0.06 (9 H, s), 3.81 (3 H, s), 6.71 (2 H, d, *J*=8.8 Hz), 6.87 (2 H, d, *J*=8.8 Hz); ^{19}F NMR (CDCl$_3$): δ 93.06 (s) from C$_6$F$_6$; IR (neat): ν 1506, 1282, 1246, 1124, 850 cm^{-1}.

Note

For the preparation of the starting material, see page 199 of this volume.

N-(4-Anisyl)-2,2,2-(trifluoroacetimidoyl)trimethylsilane as well as other silanes with different substituents on nitrogen were found to react with such various electrophiles as aldehydes, ketones, or acid chlorides in the presence of tetra-*n*-butylammonium fluoride (TBAF).   On the other hand, although it was possible to construct the same product *via* lithiation of the imidoyl iodide, this was reported to be limited only for the case of R=2,6-Me$_2$C$_6$H$_3$ due to the instability of the lithiated species under the reaction conditions.   See Uneyama, K., *et al. J. Org. Chem.* **1993**, *58*, 32.

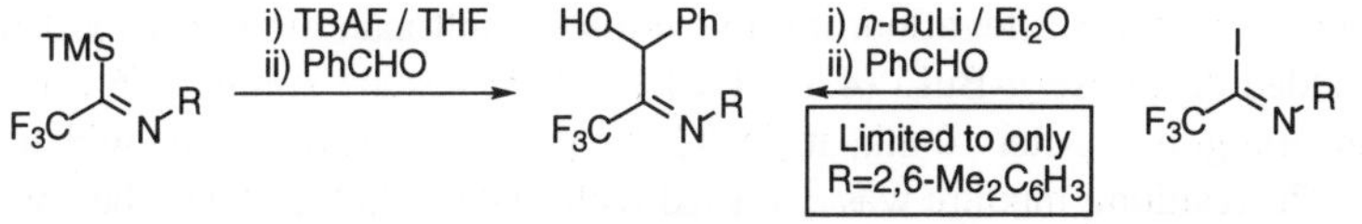

Uneyama, K., *et al. J. Org. Chem.* **1996**, *61*, 6055.

Ethyl 2-diazo-3,3,3-trifluoropropionate

$$F_3C \overset{O}{\underset{}{\parallel}} CO_2Et \quad \xrightarrow{a),\ b),\ c)} \quad F_3C \overset{N_2}{\underset{}{\parallel}} CO_2Et$$

a) Tosyl hydrazide / CH_2Cl_2,   b) pyridine,   c) $POCl_3$

Synthetic method

Ethyl 3,3,3-trifluoro-2-oxopropionate (20 g, 0.1 mol) and tosyl hydrazide (18.6 g, 0.1 mol) were mixed in CH_2Cl_2 (120 mL). The mixture was briefly refluxed then stirred at room temperature overnight. Pyridine (50 mL) was added, then with the exclusion of moisture $POCl_3$ (9.4 mL, 0.1 mol) was added at such a rate that a gentle reflux was maintained. After addition, refluxing of the reaction mixture was continued for an additional 20 min. H_2O (200 mL) was added, and the organic layer was separated. The H_2O layer was extracted with Et_2O (3x80 mL). The combined organic layer was washed with 1 N HCl aq. to remove pyridine, then washed successively with sat. $NaHCO_3$ aq. and brine, and dried over Na_2SO_4. The bulk of the solvent was removed in a rotary evaporator, and the remaining solvent was carefully distilled off under atmospheric pressure. Further distillation under reduced pressure gave 14 g of ethyl 2-diazo-3,3,3-trifluoropropionate in 77% yield as a yellow liquid.

Spectral data

bp. 60-62 °C/100 mmHg; ^1H NMR (CCl_4): δ 1.25 (3 H, t, J= 7 Hz), 4.20 (2 H, q, J= 7 Hz); ^{19}F NMR (CCl_4): δ 20.0 (s) from CF_3CO_2H; IR (neat): ν 2200, 1750 cm^{-1}.

Note

For the preparation of trifluorinated pyruvate, see page 179 of this volume.

The distillation can also be carried out safely at the atmospheric pressure, although some decomposition was noted when the final bath temperature was raised above 150 °C in order to drive out the product from the high boiling residues. Unlike some other diazo compounds, this material can be stored in a refrigerator for several months without any deterioration.

Shi, G.-Q.; Xu, Y.-Y. *J. Org. Chem.* **1990**, *55*, 3383.

2,3,3-Trifluoroprop-1-en-1-yl *p*-toluenesulfonate

$$CHF_2CF_2CH_2OTs \quad \xrightarrow{a)} \quad \overset{F_2HC}{\underset{F}{\diagdown}} C=C \overset{H}{\underset{OTs}{\diagup}}$$

a) *n*-BuLi / THF

Synthetic method

To a solution of 2,2,3,3-tetrafluoropropyl *p*-toluenesulfonate (0.858 g, 3.0 mmol) in THF (12 mL) was added dropwise *n*-BuLi (4.1 mL of a 1.6 M hexane solution, 6.6 mmol) at -78 °C over 20 min under argon. After 10 min at -78 °C, the reaction was quenched with a cold 10% HCl aq. (50 mL). The resulting mixture was extracted with Et_2O (3x30 mL) and the combined extracts were dried over anhydrous Na_2SO_4, filtered, and concentrated *in vacuo*. After the isomer distribution of the product was determined by ^{19}F NMR, the residue was purified by silica gel column chromatography (benzene) to give 0.559 g of 2,3,3-trifluoroprop-1-en-1-yl *p*-toluenesulfonate ((Z):(E)= 86:14) in 70% yield.   The (E)- and (Z)-isomers were easily separated by silica gel column chromatography (hexane:benzene=1:1).

Spectral data

(Z)-isomer

Rf 0.46 (hexane:benzene=1:1); mp. 75.0-76.0 ℃ (hexane); ^{1}H NMR (CDCl$_3$): δ 2.46 (3 H, s), 5.96 (1 H, dt, J=52.6, 7.0 Hz), 6.63 (1 H, dt, J=18.0, 2.0 Hz), 7.30 and 7.74 (4 H, AB q, J=8.8 Hz); ^{19}F NMR (CDCl$_3$): δ -45.4 (2 F, ddd, J=52.6, 17.6, 2.0 Hz), -68.3 (1 F, ddt, J=18.0, 17.6, 7.0 Hz) from CF$_3$CO$_2$H; IR (KBr): ν 1720, 1350, 1172 cm^{-1}.

(E)-isomer

Rf 0.57 (hexane:benzene=1:1); mp. 24.1-25.5 ℃ (hexane); ^{1}H NMR (CDCl$_3$): δ 2.44 (3 H, s), 6.16 (1 H, dt, J=51.2, 15.8 Hz), 6.85 (1 H, d, J=4.4 Hz), 7.30 and 7.71 (4 H, AB q, J=8.2 Hz); ^{19}F NMR (CDCl$_3$): δ -48.3 (2 F, dd, J=51.2, 16.9 Hz), -87.0 (1 F, ddt, J=16.9, 15.8, 4.4 Hz) from CF$_3$CO$_2$H; IR (KBr): ν 1720, 1379, 1172 cm^{-1}.

Note

For the preparation of the starting tosylate, see page 230 of this volume.

Funabiki, K., *et al. Chem. Lett.* **1994**, 1075.

4-Trifluoromethyl-5,5-bis(2-phenylethyl)-1,3-dioxolane-2-thione, 1,1,1-Trifluoro-3-(2-phenylethyl)-5-phenylpent-2-ene

a) 1,1'-Thiocarbonyldiimidazole / toluene,   b) (MeO)$_3$P

Synthetic method

4-Trifluoromethyl-5,5-bis(2-phenylethyl)-1,3-dioxolane-2-thione

A solution of 1,1,1-trifluoro-3-(2-phenylethyl)-5-phenylpentane-2,3-diol (119 mg, 0.35 mmol) and 1,1'-thiocarbonyldiimidazole (75 mg, 0.42 mmol) in toluene (2.0 mL) was heated at 110 ℃ for 30 min.   After concentration *in vacuo*, the residue was purified by chromatography (hexane:AcOEt=5:1) to afford 127 mg of 4-trifluoromethyl-5,5-bis(2-phenylethyl)-1,3-dioxolane-2-thione in 95% yield as a colorless solid.   Recrystallization (hexane) gave an analytical sample as colorless prisms.

Spectral data

mp. 87-88 ℃; ^{1}H NMR (CDCl$_3$): δ 2.18-2.38 (4 H, m), 2.68-2.94 (4 H, m), 4.79 (1 H, q, J=6.6 Hz), 7.18-7.37 (10 H, m); ^{19}F NMR (CDCl$_3$): δ -72.6 (d, J=6.5 Hz); IR (KBr): ν 3040, 2970, 1610, 1500, 1460, 1400, 1320, 1280, 1190, 1150, 1130, 1040 cm^{-1}.

Synthetic method

1,1,1-Trifluoro-3-(2-phenylethyl)-5-phenylpent-2-ene

A solution of 4-trifluoromethyl-5,5-bis(2-phenylethyl)-1,3-dioxolane-2-thione (57 mg, 0.15 mmol) in trimethyl phosphite (0.5 mL) was heated at 130 ℃ for 13 h.   Excess trimethyl phosphite was removed *in vacuo*, and the residue was purified by column chromatography (hexane) to afford 39 mg of 1,1,1-trifluoro-3-(2-phenylethyl)-5-phenylpent-2-ene in 86% yield as a colorless oil.

Spectral data
    ^{1}H NMR (CDCl$_3$): δ 2.40-2.45 (2 H, m), 2.54-2.59 (2 H, m), 2.74-2.79 (4 H, m), 5.46 (1 H, q, *J*=8.6 Hz), 7.158-7.32 (10 H, m); ^{19}F NMR (CDCl$_3$): δ -57.4 (d, *J*=7.7 Hz); IR (neat): ν 3040, 2950, 1670, 1500, 1460, 1280, 1120 cm^{-1}.

Note
    For the reaction of aldehydes and CF$_3$TMS, see pages 122, 123, and 126 of this volume.
    The starting material was synthesized by the TBAF-catalyzed reaction of CF$_3$TMS with the aldehyde which was, in turn, prepared by i) reaction of 1,5-diphenylpentan-3-one with TMSCN, followed by ii) reduction of the resultant nitrile group by DIBAL-H, and the crude product was desilylated by the action of 2 equiv of TBAF in THF.

                                    Terashima, S., *et al. Chem. Pharm. Bull.* **1997**, *45*, 43.

(Trifluoromethy)trimethylsilane

                        a)
    CF$_3$Br $\longrightarrow$ CF$_3$TMS

            a) TMSCl, (Me$_2$N)$_3$P / PhCN

Synthetic method
    Into a 2-L three-necked flask fitted with an efficient dry ice-acetone cold finger and a mechanical stirrer was placed 102 mL (87.3 g, 0.83 mol) of chlorotrimethylsilane dissolved in 100 mL of benzonitrile.   Stirring was started, and the solution was cooled to *ca.* -30 °C.   Bromotri-fluoromethane (261 g, 1.75 mol) was precondensed in a flask, then allowed to evaporate into the reaction flask.   The bath was cooled progressively until it was -60 °C.   To the resulting slurry was added a solution of 216 g (0.876 mol) of hexamethylphosphorus triamide in 175 mL of benzo-nitrile over a period of 2 h.   After an additional hour at -60 °C, the bath and cold finger were allowed to warm up to room temperature, then connected to a dry ice-acetone-cooled trap and subjected to an aspirator vacuum (20 mmHg) with mild warming (45 °C) to drive out all volatile material.   The liquid in the trap was washed rapidly with ice-cold H$_2$O (3x75 mL), the product (top) layer was separated and dried over MgSO$_4$, and the dry liquid was decanted into a 100-mL flask.   Fractional distillation using a 15-cm column packed with glass helices afforded 77.1 g of (trifluoromethy)trimethylsilane as a colorless liquid in 65% yield based on chlorotrimethylsilane.

Spectral data
    bp. 55-55.5 °C; ^{1}H NMR (CDCl$_3$): δ 0.25 (s); ^{13}C NMR (CDCl$_3$): δ -5.2, 131.7 (q, *J*=321.9 Hz); ^{19}F NMR (CDCl$_3$): δ -66.1 (s); ^{29}Si NMR (CDCl$_3$): δ 4.7 (q, *J*=37.9 Hz).

Note
    By a procedure similar to the one described above based on the originally reported method by Ruppert (Ruppert, I., *et al. Tetrahedron Lett.* **1984**, *25*, 2195), the corresponding (pentafluoro-ethyl)- and (*n*-heptafluoropropyl)trimethylsilanes were synthesized in 50% and 68% yield, respectively.   Very recently, Prakash and Olah's group reported the preparation of chloro- and bromodifluoromethylated trimethylsilanes as well as difluorobis(trimethylsilyl)methane and 1,1,2,2-tetrafluoro-1,2-bis(trimethylsilyl)ethane.   See Surya Prakash, G. K., *et al. J. Am. Chem. Soc.* **1997**, *119*, 1572.
    Fuchikami's group also reported the preparation of (difluoromethyl)-, (bromodifluoro-methyl)-, (1,1-difluoropropyl)-, (1,1-difluoro-2-methylpropyl)-, and (cyclohexyldifluoromethyl)-dimethylphenylsilanes and their reactions with aldehydes and ketones in the presence of a catalytic

amount of potassium fluoride in DMF.   See Hagiwara, T.; Fuchikami, T. *Synlett* **1995**, 717.

Surya Prakash, G. K., *et al. J. Org. Chem.* **1991**, *56*, 984.

2-Dibenzylamino-4,4,4-trifluoro-3-hydroxybutyric acid,
2-Amino-4,4,4-trifluoro-3-hydroxybutyric acid
(DL-4,4,4-Trifluorothreonine, DL-4,4,4-Trifluoro-*allo*-threonine)

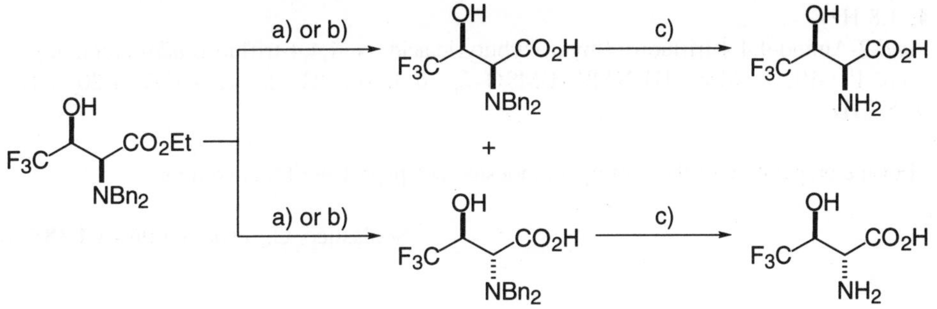

a) 0.5 N KOH / MeOH-H$_2$O,   b) 0.4 N NaOEt / EtOH-H$_2$O,   c) H$_2$, 10% Pd/C / EtOH

Synthetic method
   2-Dibenzylamino-4,4,4-trifluoro-3-hydroxybutyric acid
   Method A: A suspension of ethyl *syn*-2-dibenzylamino-4,4,4-trifluoro-3-hydroxybutyrate (100 mg, 0.262 mmol) in 0.5 N KOH in MeOH:H$_2$O=7:3 (3.15 mL, 1.57 mmol) was stirred at 20 ℃ for 5 h.   The resulting solution was treated with 0.3 M KH$_2$PO$_4$ (5.24 mL), evaporated to dryness, taken up with AcOEt (15 mL) to give, after drying with Na$_2$SO$_4$, a crude mixture of 2-dibenzylamino-4,4,4-trifluoro-3-hydroxybutyric acid in an 85:15 *syn:anti* ratio, as judged by ^{1}H NMR.   Column on silica gel chromatography (AcOEt) gave pure *syn*-2-dibenzylamino-4,4,4-trifluoro-3-hydroxybutyric acid (75 mg, 80% yield) and the corresponding *anti* form (12 mg, 13% yield), both as oils.
   Method B: A suspension of ethyl *syn*-2-dibenzylamino-4,4,4-trifluoro-3-hydroxybutyrate (100 mg, 0.262 mmol) was treated with a 0.4 N solution of NaOEt in EtOH:H$_2$O=99.8:0.2 (7.9 mL, 0.79 mmol).   The resulting solution was stirred at 20 ℃ for 4 d and then worked up as above described to give a crude mixture in a ratio of 35:65.   Silica gel column chromatography gave pure *syn*-2-dibenzylamino-4,4,4-trifluoro-3-hydroxybutyric acid (30 mg, 32% yield) and the corresponding *anti* form (58 mg, 63% yield), both as oils.
Spectral data
   syn-2-Dibenzylamino-4,4,4-trifluoro-3-hydroxybutyric acid
   Rf 0.45 (AcOEt); ^{1}H NMR (DMSO-d_6): δ 3.54 (1 H, d, *J*=8.3 Hz), 3.77 (2 H, d, *J*=13.5 Hz), 4.03 (2 H, d, *J*=13.5 Hz), 4.61 (1 H, dq, *J*=8.3, 6 Hz), 7.15-7.60 (10 H, m).
   anti-2-Dibenzylamino-4,4,4-trifluoro-3-hydroxybutyric acid
   Rf 0.21 (AcOEt); ^{1}H NMR (DMSO-d_6): δ 3.53 (1 H, d, *J*=6 Hz), 3.72 (2 H, d, *J*=13.5 Hz), 3.97 (2 H, d, *J*=13.5 Hz), 4.44 (1 H, dq, *J*=7.7, 6 Hz), 7.15-7.60 (10 H, m).

Synthetic method
   2-Amino-4,4,4-trifluoro-3-hydroxybutyric acid
   A solution of *syn*-2-dibenzylamino-4,4,4-trifluoro-3-hydroxybutyric acid (100 mg, 0.28

mmol) in 95% EtOH (10 mL) was hydrogenated over 10% palladium on carbon (30 mg) for 5 h at reflux.   After filtration of the catalyst, the solution was evaporated to dryness to give pure *syn*-2-amino-4,4,4-trifluoro-3-hydroxybutyric acid (DL-4,4,4-trifluorothreonine) by TLC and ^{1}H NMR. The yield was 90%.   The same procedure for the corresponding *anti*-2-dibenzylamino-4,4,4-trifluoro-3-hydroxybutyric acid furnished pure *anti*-2-amino-4,4,4-trifluoro-3-hydroxybutyric acid (DL-4,4,4-trifluoro-*allo*-threonine) in 90% yield.

Spectral data

   syn-2-Amino-4,4,4-trifluoro-3-hydroxybutyric acid (DL-4,4,4-trifluorothreonine)

      mp. 212-214 °C (dec); ^{1}H NMR (DMSO-d_6): δ 3.40 (1 H, d, *J*=1.8 Hz), 4.70 (1 H, dq, *J*=8.4, 1.8 Hz).

   anti-2-Amino-4,4,4-trifluoro-3-hydroxybutyric acid (DL-4,4,4-trifluoro-*allo*-threonine)

      mp. 191-193 °C (dec); ^{1}H NMR (DMSO-d_6): δ 3.36 (1 H, d, *J*=5.7 Hz), 4.20 (1 H, dq, *J*=7.0, 5.7 Hz).

Note

   For the preparation of the starting aminoester, see page 138 of this volume.

Scolastico, C., *et al. Synthesis* **1985**, 850.

2-(Trifluoromethyl)acryloyl chloride

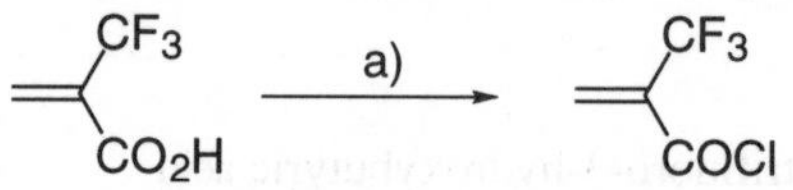

a) Phthaloyl dichloride

Synthetic method

   2-(Trifluoromethyl)acrylic acid (14.0 g, 100 mmol) was added to phthaloyl dichloride (21.6 mL, 150 mmol) and the mixture was heated at 140 °C under a nitrogen atmosphere.   The refluxing mixture was cooled after 2 h and the condenser was replaced with a distillation head.   The crude material was distilled under atmospheric pressure to furnish 14.6 g of 2-(trifluoromethyl)-acryloyl chloride (92 mmol) in 92% yield.

Spectral data

   bp. 92 °C; ^{1}H NMR (CDCl$_3$): δ 6.69 (1 H, br q), 7.12 (1 H, br q); ^{19}F NMR (CDCl$_3$): δ 11.9 (m) from CF$_3$CO$_2$H; IR (neat): ν 1770 cm^{-1}.

Note

   This method was basically applicable to the conversion of (*E*)-4,4,4-trifluorobut-2-enoic acid or 3,3,3-trifluoropropionic acid into the corresponding acid chlorides.

Kitazume, T., *et al. J. Org. Chem.* **1989**, 54, 5630.

5 Preparation of Poly- or Perfluorinated Materials

5.1 Introduction of Poly- or Perfluorinated Groups

5.1.1 Introduction of Poly- or Perfluorinated Groups (Nucleophilic)

N-Allyl-2-[(*tert*-butyldimethylsilyl)oxy]-4,4,5,5,6,6,7,7,8,8,9,9,9-tridecafluoronon-3-ylamine

$$n\text{-}C_6F_{13}I \; + \quad \underset{\text{OTBS}}{\diagup\!\!\diagdown}\!=\!N\diagup\!\!\diagup \quad \xrightarrow{\text{a), b)}} \quad \underset{\text{OTBS}}{\overset{\text{HN}\diagup\!\!\diagdown}{\diagup\!\!\diagdown}}C_6F_{13}\text{-}^n \; + \; \underset{\text{OTBS}}{\overset{\text{HN}\diagup\!\!\diagdown}{\diagup\!\!\diagdown}}C_6F_{13}\text{-}^n$$

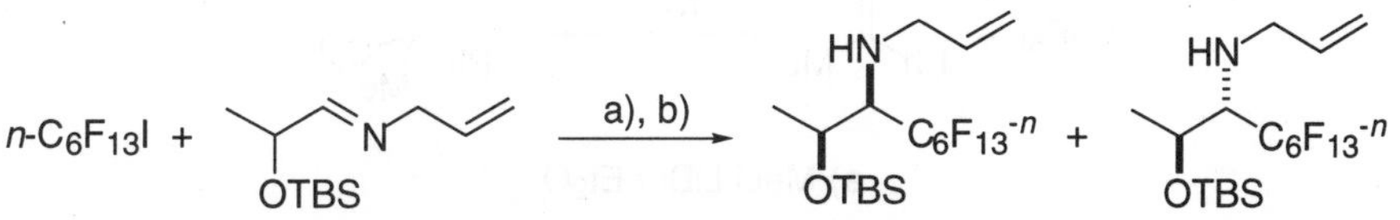

a) $BF_3 \cdot OEt_2$ / Et_2O, b) MeLi·LiBr

Synthetic method

A solution of N-{[2-(*tert*-butyldimethylsilyl)oxy]prop-1-ylidene}allylamine (1 mmol) and 1,1,2,2,3,3,4,4,5,5,6,6,6-tridecafluoro iodide (1.2 mmol) in 10 mL of dry Et_2O was stirred and cooled at -78 °C. As 1.2 mmol of $BF_3 \cdot OEt_2$ was added, precipitation of the BF_3-imine complex soon occurred. An ethereal solution of MeLi·LiBr (1.1 mmol) was added to the mixture at -78 °C over 10 min. The mixture became clear near the end of the addition. After the mixture was stirred for 1 h at -78 °C, NaHCO$_3$ aq. was added to it. The organic phase was separated, and the aqueous phase was extracted with Et_2O. The combined extracts were washed with brine and dried over Na_2SO_4. N-Allyl-2-[(*tert*-butyldimethylsilyl)oxy]-4,4,5,5,6,6,7,7,8,8,9,9,9-tridecafluoro-non-3-ylamine was obtained in 81% yield as an 85:15 *syn:anti* diastereomer mixture. Spectral data were given only for the *syn* diastereomer.

Spectral data

bp. 73 °C (bath temperature)/0.6 mmHg; ^1H NMR (CDCl$_3$): δ 0.06 (3 H, s), 0.09 (3 H, s), 0.89 (9 H, s), 1.29 (3 H, d, J=6.4 Hz), 2.72 (1 H, br s), 3.05 (1 H, dd, J=19.5, 8.2 Hz), 3.47 (2 H, m), 4.34 (1 H, m), 5.11 (1 H, m), 5.21 (1 H, m), 5.87 (1 H, m); ^{13}C NMR (CDCl$_3$): δ -5.32, -4.16, 14.09, 23.22 25.68, 52.58, 62.08 (dd, J=24, 18 Hz), 65.16 (m), 100-125 (6C), 116.05, 136.78; IR (neat): ν 3368, 2956, 1364, 1300-1100 cm^{-1}.

Note

When a methoxymethyl moiety was employed as the hydroxy protective group, the opposite diastereoselection was observed up to *syn:anti*=97:3. 1,3-Asymmetric induction was also tried and, although conditions furnishing high *syn* selectivity were not discovered, as high as 98% selectivity was attained for the construction of the product with an *anti* relationship.

$n\text{-}C_6F_{13}I$ + [structure with OMOM and N-Pr^{-n} imine]

i) BF$_3$·OEt$_2$ / Et$_2$O
ii) MeLi·LiBr

[product structures with HN-Pr^{-n}, OMOM, C$_6$F$_{13}^{-n}$]

42% yield
syn: anti = 3:97

$n\text{-}C_6F_{13}I$ + [structure with CO$_2$Me, i-Pr, N, Ph]

i) BF$_3$·OEt$_2$ / Et$_2$O
ii) MeLi·LiBr

[product structures with MeO$_2$C, Ph, i-Pr, N-H, C$_6F_{13}^{-n}$]

54% yield
syn: anti = 2:98

Uno, H., *et al. J. Org. Chem.* **1992**, *57*, 1504.

3,3,4,4,4-Pentafluoro-2-phenylbutan-2-ol

C_2F_5I + [acetophenone structure: Ph, O, Me] →(a)→ [product: Ph, OH, C$_2$F$_5$, Me]

a) MeLi·LiBr / Et$_2$O

Synthetic method

Under anhydrous conditions and a dry nitrogen atmosphere, a 100-mL, three-necked flask cooled in a dry ice-isopropyl alcohol bath was charged with anhydrous Et$_2$O (30 mL), acetophenone (1 mL, 1.03 g, 8.57 mmol), and pentafluoroethyl iodide (3.18 g, 12.93 mmol). The mixture was magnetically stirred, and a MeLi·LiBr solution (1.6 *M* in Et$_2$O, 7.5 mL, 12.0 mmol) was added. Stirring was continued for 30 min at -78 °C, after which time the contents of the flask were added to a 3 *N* HCl aq. solution (75 mL) together with 25 mL of Et$_2$O. After being shaken, the layers were separated, and the aqueous layer was extracted with an additional 25 mL of Et$_2$O. The combined ethereal extracts were dried over anhydrous Na$_2$SO$_4$ and filtered, and the solvent was removed with a rotary evaporator. Distillation of the residue gave 1.81 g of 3,3,4,4,4-penta-fluoro-2-phenylbutan-2-ol in 88% yield.

Spectral data

bp. 58-59 °C/3 mmHg; ^{1}H NMR (CDCl$_3$): δ 1.77 (3 H, s), 2.51 (1 H, s), 7.10-7.15 (5 H, m); ^{13}C NMR (CDCl$_3$): δ 24.60, 75.12 (t, *J*=24.0 Hz), 114.39 (tq, *J*=261.5, 34.6 Hz), 119.31 (tq, *J*=287.9, 36.5 Hz), 126.10, 128.24, 128.54, 138.55; IR (neat): ν 3480, 3095, 3060, 3030, 3000, 2940, 1610, 1500, 1475, 1465, 1452, 1385, 1340, 1285, 1273, 1220-1210, 1203-1180, 1134, 1108, 1076, 1063, 1030, 1007, 928, 915, 818, 759, 742, 735, 696, 663 cm^{-1}.

$n\text{-}C_8F_{17}I$ + [structure: Ph, CO$_2$Me]

MeCu, MeLi
Et$_2$O, -78 °C→rt
→ [product: Ph, Me, O, C$_8$F$_{17}^{-n}$]   75% yield

i) cat. CuI, MeLi
ii) PhMgBr, -78 °C→rt
→ [product: Ph, Ph, O, C$_8$F$_{17}^{-n}$]   65% yield

Note

Perfluoroalkyl iodides were reported to coexist in the presence of organocopper or Grignard reagents, enabling the carbonyl addition of a perfluoroalkyl group followed by the conjugate addition of other organometallic species in a one pot manner as shown in the previous page. See Uno, H., *et al. Chem. Lett.* **1987**, 1153. This article also dealt with the preparation of perfluoro-alkylated ketones from esters, attaining good to excellent yields (58-99%).

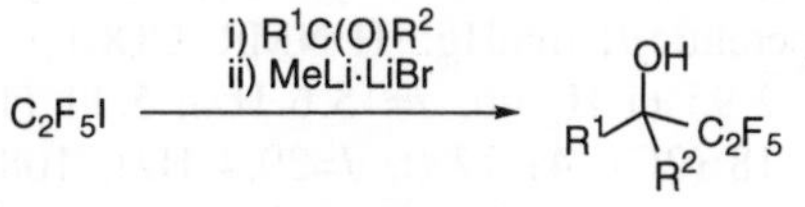

$$C_2F_5I \xrightarrow{\substack{\text{i) } R^1C(O)R^2 \\ \text{ii) MeLi·LiBr}}} R^1 \overset{OH}{\underset{R^2}{\diagup}} C_2F_5$$

R^1	R^2	Equivalent		Time (min)	Yield (%)
		C$_2$F$_5$I	MeLi		
Ph	H	1.84	1.46	30	91
-(CH$_2$)$_5$-		1.77	1.77	8	100
-(CH$_2$)$_3$-CH=CH-		1.56	1.24	30	81

(Reproduced with permission from Gassman, P. G.; O'Reilly, N. J. *J. Org. Chem.* **1987**, *52*, 2481)

Gassman, P. G.; O'Reilly, N. J. *J. Org. Chem.* **1987**, *52*, 2481.

5.1.2 Introduction of Poly- or Perfluorinated Groups (Electrophilic, Radical)

**6-(1,1,2,2,3,3,4,4,5,5,6,6,6-Tridecafluorohexyl)-
1-[(trimethylsilyl)oxy]cyclohex-1-ene,
2-(1,1,2,2,3,3,4,4,5,5,6,6,6-Tridecafluorohexyl)-
1-[(trimethylsilyl)oxy]cyclohex-1-ene,
2-(1,1,2,2,3,3,4,4,5,5,6,6,6-Tridecafluorohexyl)cyclohexanone**

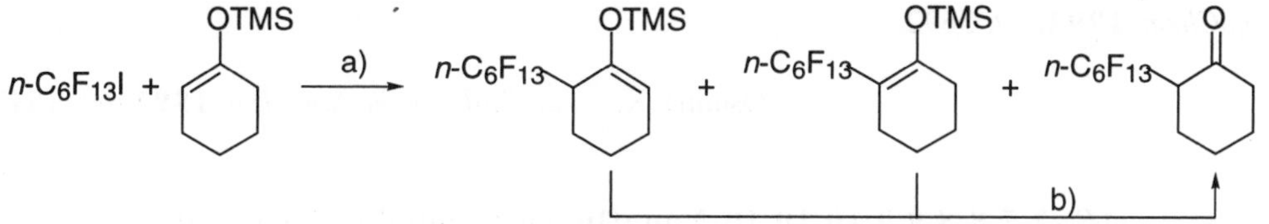

a) Et$_3$B, 2,6-Dimethylpyridine / hexane, b) conc HCl / THF

Synthetic method

Et$_3$B (a 1.0 *M* hexane solution, 0.4 mL, 0.4 mmol) was added to a solution of 1-[(trimethyl-silyl)oxy]cyclohex-1-ene (0.34 g, 2.0 mmol), 1,1,2,2,3,3,4,4,5,5,6,6,6-tridecafluoro iodide (1.19 g, 2.6 mmol) and 2,6-dimethylpyridine (0.20 g, 1,9 mmol) in hexane (5 mL) at room temperature. After stirring for 8 h at 25 °C, the resulting precipitate was filtered through Celite 545. The filtrate was concentrated *in vacuo*. The residual oil was submitted to silica gel column chromatography to give a mixture of 6- and 2-(1,1,2,2,3,3,4,4,5,5,6,6,6-tridecafluorohexyl)-1-[(trimethylsilyl)oxy]-cyclohex-1-ene (0.64 g, 65% yield, in a ratio of 68:32), and 2-(1,1,2,2,3,3,4,4,5,5,6,6,6-trideca-fluorohexyl)cyclohexanone (92 mg, 11% yield).

The residual oil after concentration of the filtrate was treated with conc HCl (35 wt%, 1 mL) in THF (5 mL) for 10 min. The reaction mixture was slowly poured into sat. NaHCO$_3$ aq. (40 mL) and extracted with AcOEt (2x30 mL). The combined organic layer was dried over anhydrous Na$_2$SO$_4$ and concentrated *in vacuo*. Purification by silica gel column chromatography gave 2-(1,1,2,2,3,3,4,4,5,5,6,6,6-tridecafluorohexyl)cyclohexanone in 74% yield as the sole product.

Spectral data

6-(1,1,2,2,3,3,4,4,5,5,6,6,6-Tridecafluorohexyl)-1-[(trimethylsilyl)oxy]cyclohex-1-ene

bp. 68-71 °C (bath temperature)/1 mmHg; ^{1}H NMR (CDCl$_3$): δ 0.20 (9 H, s), 1.48-1.88 (3 H, m), 1.96-2.16 (3 H, m), 2.95 (1 H, tm, J=15.6 Hz), 5.11 (1 H, t, J=4.1 Hz); ^{13}C NMR (CDCl$_3$): δ -0.10, 18.82, 23.18 (2C), 41.77 (t, J=20.4 Hz), 108.6, 141.1. ^{13}C signals of a perfluorohexyl group could not be observed.; ^{19}F NMR (CDCl$_3$): δ -81.39 (3 F, t, J=9.8 Hz), -108.1 (1 F, dm, J=284 Hz), -113.8 (1 F, dm, J=284 Hz), -121.2 (2 F, br s), -122.2 (2 F, br s), -123.2 (2 F, br s), -126.4 - -126.8 (2 F, m); IR (neat): ν 2958, 1669, 1324, 1239, 1203, 1146, 1120, 913, 846, 694, 660 cm^{-1}.

2-(1,1,2,2,3,3,4,4,5,5,6,6,6-Tridecafluorohexyl)-1-[(trimethylsilyl)oxy]cyclohex-1-ene

bp. 68-72 °C (bath temperature)/1 mmHg; ^{1}H NMR (CDCl$_3$): δ 0.20 (9 H, s), 1.55-1.75 (4 H, m), 2.14-2.19 (4 H, m); ^{13}C NMR (CDCl$_3$): δ 0.50, 22.10, 22.45, 23.51 (t, J=5.0 Hz), 31.38, 117.1 (t, J=32.7 Hz), 155.8 (t, J=5.5 Hz). ^{13}C signals of a perfluorohexyl group could not be observed.; ^{19}F NMR (CDCl$_3$): δ -81.39 (3 F, tt, J=9.7, 3.6 Hz), -108.5 (2 F, t, J=14.0 Hz), -122.1 - -123.2 (6 F, m), -126.4 - -126.8 (2 F, m); IR (neat): ν 2944, 1661, 1375, 1292, 1238, 1204, 1145, 1119, 1074, 1065, 933, 861, 848, 704 cm^{-1}.

2-(1,1,2,2,3,3,4,4,5,5,6,6,6-Tridecafluorohexyl)cyclohexanone

bp. 68 °C/0.5 mmHg; ^{19}F NMR (CDCl$_3$): δ -82.4, -115.3 - -127.8; IR (CCl$_4$): ν 1725, 1680 cm^{-1}.

Note

Physical data for 2-(1,1,2,2,3,3,4,4,5,5,6,6,6-tridecafluorohexyl)cyclohexanone was extracted from the following article: Cantacuzène, D.; Dorme, R. *Tetrahedron Lett.* **1975**, 2031.

2-(1,1,2,2,3,3,4,4,5,5,6,6,7,7,8,8,8-Heptadecafluorooctyl)cyclohexanone was reported to be partially dehydrofluorinated during silica gel column chromatography affording a 1:2.5 mixture of perfluoroalkyl:perfluoroalkylidene materials, while only the perfluoroalkylated material was observed by ^{19}F NMR analysis of the crude reaction mixture. See Umemoto, T.; Adachi, K. *J. Org. Chem.* **1994**, *59*, 5692.

Oshima, K., *et al. Bull. Chem. Soc. Jpn.* **1991**, *64*, 1542.

(*E*)-7,7,8,8,9,9,10,10,10-Nonafluoro-5-iododec-5-en-1-ol, (*Z*)-7,7,8,8,9,9,10,10,10-Nonafluorodec-5-en-1-ol

$$n\text{-}C_4F_9I \xrightarrow{\text{a)}} \underset{n\text{-}C_4F_9}{(CH_2)_4OH} \xrightarrow{\text{b), c)}} \underset{n\text{-}C_4F_9}{(CH_2)_4OH}$$

a) HC≡C(CH$_2$)$_4$OH, AIBN, b) *n*-BuLi / Et$_2$O, c) MeOH

Synthetic method

(*E*)-7,7,8,8,9,9,10,10,10-Nonafluoro-5-iododec-5-en-1-ol

5-Hexyn-1-ol (980 mg, 10 mmol), 1,1,2,2,3,3,4,4,4-nonafluorobutyl iodide (5.2 g, 15 mmol), and AIBN (130 mg, 0.8 mmol) were placed in a heavy walled glass tube (2.5 cm o. d., 1.8 cm i. d.x 20 cm, with a narrow neck 5 cm from the bottom) equipped with a magnetic stir bar. The mixture was frozen (liquid nitrogen, 77 K), degassed, and thawed under a nitrogen atmosphere to eliminate oxygen. This process was repeated, and the tube was sealed under vacuum while the contents were still frozen. The mixture was slowly warmed to room temperature and heated to 80 °C for 18 h. After being cooled with liquid nitrogen, the tube was broken open. The crude product was purified, and the (E)- and (Z)-isomers were separated by flash chromatography (hexane:AcOEt=10:1). Fractions with over 98% isomeric purity (GC) were combined; the remaining fractions were pooled, concentrated, and chromatographed again. After two chromatographic separations, two portions were obtained, 2.22 g of geometrically pure (E)-7,7,8,8,9,9,10,10,10-nonafluoro-5-iododec-5-en-1-ol in 50% yield and 1.46 g of a mixture of (E)- and (Z)-isomers in 33% yield. A small aliquot was analyzed by GC, which showed that before chromatography the (E):(Z) ratio was 7:1.

Spectral data
^{1}H NMR (CDCl$_3$): δ 1.5-1.8 (4 H, m), 2.59 (2 H, t, J=6.9 Hz), 3.56 (2 H, t, J=5.72 Hz), 6.26 (1 H, t, J=14.5 Hz); ^{19}F NMR (CDCl$_3$): δ -80.84 (3 F, t), -104.89 (2 F, m), -123.65 (2 F, m), -125.34 (2 F, m).

Synthetic method
(Z)-7,7,8,8,9,9,10,10,10-Nonafluorodec-5-en-1-ol
A solution of *n*-BuLi (a 1.6 *M* solution in hexane, 3 mL) was added to a solution of (E)-7,7,8,8,9,9,10,10,10-nonafluoro-5-iododec-5-en-1-ol (444 mg, 1 mmol) in 10 mL of dry Et$_2$O at -78 °C. The mixture was stirred for 30 min and quenched with precooled (-78 °C) MeOH (3 mL). After being warmed to 0 °C, the mixture was poured into 20 mL of an NH$_4$Cl solution and the product was isolated by flash chromatography (Et$_2$O) to give 270 mg of (Z)-7,7,8,8,9,9,10,10,10-nonafluorodec-5-en-1-ol in 85% yield.

Spectral data
^{1}H NMR (CDCl$_3$): δ 1.5-1.6 (4 H, m), 2.28 (2 H, br s), 3.55 (2 H, t, J=6.4 Hz), 5.53 (1 H, br q, J=14.5 Hz), 6.12 (1 H, dtt, J=12.2, 7.1, 2.4 Hz); ^{19}F NMR (CDCl$_3$): δ -80.79 (3 F, t), -104.33 (2 F, m), -123.55 (2 F, m), -125.63 (2 F, m).

Prestwich, G. D., et al. *J. Org. Chem.* **1992**, *57*, 132.

Methyl 2-(1,1,1,2,3,3,3-heptafluoroprop-2-yl)octanoate

$$i\text{-}C_3F_7I + \quad \underset{n\text{-}C_6H_{13}}{\overset{OTMS}{\diagup\!\!\!=\!\!\!<}}_{OMe} \quad \xrightarrow{a), b)} \quad n\text{-}C_6H_{13}\underset{i\text{-}C_3F_7}{\diagdown}CO_2Me$$

a) Et$_3$B / hexane, b) TBAF

Synthetic method
Et$_3$B (a 1.0 *M* hexane solution, 0.2 mL, 0.2 mmol) was added to a solution of 1-[(trimethyl-silyl)oxy]-1-methoxyoct-1-ene (2.0 mmol) and 1,1,1,2,3,3,3-haptafluoroprop-2-yl iodide (1.0 mmol) in hexane (5 mL) at 0 °C. After addition of Et$_3$B, the reaction mixture was immediately warmed to room temperature and stirred for 20 min. Sat. NaHCO$_3$ aq. (5 mL) was added to the reaction mixture. The mixture was vigorously stirred for 40 min, then poured into H$_2$O (30 mL)

and extracted with hexane (2x30 mL). The combined organic layer was dried over anhydrous Na$_2$SO$_4$ and concentrated *in vacuo*. The residual oil was purified by silica gel column chromatography to furnish methyl 2-(1,1,1,2,3,3,3-heptafluoroprop-2-yl)octanoate in 88% yield.

Spectral data

bp. 85-90 °C (bath temperature)/24 mmHg; ^{1}H NMR (CDCl$_3$): δ 0.89 (3 H, t, *J*=6.5 Hz), 1.28 (8 H, br s), 1.67-1.89 (1 H, m), 1.92-2.13 (1 H, m), 3.09-3.23 (1 H, m), 3.77 (3 H, s); ^{13}C NMR (CDCl$_3$): δ 13.77, 22.41, 25.69, 27.67, 28.64, 31.38, 46.90 (d, *J*=20.2 Hz), 52.39, 91.48 (dm, *J*=209 Hz), 120.6 (dq, *J*=287, 27.0 Hz), 168.0-168.2 (m); ^{19}F NMR (CDCl$_3$): δ -74.03 (6 F, d, *J*=6.1 Hz), -178.7 (1 F, dsept, *J*=13.4, 6.1 Hz); IR (neat): ν 2958, 2930, 2860, 1757, 1459, 1439, 1302, 1229, 1168, 1134, 1114, 1093, 976, 718 cm^{-1}.

Note

While 2,6-dimethylpyridine was required in the case of the reaction with enol silyl ethers, addition of this base to the first example of the following table resulted in the decrease of the yield of the desired material to 37% and 10% of dehydrofluorinated product was obtained.

$$R_fI + \begin{matrix} R^2 \\ \diagdown \\ R^3 \end{matrix}\begin{matrix} OSi \\ \diagup \\ OR^1 \end{matrix} \xrightarrow[\text{ii) TBAF}]{\text{i) Et}_3\text{B / hexane}} R_f\begin{matrix} CO_2R^1 \\ \diagup \\ R^2 R^3 \end{matrix}$$

R_f	R^1	R^2	R^3	Si	Time (min)	Yield (%)
n-C$_6$F$_{13}$[a)]	n-Bu	H	H	TBS	10	87
i-C$_3$F$_7$[a)]	n-Bu	H	H	TBS	30	61
n-C$_6$F$_{13}$	n-Bu	H	H	TMS	20	96
i-C$_3$F$_7$[a)]	n-Bu	H	H	TMS	20	49
i-C$_3$F$_7$	n-C$_8$H$_{17}$	H	H	TMS	20	96
n-C$_6$F$_{13}$[a)]	Me	n-Bu	H	TMS	20	83
i-C$_3$F$_7$[a)]	Me	n-Bu	H	TMS	20	40
i-C$_3$F$_7$	Me	n-C$_6$H$_{13}$	H	TMS	20	88
n-C$_6$F$_{13}$[a)]	n-C$_6$H$_{13}$	Me	Me	TMS	240	63
i-C$_3$F$_7$[a)]	n-C$_6$H$_{13}$	Me	Me	TMS	960	27

[a)] The reaction was quenched with 5 mL of sat. NaHCO$_3$ aq. (5 mL) instead of TBAF. (Reproduced with permission from Oshima, K., *et al. Bull. Chem. Soc. Jpn.* **1991**, *64*, 1542)

Oshima, K., *et al. Bull. Chem. Soc. Jpn.* **1991**, *64*, 1542.

9,9,10,10,11,11,12,12,12-Nonafluoro-7-iodododecane-1,2-diol,
9,9,10,10,11,11,12,12,12-Nonafluorododecane-1,2-diol

$$n\text{-C}_4\text{F}_9\text{I} \xrightarrow{\text{a)}} n\text{-C}_4\text{F}_9\diagup\diagdown_R \xrightarrow{\text{b)}} n\text{-C}_4\text{F}_9\diagup\diagdown\diagup_R$$

R: -(CH$_2$)$_4$CH(OH)CH$_2$OH

a) cat. Pd(PPh$_3$)$_4$, b) H$_2$, 5% Pd/C, NaHCO$_3$ / MeOH

Synthetic method

9,9,10,10,11,11,12,12,12-Nonafluoro-7-iodododecane-1,2-diol

A 500-mL flask was charged with 28.8 g (200 mmol) of oct-7-ene-1,2-diol and 69.2 g (200 mmol) of 1,1,2,2,3,3,4,4,4-nonafluorobutyl iodide.   Tetrakis(triphenylphosphine)palladium (1.7 g, 1.5 mmol) was added at room temperature with stirring.   After a few seconds an exothermic reaction occurred.   The reaction mixture was stirred at room temperature for 2 h.   Chromatography (silica gel column, 60x600 mm; CH_2Cl_2:AcOEt=2:1) gave 74.5 g of 9,9,10,10,11,11,12,12,12-nonafluoro-7-iodododecane-1,2-diol in 76% yield as an oil.

Spectral data

^1H NMR (CDCl$_3$): δ 1.37-1.58 (6 H, m), 1.82 (2 H, m), 2.17 (2 H, br s), 2.84 (2 H, m), 3.41-3.71 (3 H, m), 4.33 (1 H, m); ^{19}F NMR (CDCl$_3$): δ -81.5 (3 F, s), -113.8 (2 F, AB q, J=266 Hz), -125.0 (2 F, s), -126.4 (2 F, s); IR (CCl$_4$): ν 3410, 2937, 1229 cm^{-1}.

Synthetic method

9,9,10,10,11,11,12,12,12-Nonafluorododecane-1,2-diol

A mixture of 40 g (82 mmol) of 9,9,10,10,11,11,12,12,12-nonafluoro-7-iodododecane-1,2-diol, 7.6 g (90 mmol) of NaHCO$_3$, 3 g of 5% Pd/C, and 250 mL of MeOH was stirred at room temperature for 6 days with hydrogen passed through the reaction mixture.   After removal of Pd/C by filtration, the filtrate was concentrated and dried under vacuum (25 °C/1 mmHg) to give a residue which was further purified by chromatography (silica gel column, 60x500 mm; CH_2Cl_2:AcOEt=1:1) to afford 23.2 g of 9,9,10,10,11,11,12,12,12-nonafluorododecane-1,2-diol in 78% yield as an oil.

Spectral data

^1H NMR (CDCl$_3$): δ 1.37-1.61 (8 H, m), 2.05 (2 H, m), 3.38-3.69 (5 H, m); ^{13}C NMR (CDCl$_3$): δ 20.4, 25.8, 29.4, 29.7, 31.1 (t, J=22 Hz), 33.3, 67.1, 72.7, 105.4-123.6 (m); ^{19}F NMR (CDCl$_3$): δ -81.7 (3 F, s), -115.2 (2 F, s), -125.0 (2 F, s), -126.6 (2 F, s); IR (CCl$_4$): ν 3394, 2936, 1243 cm^{-1}.

$$R_fI + CH_2=CHR \xrightarrow{\text{cat. Pd(PPh}_3)_4} R_fCH_2CHI\text{-}R \xrightarrow[\text{NaHCO}_3\text{ / MeOH}]{\text{H}_2,\ 5\%\ \text{Pd/C,}} R_fCH_2CH_2\text{-}R$$

		Yield (%)	
R_f	R	R_f-CH$_2$CHI-R	R_f-CH$_2$CH$_2$-R
n-C$_4$F$_9$	(CH$_2$)$_4$CH(OH)CH$_2$OH	76	78
n-C$_6$F$_{13}$	(CH$_2$)$_4$CH(OH)CH$_2$OH	75	72
n-C$_8$F$_{17}$	(CH$_2$)$_2$CH(OH)CH$_2$OH	75	80
n-C$_8$F$_{17}$[a]	(CH$_2$)$_2$CH(OH)CH$_2$OH	71	_[b]
n-C$_3$F$_7$	(CH$_2$)$_4$CH(OH)CH$_2$OH	88	_[b]

[a] (PhCO$_2$)$_2$ was employed as a catalyst instead of Pd(PPh$_3$)$_4$.   [b] Not tried.
(Reproduced with permission from Qiu, W.-M.; Burton, D. J. *J. Org. Chem.* **1993**, *58*, 419)

Qiu, W.-M.; Burton, D. J. *J. Org. Chem.* **1993**, *58*, 419.

5.2 Carbon-Carbon Bond-forming Reactions

5.2.1 Aldol Type Reactions

(4R*,5R*)-5-(Pentafluoroethyl)-4-(methoxycarbonyl)-5-phenyl-2-oxazoline

$$\text{Ph} \overset{O}{\underset{}{\parallel}} \text{CF}_2\text{CF}_3 \xrightarrow{\text{a)}} \begin{array}{c} \text{CF}_3\text{CF}_2 \quad \text{CO}_2\text{Me} \\ \text{Ph}^{\prime\prime\prime\prime} \\ O \diagdown_{N} \end{array}$$

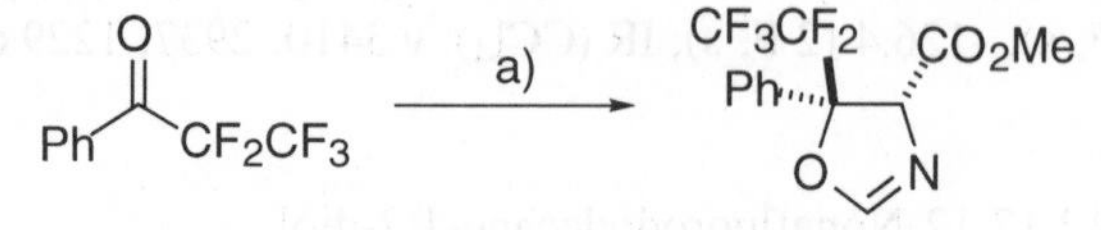

a) CN-CH₂CO₂Me, CuCl, Et₃N / CH₂ClCH₂Cl

Synthetic method

To a magnetically stirred solution of copper(I) chloride (0.1 mmol) and triethylamine (0.1-0.2 mmol) and methyl isocyanoacetate (1 mmol) in freshly distilled 1,2-dichloroethane (2 mL) was added 2,2,3,3,3-pentafluoro-1-phenylpropan-1-one (1.1 mmol) under a dry nitrogen atmosphere at room temperature (20-22 °C). Stirring was continued for 10 min, then complete consumption of the starting isocyanoacetate was confirmed. The solvent was removed *in vacuo* and (4R*,5R*)-5-(pentafluoroethyl)-4-(methoxycarbonyl)-5-phenyl-2-oxazoline was isolated *via* bulb-to-bulb distillation in 94% yield. The ratio of diastereomers ((4R*,5R*):(4R*,5S*)=>99:1) was determined by GLC analysis on the crude reaction mixture, before removal of the solvent and distillation, and by NMR (^{1}H and ^{19}F, where possible) analysis on the distilled product.

Spectral data

^{1}H NMR (CDCl₃): δ 3.27 (3 H, s), 5.41 (1 H, d, J=2.0 Hz), 7.21 (1 H, d, J=2.0 Hz), 7.36-7.41 (3 H, m), 7.43-7.46 (2 H, m); ^{19}F NMR (CDCl₃): δ -79.02 (3 F, br s), -123.01, -125.16 (2 F, AB d, J=302.3 Hz).

(4R*,5S*)-5-(pentafluoroethyl)-4-(methoxycarbonyl)-5-phenyl-2-oxazoline

^{1}H NMR (CDCl₃): δ 3.91 (3 H, s), 5.07 (1 H, d, J=2.2 Hz), 7.18 (1 H, d, J=2.2 Hz), 7.36-7.41 (3 H, m), 7.43-7.46 (2 H, m).

Note

For the preparation of the starting ketone, see page 223 of this volume.

In the present reaction, a wide range of fluorinated ketones was employed as substrates (CH₂F, CClF₂, CF₃, n-C₃F₇, n-C₄F₉ instead of C₂F₅, or C₆F₅ instead of C₆H₅ in the above case), and high yields (usually 80-99%) as well as good selectivities (up to >99:1 in favor of diastereomers with a *trans* relationship between fluoroalkyl and methoxycarbonyl groups) were generally attained.

The oxazoline ring thus formed was readily transformed into the corresponding acyclic structure as shown below.

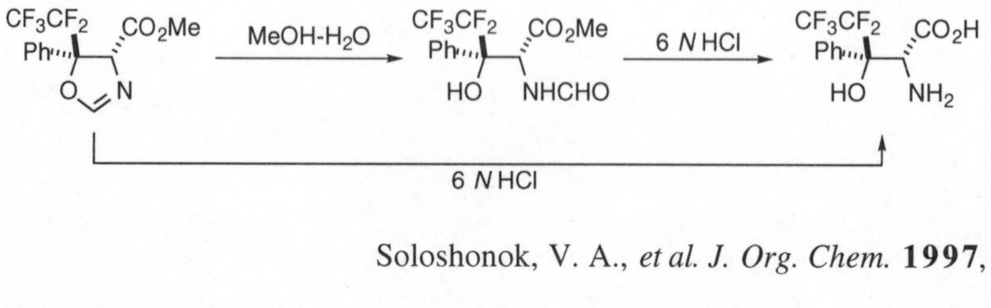

$$\begin{array}{c} \text{CF}_3\text{CF}_2 \quad \text{CO}_2\text{Me} \\ \text{Ph}^{\prime\prime\prime\prime} \\ O \diagdown_{N} \end{array} \xrightarrow{\text{MeOH-H}_2\text{O}} \begin{array}{c} \text{CF}_3\text{CF}_2 \quad \text{CO}_2\text{Me} \\ \text{Ph}^{\prime\prime\prime\prime} \\ \text{HO} \quad \text{NHCHO} \end{array} \xrightarrow{6\ N\ \text{HCl}} \begin{array}{c} \text{CF}_3\text{CF}_2 \quad \text{CO}_2\text{H} \\ \text{Ph}^{\prime\prime\prime\prime} \\ \text{HO} \quad \text{NH}_2 \end{array}$$

6 N HCl

Soloshonok, V. A., *et al. J. Org. Chem.* **1997**, *62*, 3470.

(R)-3-[(S)-4,4,4-Trifluoro-4-(trifluoromethyl)-3-hydroxy-2-methylbutyryl]-4-(prop-2-yl)oxazolidin-2-one,
(R)-4,4,4-Trifluoro-4-(trifluoromethyl)-2-methylbutane-1,3-diol

a) n-Bu₂BOTf, Et₃N / CH₂Cl₂, b) (CF₃)₂C=O, c) LiBH₄ / THF

Synthetic method

(R)-3-[(S)-4,4,4-Trifluoro-4-(trifluoromethyl)-3-hydroxy-2-methylbutyryl]-4-(prop-2-yl)-oxazolidin-2-one

To a solution of (R)-3-propionyl-4-(prop-2-yl)oxazolidin-2-one (776 mg, 4.19 mmol) in dry CH₂Cl₂ (5 mL) was added dibutylboron trifluoromethanesulfonate (1.32 g, 4.81 mmol) at -78 ℃ over 2 min. After 10 min at this temperature, Et₃N (760 mg, 5.45 mmol) was added over 10 min and the reaction mixture was warmed to 0 ℃. After 1 h at 0 ℃, the solution was cooled to -78 ℃ and gaseous hexafluoroacetone (1.5 mL at -78 ℃, 11.9 mmol) was added with a cannula. After 30 min at -78 ℃, the reaction mixture was brought to and held at -30 ℃ for 2 h, then the reaction was quenched with pH 7.0 phosphate buffer (0.1 M, 5 mL) and MeOH (10 mL), followed by the addition of 30% H₂O₂-MeOH (5 mL-25 mL). After 1 h at 0 ℃, the mixture was concentrated *in vacuo*. The residue was diluted with 10% NaHCO₃ aq. and extracted with CH₂Cl₂. The combined extracts were washed with brine, dried over MgSO₄ and filtered. After evaporation of the solvent, chromatography of the residue (hexane:AcOEt=3:1) gave 1.18 g of (R)-3-[(S)-4,4,4-trifluoro-4-(trifluoromethyl)-3-hydroxy-2-methylbutyryl]-4-(prop-2-yl)oxazolidin-2-one in 80% yield as colorless needles.

Spectral data

$[\alpha]_D^{24}$ -47.9 (c 0.98, CHCl₃); mp. 96.1-97.9 ℃ (hexane:Et₂O); ¹H NMR (CDCl₃): δ 0.88 (3 H, d, J=6.9 Hz), 0.94 (3 H, d, J=7.0 Hz), 1.43 (3 H, dq, J=7.1, 2.7 Hz), 2.37 (1 H, dsept, J=7.1, 2.8 Hz), 4.28-4.49 (3 H, m), 4.75 (1 H, q, J=7.0 Hz), 6.53 (1 H, s); ¹⁹F NMR (CDCl₃): δ -76.17 (3 F, q, J=11.5 Hz), -73.32 (3 F, dq, J=11.5, 1.5 Hz); IR (KBr): ν 3295, 1769, 1682 cm⁻¹.

Synthetic method

(R)-4,4,4-Trifluoro-4-(trifluoromethyl)-2-methylbutane-1,3-diol

A solution of (R)-3-[(S)-4,4,4-trifluoro-4-(trifluoromethyl)-3-hydroxy-2-methylbutyryl]-4-(prop-2-yl)oxazolidin-2-one (1.12 g, 3.19 mmol) in dry THF (10 mL) was treated with lithium borohydride (350 mg, 16.0 mmol) in five portions at 0 ℃. The mixture was stirred at 0 ℃ for 3.5 h, and the reaction was quenched with pH 7.0 phosphate buffer. The whole was extracted with CH₂Cl₂. The combined extracts were washed with brine, dried over MgSO₄ and filtered. After evaporation of the solvent, chromatography of the residue (hexane:Et₂O=1:1) gave 464 mg of (R)-4,4,4-trifluoro-4-(trifluoromethyl)-2-methylbutane-1,3-diol in 64% yield as a colorless oil.

Spectral data

$[\alpha]_D^{25}$ -7.75 (c 0.76, CHCl₃); mp. 96.1-97.9 ℃ (hexane:Et₂O); ¹H NMR (CDCl₃): δ 1.12 (3 H, d, J=7.3 Hz), 2.27 (1 H, s), 2.48-2.70 (1 H, m), 3.87 (1 H, dd, J=11.0, 4.1 Hz), 4.10 (1 H, t, J=11.0 Hz), 6.21 (1 H, s); ¹⁹F NMR (CDCl₃): δ -76.59 (3 F, q, J=10.0 Hz), -72.34 (3 F, q, J=10.0 Hz); IR (KBr): ν 3300 cm⁻¹.

Note

The above diol was further transformed to the corresponding mono MTPA ester in the primary hydroxy group in order to clarify that the obtained material was enantiomerically pure on the basis of ^{1}H as well as ^{19}F NMR analyses. This diol was eventually employed as the side chain of the F_6-vitamin D_2 analog.

Iseki, K., *et al. Chem. Pharm. Bull.* **1995**, *43*, 1897.

N,N-Diethyl-2,3,3,3-tetrafluoropropionamide, *N,N*-Diethyl-2-fluoro-2-(trifluoromethyl)-3-hydroxyhexanamide

a) NaHCO$_3$ / H$_2$O,　b) *n*-Bu$_2$BOTf / CH$_2$Cl$_2$,　c) *i*-Pr$_2$NEt, d) *n*-PrCHO,　e) 30% H$_2$O$_2$ aq. / phosphate buffer (pH 7)

Synthetic method

　N,N-Diethyl-2,3,3,3-tetrafluoropropionamide

　Hexafluoropropene (17 mL, 180 mmol) was allowed to react with diethylamine (11.0 g, 150 mmol) in Et$_2$O (50 mL) according to the reported procedure. The resulting crude adduct was treated with sat. NaHCO$_3$ aq. (20 mL) under cooling with an ice bath. The mixture was extracted with Et$_2$O (3x30 mL), the combined extracts were dried over anhydrous Na$_2$SO$_4$, filtered, and concentrated under reduced pressure. The residue was subjected to column chromatography on silica gel to give 24.7 g of *N,N*-diethyl-2,3,3,3-tetrafluoropropionamide in 82% yield.

Spectral data

　^{1}H NMR (CDCl$_3$): δ 1.18 (3 H, t, *J*=7.0 Hz), 1.25 (3 H, t, *J*=7.0 Hz), 3.40 (4 H, q, *J*=7.0 Hz), 5.28 (1 H, dq, *J*=46.4, 6.1 Hz); ^{19}F NMR (CDCl$_3$): δ -75.86 (3 F, dd, *J*=12.2, 6.1 Hz), -199.28 (1 F, dq, *J*=46.4, 12.2 Hz); IR (neat): ν 1671 cm^{-1}.

Synthetic method

　N,N-Diethyl-2-fluoro-2-(trifluoromethyl)-3-hydroxyhexanamide

　To a solution of *N,N*-diethyl-2,3,3,3-tetrafluoropropionamide (0.201 g, 1.0 mmol) in CH$_2$Cl$_2$ (5 mL) was added a CH$_2$Cl$_2$ solution of dibutylboryl triflate (1 *M* in CH$_2$Cl$_2$, 1.1 mL, 1.1 mmol) at 0 ℃ under argon. After 10 min, ethyldiisopropylamine (0.155 g, 1.2 mmol) was added to the reaction mixture at -10 ℃. After stirring for 15 min, butanal (0.08 g, 1.1 mmol) was added. The resultant mixture was stirred for 30 min at -10 ℃, and poured into a mixture of phosphate buffer (pH 7, 4 mL), 30% H$_2$O$_2$ aq. (1 mL), and ice. After stirring for 1 h at 0 ℃, this mixture was added to an aqueous solution of Na$_2$SO$_3$ (1.23 g) and stirred at room temperature. The mixture was extracted with Et$_2$O (3x30 mL). The combined extracts were dried over anhydrous Na$_2$SO$_4$, filtered, and concentrated. The crude residue was purified by silica gel column chromatography to give 0.240 g of *N,N*-diethyl-2-fluoro-2-(trifluoromethyl)-3-hydroxyhexan-amide in 88% yield as a 94:6 *threo: erythro* mixture.

Spectral data

　^{1}H NMR (CDCl$_3$): δ 0.94 (3 H, t, *J*=5.4 Hz), 1.15 (3 H, t, *J*=6.4 Hz), 1.23 (3 H, t, *J*=6.4 Hz), 1.4-1.8 (4 H, m), 2.6 (1 H, br s), 3.1-3.7 (4 H, m), 4.40 (1 H, br d, *J*=20.8 Hz); ^{19}F NMR

(CDCl$_3$): δ -72.69 (3 F, d, J=4.9 Hz), -184.43 (1 F, dq, J=20.8, 4.9 Hz); IR (neat): ν 3388, 1639 cm^{-1}.

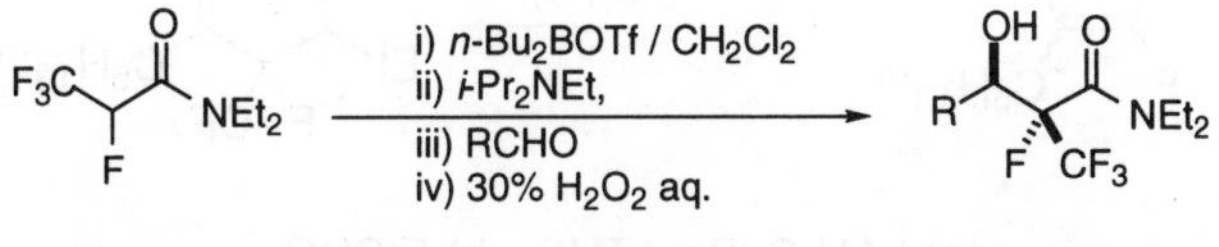

R	Yield (%)	threo:erythro		
i-Pr	88	92	:	8
t-Bu	74	85	:	15
(E)-MeCH=CH-	85	93	:	7
(E)-MeCH=C(Me)-	82	92	:	8
(E)-PhCH=CH-	86	100	:	0
Ph	84	93	:	7
4-Me-C$_6$H$_4$-	78	100	:	0
4-MeO-C$_6$H$_4$-	82	94	:	6
4-Cl-C$_6$H$_4$-	76	100	:	0

(Reproduced with permission from Kuroboshi, M.; Ishihara, T. *Bull. Chem. Soc. Jpn.* **1990**, *63*, 1191)

Note

Instead of the *in situ* preparation of *N,N*-diethyl-1,1,2,3,3,3-hexafluoropropylamine from gaseous hexafluoropropene, this reagent, obtained from a commercial supplier, can be subjected to direct hydrolysis.

Subjection of the starting tetrafluoropropionamide to an LDA or a *n*-BuLi solution at -78 °C led to the formation of complex mixtures. On the other hand, the corresponding ethyl ester afforded the aldol product in 80% yield when it was reacted with LDA in the presence of benzaldehyde, while a significantly decreased yield was obtained by the stepwise reaction (enolate formation→addition of an aldehyde) due to the formation of complex byproducts. See Qian, C.-P.; Nakai, T. *Tetrahedron Lett.* **1990**, *31*, 7043. In the authors' laboratory, such a byproduct was shown to be dimerization material constructed (obtained in 62% yield) as depicted below. bp. 102.0-102.5 °C/6.5 mmHg; ^{1}H NMR (CDCl$_3$): δ 1.37 (6 H, t, J=7.4 Hz), 4.34 (2 H, q, J=7.4 Hz), 4.43 (2 H, q, J=7.4 Hz); ^{19}F NMR (CDCl$_3$): δ 2.4 (3 F, ddd, J=10.9, 8.9, 4.7 Hz), -67.2 (1 F, ddq, J=131.8, 21.1, 10.9 Hz), -73.7 (1 F, ddq, J=131.8, 13.6, 4.7 Hz), -90.9 (1 F, ddq, J=21.1, 13.6, 8.9 Hz) from CF$_3$CO$_2$H; IR (neat): ν 1770, 1740, 1690 cm^{-1} (Yamazaki, T.; Ishikawa, N. unpublished results).

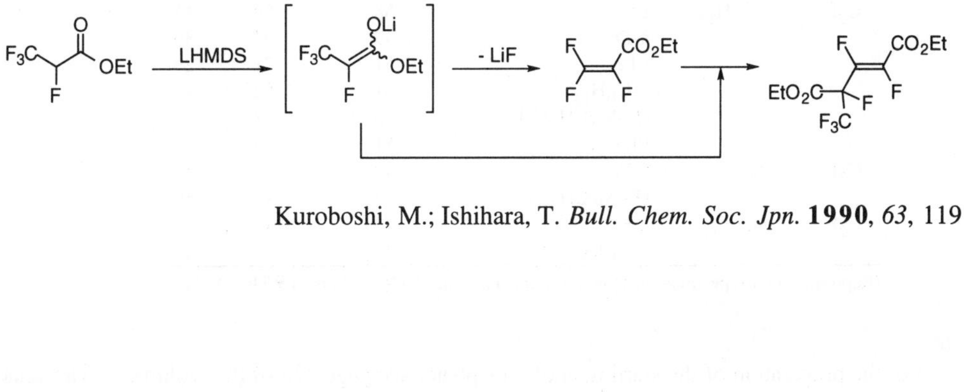

Kuroboshi, M.; Ishihara, T. *Bull. Chem. Soc. Jpn.* **1990**, *63*, 1191.

4-Fluoro-4-(trifluoromethyl)-3-hydroxydodecan-5-one

a) LAH-CuBr$_2$ / THF, b) EtCHO

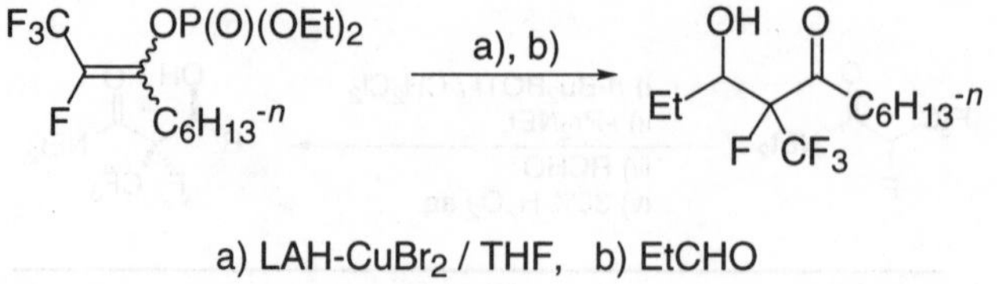

Synthetic method

Into a suspension of copper(II) bromide (0.447 g, 2.0 mmol) in THF (5 mL) was introduced *via* a syringe a THF solution (1 M) of LAH (2.0 mL, 2.0 mmol) at -30 °C, and the mixture was stirred for 30 min.   After addition of 1,1,1,2,2-pentafluorononan-3-one (0.350 g, 1.0 mmol), followed by stirring for an additional 3 h at -30 °C, propanal (0.174 g, 3.0 mmol) was added dropwise to the mixture at the same temperature.   After 10 h of stirring at -30 °C, the reaction mixture was hydrolyzed with cold sat. NH$_4$Cl aq. (15 mL) containing 6 N HCl aq. (5 mL).   The resulting mixture was extracted with Et$_2$O (3x30 mL) and the ethereal extracts were washed with H$_2$O, dried (Na$_2$SO$_4$), filtered, and concentrated under vacuum.   The ratio of *threo* to *erythro* isomer was determined to be 59:41 by ^{19}F NMR of the crude product, which was then chromatographed (hexane:AcOEt) on a column of silica gel to give 0.190 g of 4-fluoro-4-(trifluoromethyl)-3-hydroxydodecan-5-one in 70% yield.

Spectral data

^{1}H NMR (CDCl$_3$): δ 0.6-2.3 (17 H, m), 2.3-2.9 (2 H, m), 3.5-4.5 (1 H, m); IR (Nujol): ν 3426, 1734, 1279, 1200, 1112 cm^{-1}.

    erythro isomer

^{19}F NMR (CDCl$_3$): δ -74.06 (3 F, d, J=6.1 Hz), -189.79 (1 F, m).

    threo isomer

^{19}F NMR (CDCl$_3$): δ -73.64 (3 F, d, J=6.1 Hz), -189.67 (1 F, m).

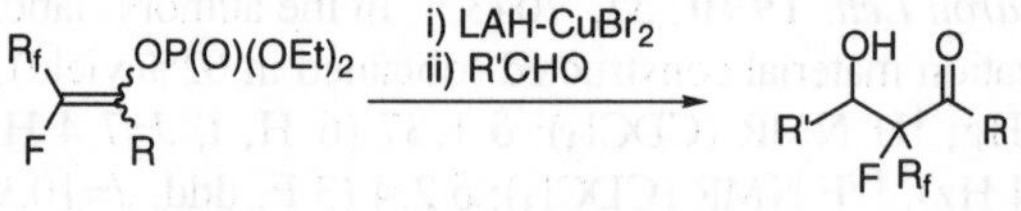

R$_f$	R	R'	Yield (%)	*threo*:*erythro*		
CF$_3$	n-C$_6$H$_{13}$	n-Pr	70	56	:	44
		(E)-MeCH=CH-	39	50	:	50
C$_2$F$_5$	n-C$_6$H$_{13}$	Et	58	59	:	41
		n-Pr	49	64	:	36
		i-Pr	51	71	:	29
		n-C$_6$H$_{13}$	51	52	:	48
		(E)-MeCH=CH-	66	57	:	43
		Ph	51	60	:	40
C$_2$F$_5$	Ph	n-Pr	49	41	:	59
		(E)-MeCH=CH-	34	41	:	59
C$_2$F$_5$	c-C$_6$H$_{11}$	n-Pr	72	41	:	59
		(E)-MeCH=CH-	38	44	:	56

(Reproduced with permission from Ishihara, T., *et al. J. Org. Chem.* **1990**, *55*, 3107)

Note

    For the preparation of the starting enol phosphate, see page 223 of this volume.   The relative stereochemical nomenclature, *threo* and *erythro*, proposed by Noyori and co-workers was employed and the *threo* isomer has an *anti* relationship between 2-F and 3-OH groups. For the definition of this nomenclature, see Noyori, R., *et al.* *J. Am. Chem. Soc.* **1983**, *105*, 1598.

Ishihara and his co-workers noted the intermediary formation of an aluminum enolate which, in the absence of an appropriate aldehyde, was found to afford 1,1,1,2-tetrafluorononan-3-one in 91% yield in the example described above.

Ishihara, T., *et al.* *J. Org. Chem.* **1990**, *55*, 3107.

5.2.2 Wittig Reactions

Ethyl 4,4,4-trifluoro-3-(trifluoromethyl)but-2-enoate

a) Molecular sieves 4 Å,   b) Ph$_3$P=CHCO$_2$Et / Et$_2$O

Synthetic method

Under a nitrogen atmosphere, a mixture of hexafluoroacetone trihydrate (32 mL, 210 mmol) and molecular sieves 4 Å (61 g) was heated at 127 ℃ to generate anhydrous hexafluoroacetone. The gaseous hexafluoroacetone was dried and introduced to an Et$_2$O (25 mL) solution of (ethoxy-carbonylmethylene)triphenylphosphorane (10 g, 29 mmol) chilled at -78 ℃. The reaction mixture was stirred at -78 ℃ for 3 h, then warmed to room temperature overnight. After addition of pentane (20 mL), the white precipitate (Ph$_3$P=O) was filtered off. Then the materials with lower boiling point (Et$_2$O and pentane) was carefully distilled off. The bulb-to-bulb distillation of the residual liquid gave 5.8 g (24 mmol) of ethyl 4,4,4-trifluoro-3-(trifluoromethyl)but-2-enoate in 84% yield (based on the phosphorane used) as a clear oil.

Spectral data

bp. 124-127 ℃; Rf 0.46 (CHCl$_3$:MeOH=19:1); ^{1}H NMR (CDCl$_3$): δ 1.32 (3 H, t, J=7.1 Hz), 4.32 (1 H, q, J=7.1 Hz), 6.88 (1 H, s); ^{19}F NMR (CDCl$_3$): δ -62.02 (3 F, q, J=6.7 Hz), -66.52 (3 F, q, J=6.7 Hz); IR (neat): ν 2980, 1780, 1670, 1395, 1295, 1220, 1175 cm^{-1}.

Note

Physical properties except for ^{1}H NMR data were extracted from a previous report by Marshall's group. See Marshall, G. R., *et al.* *J. Med. Chem.* **1981**, *24*, 1043. For the preparation of the corresponding α,β-unsaturated imines, see Haas, A., *et al.* *J. Fluorine Chem.* **1997**, *83*, 133.

For the traditional preparation of anhydrous hexafluoroacetone, see the following procedure (in *Syntheses of Fluoroorganic Compounds*; Knunyants, I. L.; Yakobson, G. G, Eds.; Springer-Verlag: Berlin, 1985; p. 40).

Anhydrous hexafluoroacetone was prepared in a 1-liter three-necked flask provided with a stirrer, a dropping funnel, and a reflux condenser connected to a gas-washing bottle with H$_2$SO$_4$, a column packed with P$_2$O$_5$, a trap for unchanged hexafluoroacetone hydrate cooled to -20 ℃, and a trap for hexafluoroacetone cooled to -78 ℃. All parts of the above apparatus were thoroughly dried. In a flask was placed 300 g of conc H$_2$SO$_4$, then 250 mL of crude hexafluoroacetone hydrate was dropped in with vigorous stirring, at such a rate that all the liberated gas was condensed in the trap.   The flask was then slowly heated to 135-140 °C for 2-3 h.   The yield of anhydrous hexafluoroacetone was 130-140 g (bp. -28 °C).

Nishino, N., *et al. Bull. Chem. Soc. Jpn.* **1996**, *69*, 1383.

Ethyl 2-fluoro-3-(trifluoromethyl)hex-2-enoate

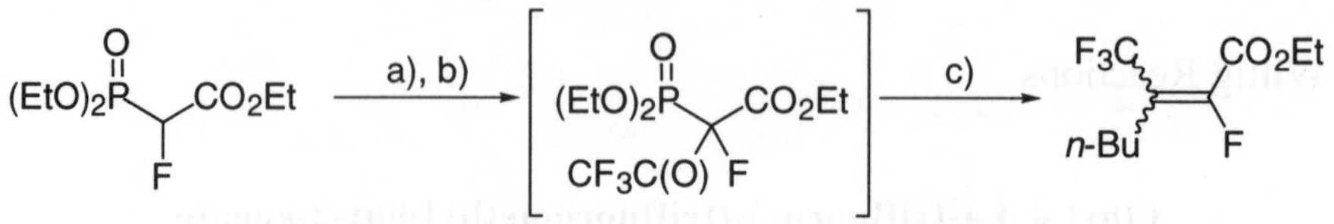

a) *n*-BuLi / THF,   b) (CF$_3$CO)$_2$O,   c) *n*-BuLi

Synthetic method

n-BuLi (2 mmol in 1.5 mL of hexane) was added dropwise over 30 min to a stirred solution of triethyl 2-fluoro-2-phosphonoacetate (2 mmol) in absolute THF (15 mL) at -78 °C under nitrogen. The mixture was stirred at -78 °C for a further 0.5 h, and trifluoroacetic anhydride (2 mmol) was added to it in one portion.   Stirring was continued at -78 °C for 1 h after which *n*-butyllithium (2 mmol) was added dropwise to the mixture, which was stirred and allowed to warm to room temperature within 4 h.   The reaction mixture was poured into H$_2$O (30 mL) and the H$_2$O layer was extracted with Et$_2$O (3x15 mL).   The combined organic layers were washed with brine (3x10 mL) and H$_2$O (3x10 mL) and dried over MgSO$_4$.   Evaporation of the solvent gave a residue which was purified by column chromatography (petroleum ether (60-90 °C):AcOEt=99:1) to give ethyl 2-fluoro-3-(trifluoromethyl)hex-2-enoate in 58% yield as an 80:20 (*E*):(*Z*) mixture.

Spectral data

bp. 50 °C/2 mmHg; ^{1}H NMR (CDCl$_3$): δ 0.92 (3 H, t, *J*=7.1 Hz), 1.20-1.54 (7 H, m), 2.26-2.44 (2 H, m; (*E*)-isomer), 2.56-2.66 (2 H, m; (*Z*)-isomer), 4.22-4.38 (2 H, m); ^{19}F NMR (CDCl$_3$): δ -18.2 (3 F, d, *J*=10 Hz; (*E*)-isomer), -15.8 (3 F, d, *J*=24 Hz; (*Z*)-isomer), 33.0-34.9 (1 F, m; (*E*), (*Z*)-mixture) from CF$_3$CO$_2$H; IR (neat): ν 2960, 1750, 1670, 1300, 1080 cm^{-1}.

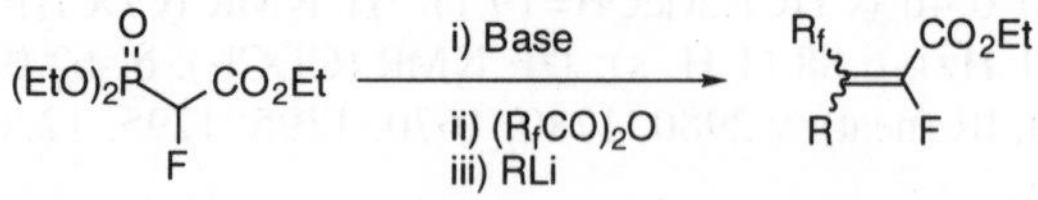

R$_f$	R	Base	Yield (%)	(*E*):(*Z*)	
CF$_3$	*n*-Bu	*n*-BuLi	58	80 :	20
CF$_3$	*n*-BuC≡C	*n*-BuLi	75	94 :	6
CF$_3$	*n*-BuC≡C	LDA	15	86 :	14
CF$_3$	*n*-BuC≡C	NaH	51	75 :	25
CF$_3$	2-thienyl	*n*-BuLi	80	0 :	100
C$_2$F$_5$	PhC≡C	*n*-BuLi	71	100 :	0
C$_2$F$_5$	2-thienyl	*n*-BuLi	82	0 :	100
n-C$_3$F$_7$	2-furyl	*n*-BuLi	61	100 :	0

(Reproduced with permission from Shen, Y.-C.; Ni, J.-H. *J. Org. Chem.* **1997**, *62*, 7260)

Shen, Y.-C.; Ni, J.-H. *J. Org. Chem.* **1997**, *62*, 7260.

5.2.3 Alkylations

3-(1,1,1,3,3,3-Hexafluoroprop-2-yloxy)-1-phenylprop-1-ene, 1,1,1,3,3-Pentafluoro-2,2-dihydroxy-6-phenylhex-5-ene

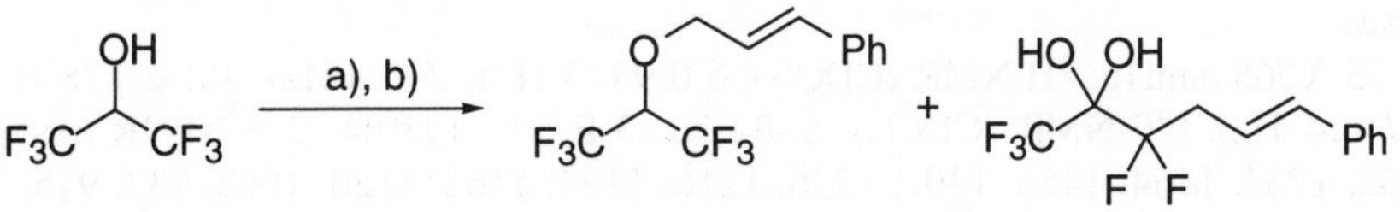

a) n-BuLi / THF,   b) (E)-PhCH=CHCH$_2$OAc, Pd(OAc)$_2$, PPh$_3$ / THF

Synthetic method

To a solution of 1,1,1,3,3,3-hexafluoropropan-2-ol (0.63 mL, 6.0 mmol) in THF, n-butyllithium (1.6 M in hexane; 7.50 mL, 12.0 mmol) was added dropwise at 0 °C and stirred for 15 min at room temperature.   The mixture was added to a boiling mixture of palladium(II) acetate (56 mg, 0.25 mmol) and triphenylphosphine (0.262 g, 1.0 mmol) in THF (25 mL).   To the resulting boiling mixture, cinnamyl acetate (0.839 mL, 5.0 mmol) was added and the mixture was refluxed for 2.5 h.   The mixture was poured into 1 N HCl and extracted with AcOEt.   The extract was washed with brine, dried over MgSO$_4$, and condensed *in vacuo*   The residue was chromatographed on silica gel to give 0.923 g of 3-(1,1,1,3,3,3-hexafluoroprop-2-yloxy)-1-phenylprop-1-ene and 0.451 g of 1,1,1,3,3-pentafluoro-2,2-dihydroxy-6-phenylhex-5-ene in 65 and 32% yields, respectively.

Spectral data

3-(1,1,1,3,3,3-Hexafluoroprop-2-yloxy)-1-phenylprop-1-ene

^{1}H NMR (CCl$_4$): δ 4.08 (1 H, sept, J=6.0 Hz), 4.45 (2 H, d, J=5.5 Hz), 6.10 (1 H, dt, J=16.0, 5.5 Hz), 6.61 (1 H, d, J=16.0 Hz), 7.26 (5 H, s).

1,1,1,3,3-Pentafluoro-2,2-dihydroxy-6-phenylhex-5-ene

^{1}H NMR (CCl$_4$): δ 2.96 (2 H, dt, J=19.0, 6.0 Hz), 4.55-5.45 (2 H, br s), 6.11 (1 H, dt, J=16.0, 6.0 Hz), 6.51 (1 H, d, J=16.0 Hz), 7.19 (5 H, s).

Shimizu, I.; Ishii, H. *Tetrahedron* **1994**, *50*, 487.

1,1,1,2,2-Pentafluorononan-3-one, 3-(Diethylphosphoryloxy)-1,1,1,2-tetrafluoronon-2-ene

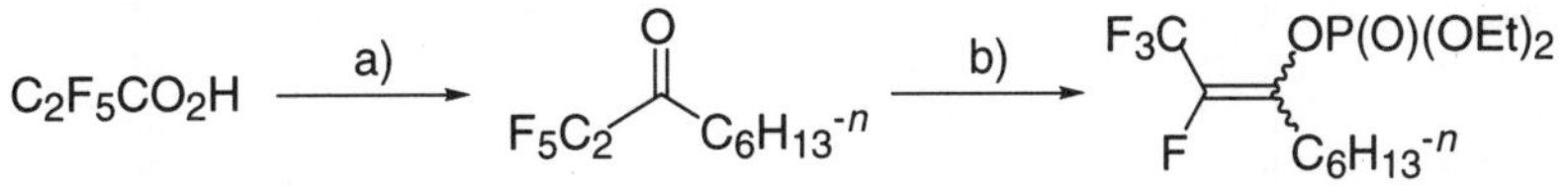

a) n-C$_6$H$_{13}$MgBr / Et$_2$O,   b) HOP(OEt)$_2$, NaH / THF

Synthetic method

1,1,1,2,2-Pentafluorononan-3-one

Hexylmagnesium bromide was prepared from magnesium tunings (3.6 g, 150 mmol) and 1-bromohexane (20.6 g, 125 mmol) in Et$_2$O (50 mL).   A solution of 2,2,3,3,3-pentafluoropropionic acid (8.2 g, 50 mmol) in Et$_2$O (15 mL) was gradually added under argon to the Grignard reagent at

such a rate that the temperature did not rise above -10 ℃.   After being stirred for 12 h at room temperature and heated to reflux for 1 h, the reaction mixture was hydrolyzed with 6 N HCl aq. (50 mL) below 0 ℃ and stirred for 1 h at room temperature.   The resultant mixture was extracted with Et$_2$O (3x50 mL) and the ethereal extracts were washed with sat. NaHCO$_3$ aq. (2x50 mL) and brine (2x50 mL), dried over anhydrous Na$_2$SO$_4$, filtered, and concentrated *in vacuo*.   The residue was distilled under reduced pressure to give 9.7 g of 1,1,1,2,2-pentafluorononan-3-one in 84% yield.
Spectral data

bp. 75 ℃/65 mmHg; ^{1}H NMR (CDCl$_3$): δ 0.94 (3 H, t, J=5.4 Hz), 1.1-2.1 (8 H, m), 2.75 (2 H, t, J=6.4 Hz); ^{19}F NMR (CDCl$_3$): δ -82.50 (3 F, s), -123.84 (2 F, s); IR (neat): ν 2952, 2930, 2852, 1744, 1464, 1456, 1402, 1336, 1216, 1194, 1162, 1120, 1062, 932, 915, 708 cm^{-1}.

Synthetic method

3-(Diethylphosphoryloxy)-1,1,1,2-tetrafluoronon-2-ene

To a suspension of NaH (60% in mineral oil, 0.48 g, 12 mmol) in THF (50 mL) was added diethyl phosphite (1.52 g, 11 mmol) under argon at 0 ℃, and stirred for 0.5 h at room temperature. To the mixture was added 1,1,1,2,2-pentafluorononan-3-one (2.82 g, 10 mmol) below 0 ℃. After stirring for 2 h at 0 ℃, the reaction mixture was poured into a mixture of ice and sat. NaHCO$_3$ aq. (20 mL), and extracted with Et$_2$O (3x20 mL).   The combined extracts were dried over anhydrous Na$_2$SO$_4$, filtered and concentrated *in vacuo*.   Purification by silica gel column chromatography gave 3.11 g of 3-(diethylphosphoryloxy)-1,1,1,2-tetrafluoronon-2-ene in 89% yield as a 73:27 mixture of (*E*)- and (*Z*)-isomers.
Spectral data

^{1}H NMR (CDCl$_3$): δ 0.84 (3 H, t, J=8.0 Hz), 1.34 (6 H, t, J=7.4 Hz), 1.1-1.9 (8 H, m), 2.2-2.7 (2 H, m), 4.16 (4 H, dq, J=6.8, 6.8 Hz); IR (neat): ν 2982, 2957, 2922, 2866, 2852, 1706, 1476, 1465, 1454, 1440, 1389, 1370, 1354, 1283, 1217, 1185, 1139, 1092, 1026, 963, 887, 821, 794, 753, 717 cm^{-1}.

(*E*)-isomer
^{19}F NMR (CDCl$_3$): δ -65.9 (3 F, d, J=11.0 Hz), -147.1 (1 F, m).
(*Z*)-isomer
^{19}F NMR (CDCl$_3$): δ -67.3 (3 F, d, J=9.8 Hz), -169.2 (1 F, m).

$$R_fCF_2CO_2H \xrightarrow{RMgX} \underset{R_fCF_2}{\overset{O}{\parallel}} R \xrightarrow{NaOP(OEt)_2} \underset{F}{\overset{R_f}{>}}=\underset{R}{\overset{OP(O)(OEt)_2}{<}}$$

R$_f$	R	Yield (%)		(*E*):(*Z*)	
		Ketone	Enol phosphate		
n-C$_3$F$_7$	n-Bu	84	92	64 :	36
n-C$_3$F$_7$	i-Bu	78	72	68 :	32
n-C$_3$F$_7$	n-C$_6$H$_{13}$	82	88	65 :	35
n-C$_3$F$_7$	c-C$_6$H$_{11}$	80	76	35 :	65
C$_2$F$_5$	Ph	87	92	15 :	85
n-C$_3$F$_7$	Ph	76	79	21 :	79
n-C$_3$F$_7$	4-MeO-C$_6$H$_4$-	84	73	18 :	82

Ishihara, T., *et al. Chem. Lett.* **1988**, 819.
The detailed experimental procedures were obtained from the following dissertation.
See Kuroboshi, M. PhD dissertation, Kyoto University (1989).

5.2.4 *Anti*-Michael Reactions

**Methyl (*S*)- and (*R*)-4,4,4-trifluoro-3-(trifluoromethyl)-
2-[(*R*)-1-phenylethylamino]butanoate,
(*R*)-2-Amino-4,4,4-trifluoro-3-(trifluoromethyl)butanoic acid (Hexafluorovaline)**

a) (*R*)-1-Phenylethylamine / MeOH
b) HCl / Et$_2$O
c) BBr$_3$ / CH$_2$Cl$_2$

Synthetic method

Methyl (*S*)- and (*R*)-4,4,4-trifluoro-3-(trifluoromethyl)-2-[(*R*)-1-phenylethylamino]butanoate
(*R*)-1-Phenylethylamine (3.884 g, 32.1 mmol) was slowly added to a solution of methyl
4,4,4-trifluoro-3-(trifluoromethyl)but-2-enoate (6.79 g, 30.56 mmol) in dry MeOH (40 mL) at -78
°C. After warming to room temperature, the mixture was stirred for 1 h and the yellow liquid,
obtained after removal of the solvent, was chromatographed (hexane:acetone=19:1) to give 8.61 g
of a 52:48 mixture of the diastereomers, methyl (*S*)- and (*R*)-4,4,4-trifluoro-3-(trifluoromethyl)-2-
[(*R*)-1-phenylethylamino]butanoate, in 81% yield.

A solution of this mixture (8.35 g, 24.3 mmol) in Et$_2$O (100 mL) was treated with Et$_2$O
saturated with HCl (15 mL). The crystalline precipitate (4.67 g, 9:1 mixture of *S:R*
diastereomers) was recrystallized (MeOH:Et$_2$O=1:4) to give 3.15 g of hydrochloride salt of methyl
(*S*)-4,4,4-trifluoro-3-(trifluoromethyl)-2-[(*R*)-1-phenylethylamino]butanoate with 99.6% de. An
additional product was isolated from the mother liquor (0.587 g, 98.6% de). The total yield of
hydrochloride salt of methyl (*S*)-4,4,4-trifluoro-3-(trifluoromethyl)-2-[(*R*)-1-phenylethylamino]-
butanoate was then 3.732 g (32.6%). The total yield of the other diastereomer was 4.33 g (37.8%,
97% purity).

Spectral data

Methyl (*S*)-4,4,4-trifluoro-3-(trifluoromethyl)-2-[(*R*)-1-phenylethylamino]butanoate
[α]$_D^{25}$ +13.68 (*c* 6.731, MeOH); ^{1}H NMR (CD$_3$OD): δ 1.81 (3 H, d, *J*=7.0 Hz), 3.56 (3 H,
s), 4.61 (1 H, d, *J*=2 Hz), 4.66 (1 H, q), 4.97 (2 H, br s), 5.07 (1 H, dsept., *J*=11, 2 Hz), 7.40-
7.48 (3 H, m), 7.57-7.60 (2 H, m); ^{13}C NMR (CD$_3$OD): δ 19.3, 54.0, 55.7, 63.3, 123.2 (q,
J=283 Hz), 123.6 (q, *J*=283 Hz), 130.1, 130.3, 131.2, 136.1, 166.0.

Methyl (*R*)-4,4,4-trifluoro-3-(trifluoromethyl)-2-[(*R*)-1-phenylethylamino]butanoate
[α]$_D^{22}$ +39.9 (*c* 6.65, MeOH); ^{1}H NMR (CD$_3$OD): δ 1.83 (3 H, d, *J*=7.0 Hz), 3.90 (3 H,
s), 4.08 (1 H, d, *J*=2 Hz), 4.58 (1 H, q), 4.98 (2 H, br s), 5.05 (1 H, dsept., *J*=8, 2 Hz), 7.49-
7.51 (3 H, m), 7.53-7.63 (2 H, m); ^{13}C NMR (CD$_3$OD): δ 22.7, 56.2, 57.5, 64.8, 125.8 (q,
J=282 Hz), 125.9 (q, *J*=282 Hz), 132.2, 133.3, 133.9, 138.7, 168.4.

Synthetic method

(R)-2-Amino-4,4,4-trifluoro-3-(trifluoromethyl)butanoic acid

Methyl (R)-4,4,4-trifluoro-3-(trifluoromethyl)-2-[(R)-1-phenylethylamino]butanoate was prepared from a cold solution of its hydrochloride salt (2.88 g) in 1 M Na$_2$CO$_3$ aq., extracted with Et$_2$O (4x50 mL) and dried over Na$_2$SO$_4$ (2.401 g).   A solution of boron tribromide (25.4 mL, 1.15 M in CH$_2$Cl$_2$, 29.2 mmol) was added to this free base (2.00 g, 5.83 mmol) dissolved in CH$_2$Cl$_2$ (50 mL) at -10 °C.   After stirring the mixture at 35 °C for 18 h it was hydrolyzed with ice (50 mL) and extracted with H$_2$O (5x30 mL).   H$_2$O was evaporated off to give a brownish residue (5.99 g).   A solution of this residue (2.42 g) in H$_2$O (20 mL) was treated with cation exchange resin (Dowex 50 WX8, H$^+$ form, 60 g).   After rinsing the column with H$_2$O (500 mL) the amino acid was eluted with 1 M HCl, the fractions positive to ninhydrin were collected and evaporated to dryness to give 0.571 g of (R)-2-amino-4,4,4-trifluoro-3-(trifluoromethyl)butanoic acid in 89.8% yield.

Spectral data

[α]$_D^{22}$ +3.22±0.05 (c 2, 5% HCl); mp. 163-165 °C (petroleum ether:EtOH); Rf 0.19 (EtOH: 25% NH$_3$ aq.:CH$_2$Cl$_2$=19:1:20); ^{1}H NMR (CD$_3$OD): δ 4.15 (1 H, br s), 4.61 (1 H, m, J=7 Hz), 4.96 (3 H, br s); ^{13}C NMR (CD$_3$OD): δ 50.9, 124.1 (q, J=279 Hz), 124.8 (q, J=279 Hz), 168.8.

Note

For the preparation of the starting α,β-unsaturated ester, see page 221 of this volume.

anti-Michael addition is classified as the reaction in which a nucleophile and an electrophile are added at the α and β positions of the carbonyl group, respectively, instead of the well-known 1,4- (Michael addition) or 1,2-addition processes (see below).   In the present case, defluorinated products were obtained possibly due to the favorable interaction of fluorine and E (usually metal) or the direct S$_N$2' type of reaction by a nucleophile (Nu).

Preference for this interesting reaction course by 4,4,4-trifluoro-3-(trifluoromethyl)but-2-enoate was supported by semiempirical molecular orbital calculations.   See Yamazaki, T., *et al.* *Rev. Heteroatom Chem.* **1996**, *14*, 165.

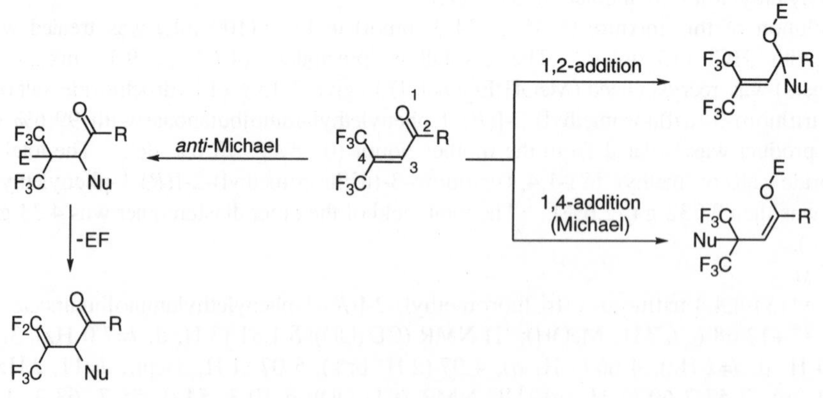

See also Eberle, M. K., *et al.* *Helv. Chim. Acta* **1998**, *81*, 182.   Keese *et al.* recently synthesized the corresponding chiral hexafluoroleucine.   See Keese, R., *et al.* *Helv. Chim. Acta* **1998**, *81*, 174.

Keese, R.; Hinderling, C. *Synlett* **1996**, 695.

Ethyl 4,4-difluoro-3-(trifluoromethyl)but-3-enoate

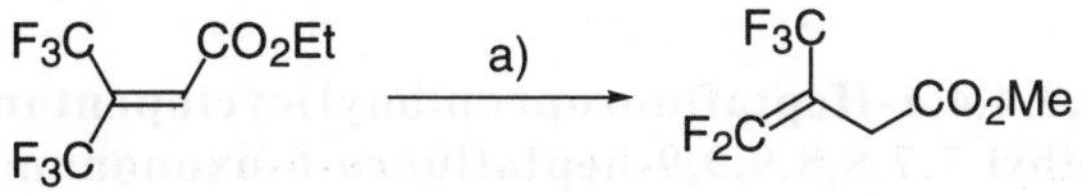

a) LiAlH(OEt)$_3$ / Et$_2$O

Synthetic method

In a flame-dried two-necked flask under an argon atmosphere a solution of EtOH (11.61 g, 252 mmol) in Et$_2$O (50 mL) was slowly added within 1 h to a suspension of lithium aluminum hydride (3.19 g, 84 mmol) in Et$_2$O (80 mL) cooled to 0 °C.   The resulting mixture was then stirred for 1 h.   29.5 mL of a lithium triethoxyaluminohydride solution thus obtained was added by portions (3 mL) to a solution of ethyl 4,4,4-trifluoro-3-(trifluoromethyl)but-2-enoate (5 g, 21 mmol) in Et$_2$O (40 mL) cooled to -78 °C.   After complete addition, the mixture was stirred at this temperature, then MeOH (1.1 mL) was added.   The mixture was allowed to warm to -10 °C, then H$_2$O (1.1 mL), 15% NaOH aq. (1.1 mL), and again H$_2$O (3.3 mL) was added successively.   The hydrolysis was complete when a white precipitate appeared.   The mixture was filtered, and the aluminum salt was washed with Et$_2$O (2x30 mL).   The ethereal solution was dried over Na$_2$SO$_4$. The solvent was distilled off and the residue was distilled under reduced pressure to give 3.56 g (16.3 mmol) of ethyl 4,4-difluoro-3-(trifluoromethyl)but-3-enoate in 77% yield.

Spectral data

bp. 80 °C/185 mmHg; ^{1}H NMR (CD$_3$OD): δ 1.33 (3 H, t, *J*=7.15 Hz), 3.22 (2 H, dd, *J*=2.1, 2.05 Hz), 4.25 (2 H, q, *J*=7.15 Hz); ^{19}F NMR (CD$_3$OD): δ -61.6 (3 F, dd, *J*=20.6, 11.5 Hz), -75.16 (1 F, dtq, *J*=20.6, 14.8, 2.05 Hz), -78 (1 F, dtq, *J*=14.8, 11.5, 2.1 Hz); IR (CCl$_4$): ν 2970-2850, 1745, 1365, 1325, 1285, 1250, 1190, 1170, 1130, 1020 cm^{-1}.

Note

Wakselman and his co-workers also tried the reaction of bistrifluoromethylated acrylate with such nucleophiles as sodium malonate, sodium thiophenoxide, methyllithium, or piperidine, and all examples but the last nucleophile were found to furnish the above type of *anti*-Michael products (a nucleophile attacked at the 2 position, followed by the elimination of fluoride from either CF$_3$ groups).   In the case of piperidine, on the other hand, an adduct was isolated as shown below. This was explained as the result of a similar *anti*-Michael addition (detectable by ^{19}F NMR; after 2 h, no *anti*-Michael adduct was observed), followed by the addition of HF from the *in situ*-generated piperidine-HF salt.

For the preparation of the starting α,β-unsaturated ester, see page 221 of this volume.

The original procedure described the usage of 2.92 g of EtOH (63 mmol), but 3 equiv of ethanol is apparently required for the formation of LiAlH(OEt)$_3$ from LiAlH$_4$.   So, the amount of EtOH in the above procedure was recalculated by the authors.

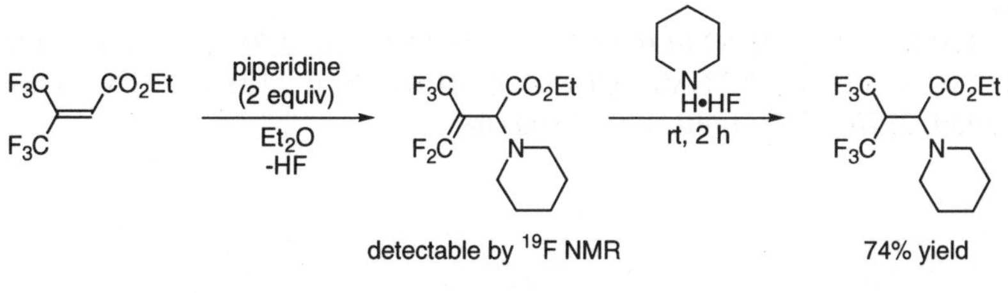

detectable by ^{19}F NMR   74% yield

Molines, H., *et al. J. Org. Chem.* **1992**, *57*, 5530.

5.2.5 Miscellaneous

2-(2,2,3,3,4,4,4-Heptafluoropropionyl)cyclopentan-1-one, Methyl 7,7,8,8,9,9,9-heptafluoro-6-oxononanate

$n\text{-}C_3F_7CO_2Me$ $\xrightarrow{\text{a), b), c), d)}}$ [2-(heptafluoropropionyl)cyclopentan-1-one structure] $C_3F_7\text{-}n$ $\xrightarrow{\text{e)}}$ $n\text{-}C_3F_7$—[chain]—CO_2Me

a) NaOMe / Et$_2$O, b) Cyclopentanone / Et$_2$O,
c) AcOH, Cu(OAc)$_2$ / H$_2$O, d) 15% H$_2$SO$_4$ aq. / Et$_2$O, e) MeOH

Synthetic method
2-(2,2,3,3,4,4,4-Heptafluoropropionyl)cyclopentan-1-one
Methyl 2,2,3,3,4,4,4-heptafluoropropionate (0.1 mol) was added dropwise over 30 min to a suspension of NaOMe (5.94 g, 0.11 mol), vigorously stirred and cooled to 0 °C by an ice bath, in dry Et$_2$O (100 mL) under nitrogen. To this mixture, maintained at 0 °C, was added dropwise, over 30 min, a solution of cyclopentanone (0.1 mol) in dry Et$_2$O (20 mL). Stirring was continued for 23 h, and the mixture was slowly allowed to reach 20 °C. The mixture was then treated with a solution of glacial AcOH (6.6 mL, 0.11 mol) in H$_2$O (30 mL), followed by the addition of copper(II) acetate (10 g) dissolved in H$_2$O (150 mL). The organic layer was separated, then washed with H$_2$O (3x20 mL) and dried (Na$_2$SO$_4$). The solvent was removed under reduced pressure and the residue was dried under vacuum. The copper chelate obtained was dissolved in Et$_2$O (100 mL), treated with 15% H$_2$SO$_4$ aq. (100 mL), and stirred at room temperature for 20 min. The organic layer was separated, washed with H$_2$O (20 mL) and dried (CaCl$_2$). The Et$_2$O layer was then concentrated under reduced pressure and the resulting oil was purified by distillation to give 2-(2,2,3,3,4,4,4-heptafluoropropionyl)cyclopentan-1-one in 57% yield.
Spectral data
bp. 45 °C/30 mmHg; ^{1}H NMR (CDCl$_3$): δ 2.01 (2 H, m), 2.53 (2 H, t, J=7.9 Hz), 2.80 (2 H, m), 13.47 (1 H, br s); ^{19}F NMR (CDCl$_3$): δ -81.3 (3 F), -119.1 (2 F, q), -122.5 (2 F, s); IR (neat): ν 1686, 1628, 1300-1100 cm^{-1}.

Synthetic method
Methyl 7,7,8,8,9,9,9-heptafluoro-6-oxononanate
A solution of 2-(2,2,3,3,4,4,4-heptafluoropropionyl)cyclopentan-1-one (0.01 mol) and MeOH (5 mL) was heated at 70 °C and stirred for 14 h. The excess MeOH was then evaporated and the crude product was distilled under vacuum in the presence of a small amount of CaCl$_2$ to afford methyl 7,7,8,8,9,9,9-heptafluoro-6-oxononanate in 93% yield.
Spectral data
bp. 57 °C/0.8 mmHg; ^{1}H NMR (CDCl$_3$): δ 1.69 (4 H, m), 2.36 (2 H, t, J=6.9 Hz), 2.78 (2 H, t, J=6 Hz), 3.68 (3 H, s); ^{19}F NMR (CDCl$_3$): δ -81.0 (3 F, t), -121.7 (2 F, q), -127.1 (2 F, s); IR (neat): ν 2959, 2870, 1757, 1740, 1300-1100 cm^{-1}.

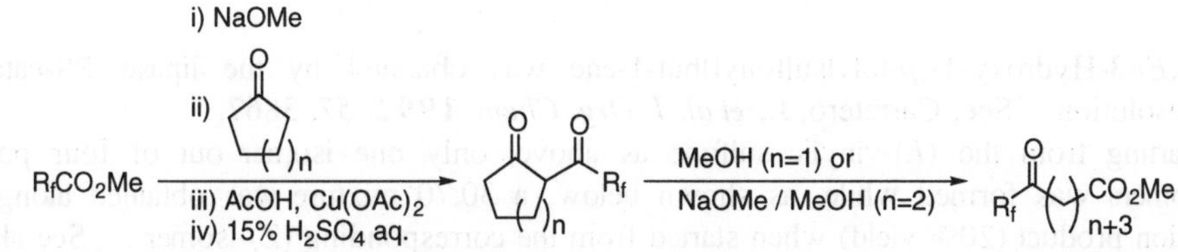

i) NaOMe

ii)

R_fCO_2Me → (via cyclopentanone) → diketone → (MeOH (n=1) or NaOMe / MeOH (n=2)) → ketoester R_f—CO$_2$Me

iii) AcOH, Cu(OAc)$_2$
iv) 15% H$_2$SO$_4$ aq.

R_f	n	Yield (%) diketone	Yield (%) ketoester	R_f	n	Yield (%) diketone	Yield (%) ketoester
n-C$_5$F$_{11}$	1	70	92	n-C$_3$F$_7$	2	56	84
n-C$_7$F$_{15}$	1	79	90	n-C$_5$F$_{11}$	2	64	85
n-C$_6$F$_{13}$	1	74	90	n-C$_6$F$_{13}$	2	64	80
n-C$_8$F$_{17}$	1	68	90	n-C$_7$F$_{15}$	2	64	76
				n-C$_8$F$_{17}$	2	65	89

(Reproduced with permission from Trabelsi, H.; Cambon, A. *Synthesis* **1992**, 315)

Trabelsi, H.; Cambon, A. *Synthesis* **1992**, 315.

(2R, 3S, 4S) -2-Fluoro-2-(trifluoromethyl)-3-(p-toluenelsulfonyl)methylpentan-4-olide

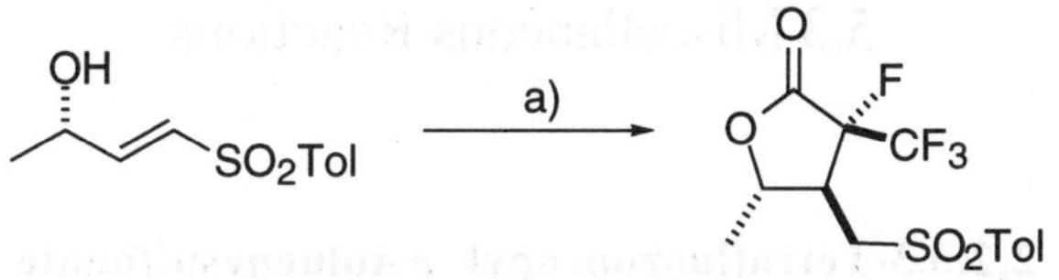

a) CF$_3$CHFCF$_2$NEt$_2$ / CH$_2$Cl$_2$

Synthetic method

(S,E)-3-Hydroxy-1-(p-toluenesulfonyl)but-1-ene (99% ee, 150 mg, 0.662 mmol) was dissolved in dry CH$_2$Cl$_2$ (4 mL) under a nitrogen atmosphere. After a solution of hexafluoro-propene-diethylamine adduct (PPDA; 0.19 mL, *ca.* 1.0 mmol) in dry CH$_2$Cl$_2$ (1 mL) was dropwise added under ice-cooling, the resulting solution was stirred at room temperature for 4.5 h and then H$_2$O (4 mL) was added to quench the reaction. The resulting mixture was extracted with CH$_2$Cl$_2$ (3x5 mL). The organic layers were combined, dried over anhydrous MgSO$_4$, and evaporated *in vacuo*. The residue was chromatographed on silica gel (hexane:AcOEt=2:1) to give 171 mg of (2R,3S,4S)-2-fluoro-2-(trifluoromethyl)-3-(p-toluenelsulfonyl)methylpentan-4-olide in 73% yield as colorless crystals. The enantiomeric excess was determined by HPLC using CHIRALPAK OD column (hexane:i-PrOH=9:1).

Spectral data

$[\alpha]_D^{19}$ 9.00 (c 1.00, CHCl$_3$); mp. 84-85 °C (103-104 °C for the corresponding racemic form); ^{1}H NMR (CDCl$_3$): δ 1.73 (3 H, d, J=6.2 Hz), 2.49 (3 H, s), 3.13-3.32 (1 H, dm, J=18.1 Hz), 3.23 (1 H, dd, J=13.1, 1.3 Hz), 3.55 (1 H, diffused d, J=10.3 Hz), 4.67 (1 H, quintet-like, J≈6.5 Hz), 7.43 (2 H, d, J=7.9 Hz), 7.82 (2 H, d, J=8.4 Hz); ^{13}C NMR (CDCl$_3$): δ 20.2, 21.7, 44.7, 52.4, 78.1, (d, J=5.9 Hz), 91.6 (dq like, J=209.5, 33.4 Hz), 120.6 (dq like, J=286.1, 31.6 Hz), 128.1, 130.4, 135.5, 146.1, 163.5 (d, J=22.0 Hz); IR (KBr): ν 1803, 1326, 1207, 1173, 1153, 1057 cm^{-1}.

Note

(S,E)-3-Hydroxy-1-(p-tolylsulfonyl)but-1-ene was obtained by the lipase PS-catalyzed optical resolution.   See, Carretero, J., *et al. J. Org. Chem.* **1992**, *57*, 3867.

Starting from the (E)-vinylic sulfone as above, only one isomer out of four possible diastereomers was formed while, as shown below, a 30:70 mixture was obtained along with fluorination product (20% yield) when started from the corresponding (Z)-isomer .   See also the following work: Ogura, K., *et al. Heterocycles* **1998**, *48*, 15.

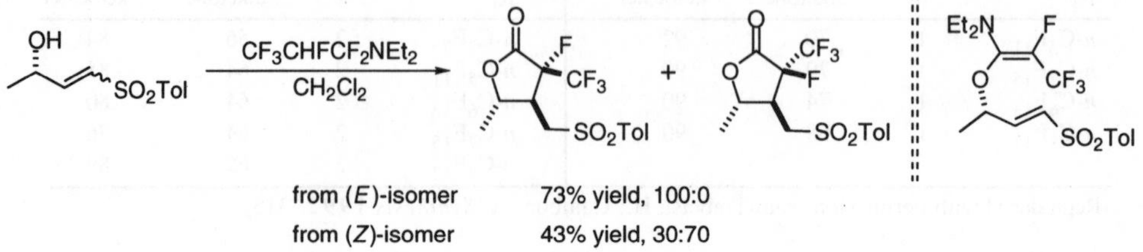

from (E)-isomer   73% yield, 100:0
from (Z)-isomer   43% yield, 30:70

Ogura, K., *et al. Tetrahedron Lett.* **1997**, *38*, 5173.

5.3 Miscellaneous Reactions

2,2,3,3-Tetrafluoropropyl *p*-toluenesulfonate

$$CHF_2CF_2CH_2OH \xrightarrow{a} CHF_2CF_2CH_2OTs$$

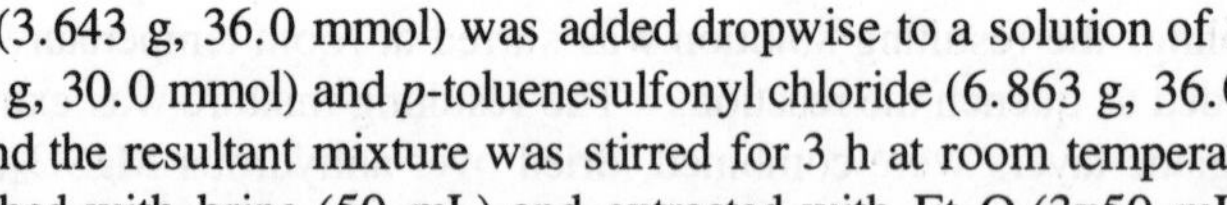

a) TsCl, Et$_3$N / CH$_2$Cl$_2$

Synthetic method

Triethylamine (3.643 g, 36.0 mmol) was added dropwise to a solution of 2,2,3,3-tetrafluoro-propan-1-ol (3.961 g, 30.0 mmol) and *p*-toluenesulfonyl chloride (6.863 g, 36.0 mmol) in CH$_2$Cl$_2$ (20 mL) at 0 °C, and the resultant mixture was stirred for 3 h at room temperature.   The reaction mixture was quenched with brine (50 mL) and extracted with Et$_2$O (3x50 mL).   The combined organic layer was washed with 25% NH$_3$ aq. (3x30 mL) and brine (3x30 mL) and dried over anhydrous Na$_2$SO$_4$.   The solvents were removed at reduced pressure to leave the residual oil, which was subjected to column chromatography (benzene) to give 2,2,3,3-tetrafluoropropyl *p*-toluenesulfonate (8.323 g) in 97% yield.

Spectral data

Rf 0.63 (benzene); ^{1}H NMR (CDCl$_3$): δ 2.39 (3 H, s), 4.22 (2 H, tt, *J*=12.0, 1.3 Hz), 5.69 (1 H, tt, *J*=52.0, 4.1 Hz), 7.18 and 7.59 (4 H, AB quartet, *J*=8.0 Hz); ^{19}F NMR (CDCl$_3$): δ -45.6 (2 F, dt, *J*=12.0, 4.1 Hz), -59.5 (2 F, dt, *J*=52.0, 1.3 Hz) from CF$_3$CO$_2$H; IR (neat): ν 1379, 1179 cm^{-1}.

Funabiki, K., *et al. Chem. Lett.* **1994**, 1075.

(E)-1,1,1,2,3-Pentafluorodec-2-en-4-ol

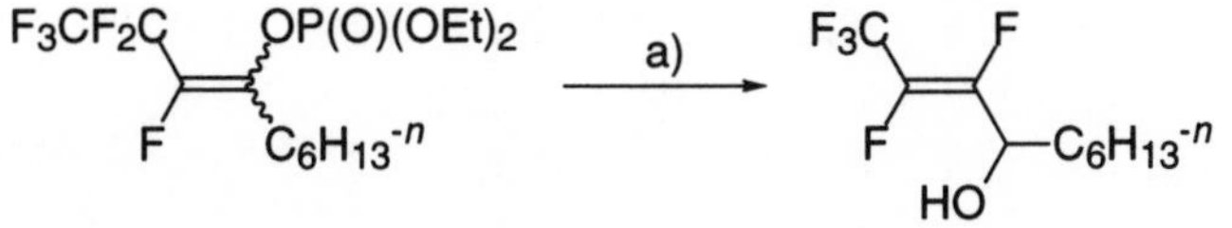

a) DIBAL-H / THF

Synthetic method

A hexane solution (1 *M*) of DIBAL-H (5.0 mL, 5.0 mmol) was added dropwise to 4-(diethyl-phosphoryloxy)-1,1,1,2,2,3-hexafluorodec-3-ene (0.400 g, 1.0 mmol) in THF (5 mL) at 0 °C. This mixture was heated to gentle reflux.   After 3 h, the reaction was quenched with a mixture of ice and 6 *M* HCl aq. (5 mL), and the resulting mixture was extracted with Et_2O (3x30 mL).   The extracts were dried (Na_2SO_4), filtered, and concentrated *in vacuo* to give a residual oil, which was chromatographed (hexane:AcOEt) on silica gel to provide 0.180 g of (E)-1,1,1,2,3-pentafluorodec-2-en-4-ol in 73% yield.

Spectral data

^1H NMR ($CDCl_3$): δ 0.87 (3 H, t, *J*=5.4 Hz), 1.1-2.0 (8 H, m), 2.59 (1 H, br s), 4.52 (1 H, ddt, *J*=24.6, 6.2, 4.9 Hz); ^{19}F NMR ($CDCl_3$): δ -68.65 (3 F, dd, *J*=21.7, 10.8 Hz), -157.65 (1 F, ddq, *J*=133.9, 24.6, 21.7 Hz), -173.15 (1 F, ddq, *J*=133.9, 10.8, 4.9 Hz); IR (neat): ν 3348, 1724, 1378, 1216, 1150 cm^{-1}.

Note

For the preparation of the starting enol phosphonate, see page 223 of this volume.

Ishihara, T., *et al. J. Org. Chem.* **1990**, *55*, 3107.

Representative Distributors of Fluorochemicals

Acros Organics [ACR]

Janssen Pharmaceuticalaan 3a, 2440 Geel, Belgium
TEL: +32-14-575211, FAX: +32-14-593434
Home page: http://web.acros.be/acros
Online catalog: http://web.acros.be/acros
Online catalog: http://www.fisher1.com/fb/itv?24..f97.1....1..

Aldrich [ALD]

1001 West Saint Paul Avenue, Milwaukee, WI 53233 USA
TEL: 414-273-3850, FAX: 414-273-4979
Home page: http://www.sigald.sial.com/aldrich/
Online catalog: http://pipeline2.sial.com/pubcatalog.nsf/

Daikin Chemicals Sales, Ltd. [DCS]

14 Kanda Higashi Matsushitacho, Chiyoda-ku, 101-0042 Tokyo, Japan
TEL: +81-3-5256-0164, FAX: +81-3-5256-0163
Home page: http://www.star-net.or.jp/daikin_chm/fusso/index.html

Fluka Chemie AG [FL]

P.O.Box 260, CH-9471 Buchs, Switzerland
TEL: +41-81-755-25-11, FAX: +41-81-756-54-49
E-mail: bucmain@msmail.sial.com
Home page: http://www.sigald.sial.com/fluka/flukindx.htm
Online catalog: http://pipeline2.sial.com/pubcatalog.nsf/

F-Tech, Inc. [FT]

Tosoh Kyobashi Building, 3-2-5 Kyobashi, Chuo-ku, Tokyo 104-0031, Japan
TEL: +81-3-3274-1301, FAX: +81-3-3274-1211

Lancaster Synthesis Inc. [LS]

P.O. Box 1000, Windham, NH 03087-9777 USA
TEL: 603-889-3306, FAX: 603-889-3326
E-mail address: sshaffer@jaxnet.com
Home page: http://www.lancaster.co.uk
Online catalog: http://www.chem.com/Lancaster.html

Oakwood Products, Inc. [OAK]

1741 Old Dunbar Road, West Columbia, SC 29169 USA
TEL: 800-467-3386 (In USA), 803-739-8800 (All other), FAX: 803-739-6957
E-mail address: oakwood@netside.com
Home page: http://www.oakwoodchemical.com

SynQuest Laboratories, Inc. [SQL]

PO Box 309, Alachua, FL 32616-0309, USA
TEL: 904-462-0788, FAX: 904-462-7097
E- mail: info@synquestlabs.com
Home page: http://reagents@synquestlabs.com
Online catalog: http://reagents@synquestlabs.com/catalog/catalog.html

Tokyo Kasei Kogyo, Co., Ltd. [TCI]

1-13-6 Nihonbashi-Muromachi, Chuo-ku, Tokyo 103-0022, Japan
TEL: +81-3-3278-8153, FAX: +81-3-3278-8008
Home page: http://www.tokyokasei.co.jp/HTML/index_e.html (English)
Home page: http://www.tokyokasei.co.jp/HTML/index_j.html (Japanese)
See also http://www.tciamerica.com
Online catalog: http://tciamerica.com/cgi-bin/tkkgl-catalog.cgi (English)
Online catalog: http://tciamerica.com/cgi-bin/tkkjp-catalog.cgi (Japanese)

Availability of Fluorinated Reagents

Benzoyl fluoride	PhC(O)F	ACR, ALD, LS, OAK, SQL, TCI
1,1-Bis(dimethylamino)trifluoroethane	$CF_3CH(NMe_2)_2$	SQL
Bromotrifluoromethane	CF_3Br	ALD, OAK
3-Bromo-1,1,1-trifluoropropan-2-ol	$CF_3CH(OH)CH_2Br$	ACR, LS, OAK, SQL
3-Bromo-1,1,1-trifluoropropan-2-one (3-Bromo-1,1,1-trifluoroacetone)	$CF_3C(O)CH_2Br$	ACR, ALD, FL, LS, OAK, SQL, TCI
2-Bromo-3,3,3-trifluoropropene	$F_3CCBr=CH_2$	FT, LS, OAK, SQL, TCI
Chlorodifluoroacetaldehyde hydrate	$CClF_2CH(OH)_2$	OAK
Chlorodifluoroacetic acid	$CClF_2CO_2H$	ACR, ALD, FL, LS, OAK, SQL, TCI
Chlorodifluoroacetic anhydride	$(CClF_2CO)_2O$	ACR, ALD, OAK, LS, SQL, TCI
Chlorodifluoromethane	$CHClF_2$	ALD, OAK, SQL, TCI
1-(Chloromethyl)-4-fluoro-1,4-diazabicyclo-[2.2.2]octane bis(tetrafluoroborate) (F-TEDA-BF$_4$ or SELECTFLUOR™)		ACR, FL, LS, OAK, TCI
Chlorotrifluoroethene	$CF_2=CFCl$	ALD, DCS, LS, OAK, SQL
Dibromodifluoromethane	CBr_2F_2	ACR, ALD, FL, FT, LS, OAK, SQL
Dichlorofluoromethane	$CHCl_2F$	ALD, OAK, SQL
1,1-Dichloro-2,2,2-trifluoroethane	CF_3CHCl_2	DCS, LS, OAK, SQL, TCI
Diethylaminosulfur trifluoride (DAST)	Et_2NSF_3	ACR, ALD, FL, LS, OAK, SQL, TCI
Diethyl (bromodifluoromethyl)phosphonate	$(EtO)_2P(O)CBrF_2$	LS, OAK
N,N-Diethyl-(1,1,2,3,3,3-hexafluoropropyl)-amine (Hexafluoropropene Diethylamine; PPDA; Ishikawa's reagent)	$Et_2NCF_2CHFCF_3$	DCS, OAK, SQL, TCI
Difluoroacetic acid	CHF_2CO_2H	ACR, ALD, FL, LS, OAK, SQL, TCI

1,1-Difluoroethylene	$CF_2=CH_2$	ALD, FL, LS, OAK, SQL
Dimethyl fluoromalonate	$CHF(CO_2Me)_2$	ACR,[a] ALD,[a] FL,[a] LS, OAK, TCI[a]
Ethyl bromodifluoroacetate	$CBrF_2CO_2Et$	ACR, ALD, FL, LS, OAK, SQL, TCI
Ethyl bromofluoroacetate	$CHBrFCO_2Et$	ACR, ALD, LS, OAK, SQL, TCI
Ethyl chlorodifluoroacetate	$CClF_2CO_2Et$	LS, OAK, SQL, TCI
Ethyl fluoroacetate	CH_2FCO_2Et	ACR, ALD, FL, LS, OAK, TCI
Ethyl trifluoroacetate	CF_3CO_2Et	ACR, ALD, FL, LS, OAK, SQL, TCI
Ethyl 4,4,4-trifluorobut-2-enoate (Ethyl 4,4,4-trifluorocrotonate)	$CF_3CH=CHCO_2Et$	ACR, ALD, LS, OAK, SQL
Ethyl 4,4,4-trifluoro-3-hydroxybutanoate	$CF_3CH(OH)CH_2CO_2Et$	ALD, FT, LS, OAK
Ethyl 4,4,4-trifluoro-3-oxobutanoate (Ethyl 4,4,4-trifluoroacetoacetate)	$CF_3C(O)CH_2CO_2Et$	ACR, ALD, FL, LS, OAK, TCI
Ethyl 3,3,3-trifluoropyruvate	$CF_3C(O)CO_2Et$	LS, OAK, SQL
Ethyl 4,4,4-trifluoro-3-(trifluoromethyl)but-2-enoate	$(CF_3)_2C=CHCO_2Et$	LS, OAK
N-Fluorobenzenesulfonimide	$(PhSO_2)_2NF$	ACR, ALD, FL, LS, OAK, SQL, TCI
N-Fluoro-4,6-dimethylpyridinium-2-sulfonate		DCS, OAK
2-Fluoroethanol	CH_2FCH_2OH	ALD, DCS, FL, LS, OAK
Heptadecafluorooctyl iodide	$CF_3(CF_2)_7I$	ACR, ALD, DCS, FL, FT, LS, OAK, SQL, TCI
Heptafluorobutyric acid	$CF_3CF_2CF_2CO_2H$	ACR, ALD, DCS, FL, LS, OAK, SQL, TCI
Heptafluorobutyric anhydride	$(CF_3CF_2CF_2CO)_2O$	ACR, ALD, FL, LS, OAK, SQL, TCI
Heptafluoroprop-2-yl iodide	$(CF_3)_2CFI$	ALD, FL, DCS, LS, OAK, SQL, TCI
Heptafluoropropyl iodide	$CF_3(CF_2)_2I$	ACR, ALD, DCS, FL, LS, OAK, SQL, TCI
1,1,1,3,3,3-Hexafluoropropan-2-one (1,1,1,3,3,3-Hexafluoroacetone)	$CF_3C(O)CF_3$	ALD, OAK
1,1,1,3,3,3-Hexafluoropropan-2-one trihydrate	$CF_3C(O)CF_3 \cdot 3H_2O$	ACR,[b] ALD, FL, LS, OAK, TCI
1,1,1,3,3,3-Hexafluoropropan-2-ol	$(CF_3)_2CHOH$	ACR, ALD, DCS, FL, LS, OAK, TCI

Hexafluoropropene oxide		ALD, DCS, LS, OAK
Methyl chlorodifluoroacetate	CClF$_2$CO$_2$Me	ACR, ALD, FL, LS, OAK, SQL
Methyl heptafluorobutyrate	CF$_3$CF$_2$CF$_2$CO$_2$Me	ACR, ALD, DCS, LS, OAK, SQL
Methyl 4,4,4-trifluoro-3-oxobutanoate	CF$_3$C(O)CH$_2$CO$_2$Me	FL, LS, OAK, TCI
Nonafluorobutyl iodide	CF$_3$(CF$_2$)$_3$I	ACR, ALD, DCS, FL, FT, LS, OAK, SQL, TCI
Nonafluoropentanoic acid	CF$_3$(CF$_2$)$_3$CO$_2$H	ACR, ALD, DCS, FL, LS, OAK, SQL, TCI
Pentafluoroethyl iodide	CF$_3$CF$_2$I	ALD, DCS, FL, FT, LS, OAK, SQL
Pentafluoropropionic acid	CF$_3$CF$_2$CO$_2$H	ACR, ALD, DCS, FL, LS, OAK, SQL, TCI
Pentafluoropropionic anhydride	(CF$_3$CF$_2$CO)$_2$O	ACR, ALD, FL, LS, OAK, SQL, TCI
2,2,3,3,3-Pentafluoro-1-phenylpropan-1-one (Pentafluoropropiophenone)	PhC(O)CF$_2$CF$_3$	OAK
Pyridine-(HF)$_n$ (Pyridinium poly(hydrogen fluoride) or PPHF)		ALD, FL, OAK, SQL, TCI
Sodium fluoroacetate	CH$_2$FCO$_2$Na	FL, TCI
Tetrabutylammonium fluoride	n-Bu$_4$NF	ACR, ALD, FL, LS, OAK[c), TCI
2,2,3,3-Tetrafluoropropan-1-ol	CHF$_2$CF$_2$CH$_2$OH	ACR, ALD, FL, DCS, LS, OAK, SQL, TCI
2,2,3,3-Tetrafluoropropan-1-yl p-toluenesulfonate	CHF$_2$CF$_2$CH$_2$O-SO$_2$C$_6$H$_4$CH$_3$	SQL
Tribromofluoromethane	CBr$_3$F	ACR, ALD, DCS, FL
1,1,1-Trichloro-2,2,2-trifluoroethane	CF$_3$CCl$_3$	ACR, ALD, LS, OAK, SQL, TCI
Tridecafluorohexyl iodide	CF$_3$(CF$_2$)$_5$I	ACR, ALD, DCS, FL, FT, LS, OAK, SQL, TCI
Triethylamine trihydrofluoride	Et$_3$N·3HF	ALD, DCS, FL, LS, OAK
Triethyl 2-fluoro-2-phosphonoacetate	(EtO)$_2$P(O)CHFCO$_2$Et	ALD, FL, OAK, TCI
Trifluoroacetaldehyde ethyl hemiacetal	CF$_3$CHOH(OEt)	ACR, ALD, FL, LS,[d) OAK, TCI
Trifluoroacetic acid	CF$_3$CO$_2$H	ACR, ALD, FL, LS, OAK, SQL, TCI

Trifluoroacetic anhydride	$(CF_3CO)_2O$	ACR, ALD, FL, LS, OAK, SQL, TCI
2,2,2-Trifluoroacetophenone	$CF_3C(O)C_6H_5$	ACR, ALD, FL, LS, OAK, SQL, TCI
(E)-4,4,4-Trifluorobut-2-en-1-ol 3,3,3-Trifluoro-1,2-epoxypropane	$CF_3CH=CHCH_2OH$	OAK, LS, OAK
2,2,2-Trifluoroethanol	CF_3CH_2OH	ACR,ALD, DCS, FL, FT, LS, OAK, SQL, TCI
2,2,2-Trifluoroethyl iodide	CF_3CH_2I	ACR,ALD, FL, FT, LS, OAK, SQL, TCI
2,2,2-Trifluoroethyl p-toluenesulfonate	CF_3CH_2O- $SO_2C_6H_4\text{-}p\text{-}CH_3$	ACR, ALD, FT, LS, OAK, SQL, TCI
Trifluoromethanesulfonyl chloride	CF_3SO_2Cl	ALD, FL, LS, OAK, TCI
2-(Trifluoromethyl)acrylic acid	$CH_2=C(CF_3)CO_2H$	ACR, ALD, FT, OAK, SQL, TCI
Trifluoromethyl iodide	CF_3I	ALD, FL, FT , LS, OAK, SQL
2-(Trifluoromethyl)propanal	$CH_3CH(CF_3)CHO$	OAK, SQL
(Trifluoromethyl)trimethylsilane	CF_3TMS	ACR, ALD, FL, FT, LS, OAK, TCI
1,1,1-Trifluoropentane-2,4-dione (1,1,1-Trifluoroacetylacetone)	$CF_3C(O)CH_2C(O)CH_3$	ACR, ALD, FL, LS, OAK, TCI
1,1,1-Trifluoropropan-2-one (1,1,1-Trifluoroacetone)	$CF_3C(O)CH_3$	ACR, ALD, FL, LS, OAK, SQL
3,3,3-Trifluoropropene	$F_3CCH=CH_2$	ALD, FT, LS, OAK, SQL

a) The corresponding ethyl ester is available.   b) The corresponding sesquihydrate ($C_3F_6O\cdot1.5H_2O$) is available.   c) The corresponding trihydrate is available.   d) The corresponding methyl hemiacetal is available.

Compound Index

This is an alphabetical index of the names of all compounds
found in each heading.

Reagent Index

This is an index of the fluorine-containing starting materials and fluorinating reagents. They are first classified by function (bold print), followed by the name of each compound in italics.

Acid anhydrides

Alcohols

Aldehydes and ketones

Fluorinating reagents

Fluoroalkanes

Product Index

This is an index of the product names. The terms in bold print indicate the function of these compounds, and additional functional groups are shown in italics.

Aldehydes and ketones

Alkenes and alkynes

Ethers

Halogene-containing compounds

sulfur-containing compounds

Lactones and lactams

Other nitrogen-containing compounds

Silicon-containing compounds

Sulfur-containing compounds

Author Index

This is an alphabetical index of the names of the author(s)
whose method is noted for the preparation and/or reaction of
each compound.